International Law and Agroecological Husbandry

Remarkable advances are being made in life science and agricultural research to reform the methods of food production, particularly with regard to staple grain and legume crops, in ways that will better reflect ecological realities. However, advances in science may be insufficient to ensure that these possibilities for agricultural reform are realized in practice and in a sustainable way. This book shows how these can only be achieved through changes in legal norms and institutions at the global level.

Interdisciplinary in character, the book draws from a range of issues involving agricultural innovation, international legal history and principles, treaty commitments, global institutions, and environmental challenges, such as climate change, to propose broad legal changes for transforming global agriculture. It first shows how modern extractive agriculture is unsustainable on economic, environmental, and social grounds. It then examines the potential for natural-systems agriculture (especially perennial-polyculture systems) for overcoming the deficiencies of modern extractive agriculture, especially to offset climate change. Finally it analyses closely the legal innovations that can be adopted at national and international levels to facilitate a transition from modern extractive agriculture to a system based more on ecological principles. In particular the author argues for the creation of a Global Convention on Agroecology.

John W. Head is Wagstaff Distinguished Professor of Law at the University of Kansas School of Law, Lawrence, Kansas, USA. He is the author of several books on international law.

Other books in the Earthscan Food and Agriculture Series

Agricultural Markets Instability
Revisiting the recent food crises
Edited by Alberto Garrido, Bernhard Brümmer, Robert M'Barek, Miranda P.M. Meuwissen and Cristian Morales-Opazo

The Sociology of Food and Agriculture
Second edition
Michael Carolan

Food Security, Gender and Resilience
Improving smallholder and subsistence farming
Edited by Leigh Brownhill, Esther M. Njuguna, Kimberly L. Bothi, Bernard Pelletier, Lutta W. Muhammad and Gordon M. Hickey

Climate Change and Agricultural Development
Improving resilience through climate smart agriculture, agroecology and conservation
Edited by Udaya Sekhar Nagothu

Forgotten Agricultural Heritage
Reconnecting food systems and sustainable development
Parviz Koohafkan and Miguel A. Altieri

International Law and Agroecological Husbandry
Building legal foundations for a new agriculture
John W. Head

Food Production and Nature Conservation
Conflicts and Solutions
Edited by Iain J. Gordon, Herbert H.T. Prins and Geoff R. Squire

For further details please visit the series page on the Routledge website:
www.routledge.com/books/series/ECEFA/

International Law and Agroecological Husbandry

Building legal foundations for
a new agriculture

John W. Head

LONDON AND NEW YORK

from Routledge

First published 2017
by Routledge

2 Park Square, Milton Park, Abingdon, Oxfordshire OX14 4RN
711 Third Avenue, New York, NY 10017

Routledge is an imprint of the Taylor & Francis Group, an informa business

First issued in paperback 2018

British Library Cataloguing-in-Publication Data
A catalogue record for this book is available from the British Library

Library of Congress Cataloging in Publication Data
Names: Head, John W. (John Warren), 1953- author.
Title: International law and agroecological husbandry : building legal
foundations for a new agriculture / John W. Head.
Description: London; New York : Routledge, 2017. | Includes
bibliographical references and index.
Identifiers: LCCN 2016025179 | ISBN 9781138213920 (hbk) |
ISBN 9781315446523 (ebk)
Subjects: LCSH: Agricultural ecology—Law and legislation.
Classification: LCC K3870. H43 2017 | DDC 343.07/6—dc23
LC record available at https://lccn.loc.gov/2016025179

ISBN: 978-1-138-21392-0 (hbk)
ISBN: 978-0-367-02987-6 (pbk)

Typeset in Bembo
by Keystroke, Wolverhampton

Contents

About the author

John W. Head holds the Robert W. Wagstaff Distinguished Professorship at the University of Kansas, where he concentrates on international and comparative law. He earned his undergraduate degree from the University of Missouri at Columbia, an English law degree from Oxford University (1977), and his US law degree from the University of Virginia (1979). Before starting an academic career, he worked in the Washington, DC office of Cleary, Gottlieb, Steen & Hamilton (1980–1983), at the Asian Development Bank in Manila (1983–1988), and at the International Monetary Fund in Washington (1988–1990).

Both his teaching and his published works concentrate in the areas of international law, international business, and comparative law, with a special focus on Chinese law. Mr. Head's principal books include *Legal Transparency in Dynastic China: The Legalist-Confucianist Debate and Good Governance in Chinese Tradition* (2013, with Xing Lijuan), *Global Legal Regimes to Protect the World's Grasslands* (2012), *Global Business Law: Principles and Practice of International Commerce and Investment* (3rd ed. 2012), *Great Legal Traditions: Civil Law, Common Law, and Chinese Law in Historical and Operational Perspective* (2011, reprinted 2014), *China's Legal Soul* (2009), *Losing the Global Development War: A Contemporary Critique of the IMF, the World Bank, and the WTO* (2008), *General Principles of Business and Economic Law* (2008), *The Future of the Global Economic Organizations* (2005), *Law Codes in Dynastic China: A Synopsis of Chinese Legal History in the Thirty Centuries from Zhou to Qing* (2005, with Yanping Wang), and *The Asian Development Bank* (multiple editions, most recently co-authored with Xing Lijuan). He has also written numerous monographs, articles and other works relating to international law, some of which have been published in Chinese and Indonesian.

Mr. Head has been awarded Fulbright teaching and research fellowships to China, Italy, and Canada, and has also taught in Austria, Hong Kong, Jordan, Mexico, Mongolia, Turkey, and the United Kingdom and has undertaken special assignments in numerous locations for international financial institutions and development agencies.

Mr. Head is married to Lucia Orth, who is a lawyer, teacher, and novelist. They live southwest of Lawrence, Kansas, on a farm where they have recently completed a four-year prairie-restoration project and are now embarked on a forest-restoration project.

Preface

About this book

My overall objective in writing this book is to contribute to important ongoing interdisciplinary efforts aimed at evaluating and reforming global agriculture. Several agricultural research institutions around the world are making remarkable strides as a *scientific* matter toward making such agricultural reform possible – with special attention to grains and legumes, which comprise two-thirds of the modern human diet worldwide. These developments will, I am convinced, bring revolutionary changes to how global food production *can be* carried out. But the advances in science are not enough to change how global food production actually *will be* carried out effectively in the years ahead. I am eager to help promote that sort of change in ways that my own background allows: through examining certain key *legal and institutional* questions that must be addressed in order to transform agriculture – and then by proposing some detailed initiatives to facilitate that transformation.

This book is the first of three volumes to emerge from my work on this topic. Whereas this volume concentrates on issues of law and policy that arise within the context of the world's legal and institutional framework as it exists *today*, the second and third volumes[1] will explore prospects for *new* structures at the global and regional levels based on a refashioned concept of sovereignty as suggested only in preliminary form at the end of this book.

In writing this book I have drawn on roughly 35 years of international practice and scholarship concentrating on law, institutions, governance, environmental protection, and economic development.[2] In directing my

---

1 See John W. Head, A GLOBAL CORPORATE TRUST FOR AGROECOLOGICAL INTEGRITY: MANAGING A NEW AGRICULTURE IN A WORLD OF ECO-STATES (forthcoming) and John W. Head, ECO-CRISIS IN THE MEDITERRANEAN BASIN: A LEGAL AND INSTITUTIONAL FRAMEWORK FOR REGIONAL AGROECOLOGICAL INTEGRITY (forthcoming).

2 As noted more fully at the beginning of this book ("About the author"), the earlier stages of my career involved legal practice with an international law firm and two international financial institutions, and more recently I have authored numerous books and other works on international economic development, transnational business law, comparative legal history, and global institutions.

attention to how these matters bear specifically on agriculture, my professional background merges with my personal background. As I explain in the beginning pages of Chapter 1, I grew up on a grain-and-livestock farm that has been in our family for well over a century.

Still, because this book addresses such a broad range of issues and a blend of disciplines – ecology, plant science, economics, health, public policy, and others – I have relied heavily on comments and contributions made by numerous colleagues. One of them is Caleb Hall. As explained more fully in the Acknowledgements, he has served as primary drafting author for a few specific portions of this book and has contributed generally to the entire project.

The broadly interdisciplinary character of this book – and my reliance on a variety of sources in writing it – can be seen in the way I have formatted its text. Instead of presenting several hundred pages of typing with straight margins and a single font, I have opted for indented passages quoted from many other writers, inserted numerous figures and "boxes" to illustrate or explore specific points, and used a heavily footnoted style. The footnotes are designed, in fact, to serve several purposes: they cite the particular sources I draw from (including experts in those disciplines that do not lie in my main areas of experience), thereby allowing readers to see the specific grounds for my assertions and analysis; they provide further research material for a reader eager to study certain points more fully; and they allow me to convey a "sub-text" of important points while keeping a clean story line in the main text.

Perhaps most important, my use of footnotes, quoted passages, and other formatting features reflects how seriously I take the topics at issue in this book – international law and agriculture. This is not an essay or a report or a novel that I have written, nor an introduction to environmental or international law. Instead, I am attempting to provide a detailed but broadly-understandable cross-disciplinary analysis of (i) some fundamental flaws in today's agriculture, (ii) some extraordinarily innovative alternative forms of food production that show strong promise as a scientific matter, and (iii) some corresponding reforms in law that would be aimed at (in the words of the book's subtitle) "building legal foundations for a new agriculture".

About natural systems agriculture

The advances that have been made recently in agricultural research might be considered part of a "natural systems agriculture" or "climate-smart agriculture" movement.[3] These advances aim at improving sustainability, soil

---

3 Information about natural systems agriculture is widely available. See, e.g., Wes Jackson, *Natural Systems Agriculture: A Truly Radical Alternative*, 88 AGRICULTURE, ECOSYSTEMS & ENVIRONMENT (Feb. 2002), at 111–117 (abstract available at www.sciencedirect.com/science/article/pii/

health, biodiversity, carbon sequestration, and food security. The innovation that I believe best exemplifies the movement – and which offers the most immediate prospect for fundamentally changing humanity's relationship to the rest of the ecosphere in which we exist – is that of *perennial polycultures*. That innovation centers on developing perennial grains grown in polycultures, instead of annual grains grown in monocultures.

In this new approach, agriculture would mimic nature, not rebuke nature. Already, a perennial intermediate wheatgrass called Kernza® has been created, and several strains of perennial rice have been developed as well. Similar successes might soon be achieved in creating perennial forms of sorghum and wheat, and efforts to develop sunflowers, rosinweed, and other grain crops in perennial form are likewise accelerating rapidly now because of recent advances in genetics, plant breeding, evolutionary biology, and ecology.[4] Significantly, these new crops show great promise of being appropriate for use in a wide range of soil and climatic conditions around the world. Expressed in today's terminology, the grains under development show great promise of being "scaled up" to actually displace the harmful form of agriculture that currently prevails in the world, based on annuals grown in monocultures.

Why would we wish to displace the existing form of agriculture? Because modern agricultural production has become both a marvel and a menace. Although what I describe in this book as the "industrialization" of agriculture – of which the Green Revolution is a prominent component – has dramatically increased crop yields over recent decades, it has also dramatically increased the pace of ecological degradation, particularly through conversion of grassland areas to agricultural use.[5] That conversion has amounted to a replacement of perennial grasses with annual crops such as corn, wheat, and soybeans. This concentration on *annuals* constitutes a departure from the natural world, as does the planting of crops in *monocultures* as opposed to polycultures (mixtures of species). Moreover, modern agriculture also displays other shortcomings from an economic standpoint and a social standpoint – all of which I explore in this book.

In short, the "natural-systems" innovations in agriculture promise to bring us the *technological* capacity to reorient food production in ways that

---

S016788090100247X), and the Natural Systems Agriculture website associated with the University of Manitoba (www.umanitoba.ca/outreach/naturalagriculture/). For references to climate-smart agriculture and "sustainable intensification", see notes 137–139 and accompanying text in Chapter 5, *infra*.

4 See John W. Head, GLOBAL LEGAL REGIMES TO PROTECT THE WORLD'S GRASSLANDS 213 (2012) [hereinafter GRASSLANDS], citing reports from The Land Institute, Washington State University, and the USDA. These developments are summarized at the end of Chapter 1 and discussed in detail in Chapter 5 of this book.

5 For a detailed account of ecological degradation attributable to modern extractive agriculture, as embraced and accelerated by the Green Revolution, see section I of Chapter 3. For a study of how agricultural conversion has adversely affected the world's grasslands, see chapter 2 of GRASSLANDS, *supra* note 4.

will help address the world's food crisis, global climate change, and profound environmental degradation while also bringing other important economic and social benefits.

However, such a reorientation cannot occur unless we build the necessary *legal and institutional* capacity worldwide (that is, both at the level of national governments and at the global level) to reorient food production in a way that responds to both the urgency and the danger of the ecological, economic, and social crises that currently face the planet. In other words, legal and institutional structures should be created over the coming few years in order to ensure that the "natural systems" innovations in agriculture can (i) be hastened along and (ii) be implemented once they reach maturity for full introduction into the global food system. I am eager to contribute to that urgent project in this book by proposing specific legal reforms.

About my legal proposals

I have structured this book to culminate in a set of proposed legal reforms to facilitate a transition from industrial agriculture to a "natural systems" agriculture. The book follows this pattern:

- Part I, consisting of a single chapter, surveys the character of today's industrial agriculture – which I refer to as "modern extractive agriculture" for reasons explained there – from a global and "deep-history" perspective, thus setting the stage for a critical evaluation of it.
- Part II undertakes that critical evaluation, over the course of two chapters. Chapter 2 asserts that modern extractive agriculture is economically unsustainable. Chapter 3 asserts that it is ecologically and socially unsustainable. While some elements of those criticisms will be familiar to some readers, my aim is to offer a concise, definitive, and substantiated assessment of why alternatives to modern extractive agriculture must be sought.
- Part III, also consisting of two chapters, then describes the particular form of "natural systems" agriculture that I have noted briefly in the foregoing paragraphs – namely, a form of "agroecological husbandry" that centers on using a prairie ecology as a model, thereby relying on perennial plants grown in polycultures rather than annual plants grown in monocultures. In particular, the last portion of Chapter 5 specifies the progress that has been made to date in developing perennial versions of wheat, rice, sorghum, and other crops. In this sense, Part III explains a "scientific response" to the problems summarized in Part II by offering an attractive alternative to modern extractive agriculture.
- Part IV focuses, in turn, on a "legal response" to those problems. That is, it details key proposed legal reforms designed to help revolutionize global agriculture. (I offer some highlights in the paragraph just following this bullet-point list.) Chapter 6 focuses on *national* legal reforms; Chapter 7 focuses on *international* legal reforms.

- A short Part V, consisting of a single chapter, outlines some larger institutional and conceptual changes – including a modernization of the doctrine of sovereignty – that the legal reforms in Part IV will both require and facilitate, and it offers some closing observations regarding the transition to a "natural systems" form of agriculture.

The specific legal reforms that I enumerate in Part IV and Part V involve general principles as well as detailed rules and regulations. Several of them draw from, or elaborate on, ideas forwarded by other observers. A few of them will require substantial (but not unprecedented) departures from existing rules that have become firmly woven into most legal systems at the national and the international level. All of them are proposed for the purpose of helping facilitate the transition from modern extractive agriculture to agroecological husbandry. They include the following[6]:

- To address the ways in which modern extractive agriculture is unsustainable as a matter of *economics*, national legislatures and parliaments should adopt legal and regulatory changes at the national level that will (i) facilitate an increase in the diversity of farms and in the size of rural populations, through enhanced infrastructure support of several types, and (ii) usher in a dramatic increase in the diversity of crops – to reverse the trend that has resulted in corn, wheat, and rice accounting for 89 percent of all cereal grain production – through a wholesale reorientation of agricultural subsidies.
- To address the ways in which modern extractive agriculture is unsustainable as an *ecological* matter, national legislatures and parliaments should adopt legal and regulatory changes at the national level that will (i) expand dramatically the ongoing scientific research into perennial species of food grains and legumes that can gradually supplant the annual crops that dominate today's agriculture, (ii) expand also the ongoing scientific research into foodcrop polycultures, (iii) reorient agricultural subsidies, tying them to ecological performance, and (iv) remove fossil-carbon subsidies as gradually as necessary to avoid chaos but as quickly as necessary to blunt the worst effects of climate change.
- *In the USA in particular*, Congress should enact and implement a "50-year farm bill" that will (i) adopt a national agricultural policy based on ecological principles, with the aim of enhancing biodiversity and ecospheric health, (ii) commit adequate funds to boost research and development into "natural systems" agriculture in general, and perennial polycultures in particular, and (iii) prescribe an implementation schedule for these and other reforms that will shift acreages devoted to food crops from annual monocultures to perennial polycultures as quickly as

---

6 Full details appear in Chapters 6 and 7. For a presentation of the proposed legal reforms in the form of a "bare-bones brief", see the Appendix to this book.

possible, with a goal of having 60 percent of US cropland devoted to perennials by 40 years after the enactment of the legislation.

- In order to facilitate a change in agriculture *at the global level*, national and international advocates and organizations should put in place a new (but not unprecedented) treaty system that would (i) announce binding principles emphasizing the need for a reintegration of humans with the rest of the natural world, (ii) unambiguously adopt the Precautionary Principle, (iii) note the special urgency of issues of climate change, ecological preservation, and food security, (iv) assert that certain "safeguard rights" of all people will be protected during the transition to a new form of agriculture, (v) commit participating states to adopt and implement legal changes similar in character to those referred to above for the USA, (vi) give "special and differential treatment" to economically less developed countries, (vii) ensure that the interests of a wide range of non-state actors are also well represented, and (viii) set in motion a process to adopt a revised concept of sovereignty – one that will reflect the overriding planetary importance of long-term ecological sustainability and resilience.

At first glance, this array of legal reforms might seem impossibly ambitious, even audacious, especially in an age marked by political gridlock, international conflict, and institutional sclerosis. However, at various junctures in the following pages I will explain why we should *not* assume an attitude of despair or resignation. After all, change will inevitably occur – in agriculture, in law, in worldviews – and what really is at issue is how the change will occur and specifically how we can influence its content, pace, and direction.

Moreover, not one of the legal reforms summarized above and elaborated in this book is truly unprecedented. Indeed, today's modern extractive agriculture can itself be traced to some specific decisions and innovations made in earlier years that were just as significant and ambitious in their day as the ones now required in ours. The same can be said of law and institutions at both the national and the international level: the past century alone has seen changes that are just as substantial as the ones required today to build the global legal foundations for a new agriculture.

In my view, then, it is shortsighted to plead that the initiatives outlined here are *too* ambitious, or that they are *too* inconsistent with existing legal doctrines or entrenched political interests, or that a "Plan B" should be designed and held in reserve in case these initiatives cannot prevail over the status quo. Indeed, a more potent critique of the proposals I develop here might be that they are not innovative *enough* to secure fully the changes that agriculture needs.

About the larger context of crisis

In referring to the crisis that the world faces in agriculture, food production, and rural life more generally, I do not mean to suggest that this crisis is the only one, or the largest one, that the world faces. As will become evident in the pages that follow, I see four ecological crises converging simultaneously in a way that threatens all life on Earth – not just human life but the entire system and fabric of our shared biosphere.

Those four crises relate to the Earth's soil, its air, its water, and its biodiversity. While I focus most intensely in this book on the first of these – the crisis that traces its roots literally to the soil – this one crisis is in fact inseparable from the other three.[7] For each of these aspects of the ecosystem that serves as our shared home – soil, air, water, and the still-astonishing diversity of other species – our status as trustees for the benefit of future generations (both human and non-human) demands that we develop innovative approaches to address or mitigate the crises that have become most manifest in our own generation.

Fortunately, we have probably never been better positioned than we are now, as a species and a civilization, to address these challenges. Rapid global communications, crowdsourced knowledge, cascading scientific advancements, a growing acknowledgment of the gathering storm of ecological crises . . . all these are resources we can use to build the new policies, new laws, and new institutions that must be created and nurtured.

It is against that backdrop that I harbor some hope that the crisis the world now faces as a result of modern extractive agriculture *can* in fact be addressed. I explore in this book how that might be done – how we can gain *traction* to get it done – especially from an international legal perspective. Whether we *will* in fact address the crisis in agriculture in a satisfactory way depends importantly on political will. I would like to think that this book can help mold that political will by suggesting how to put, in Thoreau's words, *foundations under our castles in the air.*[8]

---

7 In one of the other two forthcoming volumes emerging from my work, I take a different approach: instead of concentrating attention as a *global* matter on *one* aspect of the ecosphere – soil – as this book does, that other publication focuses on all *four* aspects in a single *region*. See John W. Head, Eco-Crisis in the Mediterranean Basin: A Legal and Institutional Framework for Regional Agroecological Integrity (forthcoming).

8 "If you have built castles in the air, your work need not be lost; that is where they should be. Now put the foundations under them." Henry David Thoreau, Walden 323–324 (1854, J. Lyndon Shanley, ed., 1971).

Illustrations

Figures

Boxes

Notes on usages, citations, and abbreviations

In this book I have followed certain conventions on spelling, punctuation, and usage that might be unfamiliar to some readers. These conventions include the following:

- *Citations to law-oriented materials.* Citations to books, articles, and other legal materials appear in a less abbreviated style than that used by many law journals and books. I believe the heavily abbreviated style used in US legal texts in particular can be so unfamiliar to a general audience as to create confusion or uncertainty. Moreover, in the case of books, I have departed from the practice of putting the authors' names in all capital letters. Instead, authors' names for all works – books and articles and other items – appear in regular upper case and lower case letters; then titles of books appear in large and small capitals and titles of other works appear in italics or, in a few cases depending on the nature of the work, in regular font with quotation marks.
- *Citations to science-oriented materials.* For purposes of consistency in style, I have cited scientific articles and books in a manner that more closely resembles the generalized legal-citation style referred to above. As a consequence, those citations should be easily readable by non-science readers. In keeping with scientific-journal-article citation style, footnote references here to some articles appearing in science journals do not include specific page-number references.
- *Internet citations.* Many journal and newspaper articles I have cited in this book are available online, but I typically have not included the webpage citations for them because they can be easily found through search engines. However, for less mainstream (or non-print) resources, I typically have included webpage addresses in my citations. I have dispensed, though, with details of "last updated" and "last visited", on grounds that such information is likely to be of little use. Most of the citations to such sources were operational as of mid-2015, many of them through early 2016. For my internet citations, I have also largely dispensed with the introductory phrase "available at".

- *Internal citations.* Many of the passages that I have quoted from other authors included, in their original publication, citations to authority in the form of footnotes or endnotes. Throughout this book, unless noted otherwise, I have omitted these citations without expressly indicating "(citations omitted)" or "(footnote omitted)".
- *Specific usages.*

  - As noted in Chapter 1, the term "corn" outside North America, Australia, and New Zealand means any cereal crop or any staple crop, with its meaning varying geographically. In the United States, Canada, Australia, and New Zealand, "corn" primarily means maize. In this book I have used the terms "corn" and "maize" interchangeably.
  - Throughout this book, the term "state" typically carries the meaning it has in international law – that is, as a nation-state and not as a subsidiary political unit such as the individual domestic states that make up federal states such as India or the USA or Mexico.
  - The acronym noun "USA" is often used in this book instead of the commonly-used noun "United States", inasmuch as there are other countries (such as Mexico) with the title "United States" in their official names. However, the term "US" has been retained for use as an adjective referring to something of or from the USA, such as "US legislation" or "US states".
  - I have opted for the use of "US" and "USA" without periods, as this seems to be the more modern trend and also follows the usage found in acronyms for other political entities such as the United Nations (UN) and the European Union (EU). Naturally, I have not changed "U.S." to "US" in any quoted material or official citations.
  - The possessive form of words ending in the letter "s" have not had another letter "s" added to them. I have referred, for instance, to "the Great Plains' environment" rather than to "the Great Plains's environment".
  - I have used the abbreviation "CE", for Current Era (or Common Era), to carry the same meaning as the more outdated abbreviation "AD", for *Anno Domini*; and I have used the corresponding abbreviation "BCE", for "before Current Era", instead of "BC", for Before Christ.
  - The symbol "$" refers to US dollars. Where amounts are expressed in US dollars but would be paid partly or wholly in other currencies, I have sometimes inserted the word "equivalent", such as in the "$ equivalent" paid in EU agricultural subsidies.
  - In the interest of brevity, I have usually opted for the symbol "%" in the footnotes instead of the word "percent" – except of course in quoted material using "percent" or where the context requires it.
  - As a surrender to modern "relaxed" usage, I have used the word "will" instead of the word "shall" with first person subjects in declarative clauses and sentences. In my view, proper English usage

still calls for "shall" in declarative clauses and "will" in imperative clauses (calling for special emphasis or certainty) when the subject is first person ("I" or "we") and for the opposite when the subject is second or third person ("you", "he", "it", etc.). However, I recognize that the repeated use of "I shall" in a text can seem stilted and antiquated, so I have abandoned it here.

- *Punctuation with quotation marks.* I have followed the less-used, but more logical convention of placing quotation marks inside all punctuation, unless of course the punctuation itself appeared in the material being quoted. Doing so allows the text to reflect more faithfully how the original material reads.
- *Italicization.* I have used italicization in four circumstances: (i) where I wish to add emphasis; (ii) in textual references to titles of books, such as references to Rachel Carson's *Silent Spring* and Thomas Hobbes' masterwork *Leviathan*; (iii) to signify words or terms from languages other than English (mainly Latin and French); and (iv) in certain "levels" of subsection headings, as a navigational aid to the reader.
- *Acronyms and Abbreviations.* Although I have kept acronyms and abbreviations to a minimum, I have in the interest of brevity used the following ones (in addition to US and USA, mentioned above) in various passages throughout the text:

GCA	Global Convention on Agroecology
CGIAR	Consultative Group on International Agricultural Research
CH_4	methane
CO_2	carbon dioxide
EU	European Union
FAO	Food and Agriculture Organization of the UN
GE	genetic engineering, or genetically engineered
GHG	greenhouse gas
GM	genetically modified
GMO	genetically modified organism
IFAD	International Fund for Agricultural Development
IPCC	Intergovernmental Panel on Climate Change
N_2O	nitrous oxide
NSAC	National Sustainable Agriculture Coalition
SDT	special and differentiated treatment
UN	United Nations
UNEP	United Nations Environment Programme
USDA	US Department of Agriculture
WRI	World Resources Institute
WWF	World Wildlife Fund

Acknowledgements

I wish to thank numerous people for their help in my work on this book. I should begin, as I did in my recent book on grasslands protection, with certain members of my extended family who set the stage over the course of several generations for me to grow up on a farm. These include in particular J. Warren Head (1910–2009), John Wallace Head (1871–1955), Joseph Warren Mackey (1839–1915), and John B. White (1793–1889). These fathers of mine acquired and consolidated the farm where I was raised, and their wives supported them in managing the farm operations and raising their families.

More recently – and particularly in the last three years – my work has been sustained and enriched by the contributions of several research assistants. Some of them were students or former students; others pitched in from greater distance but with deep commitment to grappling with the issues this book addresses. These assistants include Lindsey Collins, Miriam Friesen, Clarissa Howley, Britt Lageman, Jade Martin, Jacque Patton, Hannah Sandal, and Isabel Segarra Trevino. As I noted at the beginning of the Preface, the work that Caleb Hall has done figures especially prominently in several portions of the book, which have been noted accordingly.

In addition, valuable assistance also came from legal and scientific experts in various countries and institutions, including the University of Kansas, The Land Institute, The World Bank, and the National Sustainable Agriculture Coalition. Some of these experts' contributions will be more specifically identified in pertinent portions of the text, but I would offer a blanket nod of appreciation to Chris Brown, Mark Cackler, Stan Cox, Tim Crews, Lee DeHaan, Warren Evans, Rob Glicksman, Tim Hardwick, Wallace Head, Ferd Hoefner, Wes Jackson, Rick Levy, Chuck Myers, Shuwen Wang, David Van Tassel, Don Worster, and Angus Wright. Support from the University of Kansas General Research Fund is also gratefully acknowledged – as are the patient secretarial services applied to this project by Michele Rutledge, Yolanda Huggins, and Lori Farmer, as well as the technical and editing assistance provided by various professionals at Routledge and Keystroke in preparing a rather complicated text for publication.

I wish to draw special attention to an issue of attribution and authorship. Over recent years I have benefited greatly from collaborating with various colleagues – particularly at The Land Institute – in preparing a range of papers, briefs, compilations, and other documents relating to numerous topics that I address in this book. These were collective efforts, and in many cases I have forgotten (or it was unclear nearly from the outset) just who wrote which phrases and passages. Not surprisingly, some of those written works, and many of their ideas, are reflected in the following pages. In nearly all cases I have altered them in one or more ways to fit substantively or presentationally in this book. Still, it is likely that some passages of these collectively-prepared documents appear here with little alteration. Where it has been practicable to do so, I have noted their derivation, but I imagine that I have failed to do so in some places – and for this I can only offer a blanket apology in advance.

What I hope is both original and useful about the book can be found in the manner in which I have linked a wide range of ideas and highlighted the significance of those linkages. For instance, Wes Jackson will always have thought more clearly and written more insightfully about perennial polycultures than I can; Hugo Grotius and Jean Bodin were smarter about sovereignty than I will ever be; likewise with Don Worster about environmental history, Wendell Berry about an agrarian ethic, Tim Crews about carbon sequestration, and so forth. I thank them for their contributions to my work, and I hope that I have drawn effectively from them.

Lastly, as always, I offer a special thanks to my wife Lucia for her efforts to bring both order and spark to the text by her critical review of it.

Part I
Orientation and preview

1 Reconceptualizing agriculture and its legal framework

I. Three theses

As noted in the Preface to this book, my aim here is to explore three theses. The first is that modern agricultural production – particularly the form of grain production that predominates in today's world – has become both a marvel and a menace, and that its menacing characteristics are now so unsustainable as to present an existential threat to the human species. The second thesis is that an alternative system of food production – replacing the system that now supplies nearly two-thirds of all human caloric intake – is possible through advanced crop science and ecological research. The third thesis is that this alternative system of food production is also possible as a *legal* matter; that is, changes in law can be made, based on precedents in our recent past, which will facilitate a transition from the current system to a new system of grain production that will bring a range of benefits.

In these introductory pages, I explain how I intend to develop each of these theses. I start by describing (immediately following) the system of agriculture that dominates the world today, which I call "modern extractive agriculture". My account of this contemporary form of agriculture draws heavily on "deep history", including geologic time, in order to present a broader view than is commonly perceived, especially by those of us who do not spend most of our time in farming or in focusing on ecology and evolutionary biology.

II. What is "modern extractive agriculture"?

IIA. My own competing interests

Moving thunderously toward me, laboring against the slope of the field, the mechanical monster came into view. I watched it as I stood under the vast Chinese elm that has presided over the back yard of my family farm for well over a hundred years. Against a darkening November eastern sky an hour before sundown, the corn harvester reached the end of a series of rows and stopped. The plexiglass door to the cab swung open; down the metal stairs came Will Bier. "Do you want to ride along?"

Will Bier farms roughly half of the nearly 500 acres of land that I own with various other members of my immediate family. The acreage that Will farms for us borders on land that his family owns. It all lies in an extraordinarily soil-rich area of northeast Missouri, just up from the Mississippi River bottom. I grew up there. I do not claim to be a farmer myself, but my upbringing gave me enough grounding to know the basics of running a corn-picker – and also, for that matter, a planter, a cultivator, a baler, and various other farm implements.

As soon as I joined Will in the cabin of the harvester, I realized that my basic grounding had not prepared me for this ride. Over the next hour, Will briefed me on the brains of the harvester and the techniques that it used – through sophisticated electronic and GPS equipment communicating remotely with other computer equipment and databases – to pick and shell corn that had been planted with equipment just as sophisticated as this monster.

I am involved to some degree – much more than most people – in the system of modern agriculture. Like Will Bier, I have a financial stake in that system. From my family's roots in farm country and farm culture, I also have something of a sentimental stake in that system. It is with a combination of that personal interest and personal background that I have embarked in recent years on an inquiry into the *failure* of that system and what might be done about it.

But why would I be drawn to this sort of inquiry, focusing on the *failure* of a system in whose *success* I have a financial and personal interest? Because I have a different perspective as well. For the past 30 years I have participated in efforts to advance international law, economic development, environmental protection, and cross-cultural cooperation through global institutions. That background naturally colors my view of agriculture. From my perspective, many of those efforts at the international level have failed to bring durable improvements to agriculture and to farmers.

Accordingly, with this mixture of competing interests or "stakes" in the subject, I wish to explore in this book the general contours of how agriculture in the USA and other highly developed economies came to be what it is today, why it has come to dominate world production of food, feed, and fiber, what alternatives to that system might be available so to permit a "re-boot" of agriculture, and specifically how at a *global* level a new legal and institutional framework might be built to facilitate a more sustainable future for farms, farmers, and those who rely on them – that is, all of us, all of humanity.

What do I mean in referring to "modern agriculture", or more provocatively (as in most of this book) to "modern *extractive* agriculture"? Even more fundamentally, how broad a definition of "agriculture" do I use? I summarize these definitions in the following paragraphs. However, a more comprehensive explanation of the contemporary form of agriculture that dominates world food production today will appear in Chapters 2 through 5 of this book as I proceed with my inquiry and analysis.

IIB. *Traditional, extractive, fossil-carbon-based, industrial, modern?*

In a recent book, I explored the need to protect the world's grasslands from further degradation.[1] As I pointed out there, most of the world's grassland areas – which in earlier eras covered vast, sweeping portions of the Earth's land surface – have now been destroyed or profoundly degraded. The chief cause of this has been agriculture. (Throughout this book, I will usually include livestock production as falling within the ambit of "agriculture", even though there are reasons for treating them separately for several purposes.)

More specifically, I explained in that earlier book that two of the ingredients involved most directly in the degradation of grasslands are agricultural conversion[2] and inappropriate grazing. The point of most agricultural production is either (i) to produce food for humans or (ii) to produce feed for livestock, most of which is then slaughtered or milked for human consumption. Likewise, the point of most livestock grazing is to produce food for humans – again, mainly through slaughtering.[3]

In that same earlier book, I explained that agriculture figures indirectly in some of the other main factors that cause grasslands degradation. These factors include habitat fragmentation, fire suppression, water mismanagement, and global warming. Indeed, in large measure, *all* of these negative factors also trace their roots to farming. For instance, the principal reason for habitat fragmentation and fire suppression is to accommodate agricultural production and livestock grazing. A very large use for water, including "fossil water" of the type represented by the Ogallala aquifer in the American Great

---

1 See John W. Head, GLOBAL LEGAL REGIMES TO PROTECT THE WORLD'S GRASSLANDS (2012) [hereinafter GRASSLANDS].

2 By "agricultural conversion", I mean the conversion of land from one form of land cover and/ or land use – in this case, grasslands – to agricultural use, such as by row-cropping.

3 Although the discussion in this section does not focus on livestock production, some issues relating to livestock production for meat (for human consumption) are raised in Chapter 3.

Plains, is irrigation, to keep crops alive during hot summer growing seasons. Agriculture and livestock are also major contributors to the buildup of atmospheric greenhouse gas.[4]

In my book on grasslands protection, I then offered this comment:

> A common reaction to these facts would be to say that we *need* agriculture, and that to suggest otherwise is to fly in the face of reality. However, it is becoming increasingly evident and well-accepted, especially among various types of environmental scientists, that we do *not* need – or, more precisely, we cannot sustain and must not continue – traditional extractive agriculture of the sort that humans have practiced globally for the past several thousand years.[5]

Why did I use the term "traditional extractive agriculture" in that context, and why do I use the term "modern extractive agriculture" in this book – and also include the terms "fossil-carbon-based" and "industrial" in the heading to this subsection? These choices of language reflect in part my motivation for writing this book, so let me expand some on why I use these terms.[6]

Today's predominant form of agriculture is "traditional" in the sense that it follows the general patterns that were already evident in the ancient city of Jarmo (in modern-day Iraq) roughly 5,000 to 7,000 years ago. Farmers around that city cultivated several cereal crops that were similar to today's wheat, barley, and rye. Those so-called *cultivars* were annuals with large enough seed heads to make it worthwhile to plant them each year, typically in designated plots of land in order to produce food for the residents and feed for their (non-human) animals.[7]

---

4 As noted in the following text, one principal way in which agriculture contributes to the buildup of greenhouse gases in the atmosphere is through the use of nitrogen-based fertilizers, creating nitrous oxide (N_2O). A principal way in which livestock production (especially through concentrated cattle operations) contributes to the buildup of greenhouse gases is through the release of methane (CH_4). See text accompanying note 79 in Chapter 3, *infra*, as well as note 47 in Chapter 6. For a visual representation of how agriculture bears on greenhouse gas emission, see World Resources Institute, *World GHG Emissions Flowchart*, http://pdf.wri.org/world_greenhouse_gas_emissions_flowchart.pdf.

5 GRASSLANDS, *supra* note 1, at 204 (emphasis added).

6 Several of these terms also appear in the form of a chart in Figure 1.1 on page 14.

7 One source has offered the following explanation of how this development occurred:

> Although we have no preserved evidence, these crops [resembling wheat, barley, and rye] must have been evolving for many thousands of years . . . Some archaeologists have suggested that these agricultural crops, known as *cultivars*, may have evolved purely by chance, growing on the edges of the first clearings made by humanoids, possibly finding the midden heaps that surrounded the encampments to be highly fertile ground that encouraged the grasses to undergo spectacular evolution from the small wild form to those which produced large seed heads. The large seed heads would have proved a natural food source. But how the transition

If these patterns – involving the use of cultivars for agricultural production – were, as I just indicated, "evident in the ancient city of Jarmo" roughly 5,000 to 7,000 years ago, then what happened *before* that time? Although the story is still being pieced together, it features (i) an extremely long period of hunting and gathering, followed by (ii) a period of several thousand years of "protoagriculture" culminating in the firm entrenchment of settled agriculture that eventually eclipsed hunting and gathering. William Ruddiman offers these observations about the first part of that story:

> [P]erhaps the most amazing fact about the [long] history of our ancestors is how remarkably little things actually changed over an incredibly long period of time. Back 3.6 million years ago, our australopithecine predecessors were already acquiring their food from scavenging, gathering, and hunting small animals. Yet 3.59 million years later, just 12,000 years ago, our immediate (and by then fully human) ancestors were still subsisting by hunting, gathering, and scavenging, along with some fishing.[8]

With such momentum of hunting and gathering as a form of keeping ourselves fed, how did the shift to agriculture occur? One possible answer centers on a "Neolithic revolution"; another emphasizes the likelihood of a longer process involving "protoagriculture". Consider this explanation:

> Western ideas about agricultural origins began when Europeans encountered "primitive" peoples, who were often foragers and knew little or nothing about farming. Other investigations found that humans and their societies and technologies had evolved over long evolutionary periods that came to be called Paleolithic and Neolithic, and that crop plants and domesticated animals of the world's agricultural systems had definite geographical and temporal origins.
>
> These findings led in the 1930s to the idea that early humans had developed agriculture in a "Neolithic Revolution" approximately 10,000

---

was made from a human society based on the gathering of seed heads to one in which grass seeds were deliberately planted to produce a crop is not known.

See Grasslands Conservation Council of British Columbia, *Overview of the World's Grasslands*, www.bcgrasslands.org/index.php/learn-more/152-overview-of-the-worlds-grasslands [hereinafter *Grasslands Overview*] (attributing the quoted text to the Geography Department at the University of Strathclyde in Glasgow, Scotland).

8 William F. Ruddiman, PLOWS, PLAGUES, AND PETROLEUM: HOW HUMANS TOOK CONTROL OF CLIMATE 17 (2005). Ruddiman is a paleoclimatologist and professor emeritus at the University of Virginia with a PhD in marine geology from Columbia University. For another perspective on the same phenomenon, see Barbara Bender, FARMING IN PREHISTORY: FROM HUNTER-GATHERER TO FOOD-PRODUCER 158 (1975): "Food-production is a recent development. Tool-using hunter-gatherers have existed for four million years; food-producers for eleven thousand – a mere four hundred generations."

years ago, in response to a drying climate after the end of the last Ice Age. This shift to agriculture led to the development of cities and civilization some 5,000 years later. In this view, farming first developed in the "fertile crescent" of Mesopotamia, where the local flora and fauna included the wild progenitors of the main domesticated food crops and animals.

New archaeological research has qualified this conception of the first "agricultural revolution." Several scholars argued that the shift to farming was so rapid that it must have been preceded by "protoagriculture" for thousands of years before the Neolithic period. A cool and dry period about 11,000 years ago, the Younger Dryas, was followed by a warmer period favorable for the spread of plants and animals in the Near East. New research and rethinking of the evidence have shown that some of the presumed centers of agricultural development actually acquired the idea and techniques of farming from one or more of the smaller number of earlier centers.[9]

In short, the notion that there was a brief period in which an agricultural "revolution" rather suddenly emerged has more recently been challenged by the view that a mixed system – some farming, some hunting and gathering – probably existed for a long time:

> Everywhere the shift to farming was gradual and ambivalent. Before people began to farm, they had established a pattern of settling for seasons or years at a time. They expanded their foraging and hunting, apparently driving some larger species of animals into extinction. People continued to hunt and forage even while they settled and raised crops and livestock.[10]

Perhaps one reason for such a "gradual and ambivalent" shift to farming is found in the notion of "coevolution", in which humans and certain plant

---

9 Mark B. Tauger, Agriculture in World History 3 (2010). Tauger is a professor of history at the University of West Virginia with a PhD from UCLA. For other accounts of the early development of agriculture, and of the forms of institutionalized power that accompanied it, see various works of David Christian, including his Maps of Time: An Introduction to Big History (2011) and his This Fleeting World: A Short History of Humanity (2009). Christian emphasizes the fact the first "early agrarian era" triggered the development of institutionalized power; there was an increased need for leadership, largely because there were more materials goods and more people over which control could be exercised. It is noteworthy that Christian's *historic* observation has *normative* implications for the present age: if the rise of agriculture contributed so importantly to the emergence of the forms of institutional power that we have today, then if we make *changes* to agriculture that are so fundamental as to alter how it affects social relations and the relationship between humans and the rest of the ecosphere, these changes will permit – and in fact require – fundamental modifications, also in the forms of institutional power, so that institutional power will be dramatically different under a natural-systems agriculture from what it is under the form of agriculture that we have inherited from the ancient past.

10 Tauger, *supra* note 9, at 10.

and animal species changed together. With human use, certain plants and animals became more productive and still more amenable to domestication. This made people in turn more dependent on their crop plants and domesticated animals. Tauger offers this assessment:

> These findings and arguments suggest broader interpretations of the human shift to agriculture. The economist Colin Tudge argued from the concept of protoagriculture that basic farming skills gave farmers advantages over neighboring non-farmers. Farmers could survive temporary declines in foraging and hunting food sources, which may have encouraged them to hunt more intensively, and might explain the Pleistocene extinctions of large animals; the hunters may have calculated they could fall back on raising plants if the animals disappeared.
>
> This fall-back plan, in Tudge's view, turned out to be a trap. Farming increased food production and enabled more children to survive, which in turn obliged people to rely on farming to support the growing population. He notes that early writings such as Genesis in the Bible as well as archaeological findings emphasize how strenuous and difficult farming was and how people hated it. Some ancient myths, from the Epic of Gilgamesh to the Biblical conflict between Jacob and Esau, contrast the free but animal-like (e.g., hairy) hunter with the more civilized farmer, and in both stories show the farmer tricking the hunter into giving up his freedom for the attractions of civilized life.[11]

Although the details of this narrative – and specifically the processes and influences that led to protoagriculture, coevolution, and domestication – remain somewhat unclear,[12] we do know the broad contours of how the ages-old patterns of hunting and gathering were displaced by agriculture, so that by the time the city of Jarmo was in its prime several thousand years ago, agriculture had gathered the momentum necessary to put in place a new tradition. Jason Clay offers this synopsis:

> For more than 99 percent of human history, people obtained their food by hunting, fishing, and gathering. Over the past 7,000 years that has

---

11 *Id.* at 10–11.

12 For a widely-cited definition of coevolution (in a biology context), see Kevin Y. Yip *et al.*, *An Integrated System for Studying Residue Coevolution in Proteins*, 24 BIOINFORMATICS 290–292 (2008) (defining it as "the change of a biological object triggered by the change of a related object"). In addition to the notion of coevolution, and related to it, is Mark Tauger's notion of "dual subordination". According to Tauger, farmers serve "as the interface between civilization and the environment" and have been subordinate to both. Tauger, *supra* note 9, at 1. They are, he says, "dependent on the natural environment" because "the changing circumstances of water, soil, and weather, and the actions of animals, plants, and other life forms . . . can threaten farm production"; similarly, farmers are also "subject to the rule of agencies outside their villages, usually urban authorities such as kings, armies, tax collectors, banks, and markets." *Id.* at 12–13.

changed remarkably. Today only 2 percent of all human food energy and only 7 percent of all protein is captured from the wild, and most of this is from water. The rest is produced by agriculture and aquaculture on land.

As a result, agriculture is the largest industry on the planet. It employs an estimated 1.3 billion people and each year produces [the equivalent of] $1.3 trillion worth of goods at the farm gate.[13]

I offer the above observations to explain why I refer to today's form of agriculture as "traditional". Surely a period of roughly ten or eleven thousand years – since the time when, following the last Ice Age, some forms of proto-agriculture and agriculture started developing – is long enough to create a *tradition*. If so, then the form of agriculture that we have today qualifies as "*traditional* agriculture" in the sense that it traces its roots (literally) to a time in the very deep past, even though the much deeper past (before ten or eleven thousand years ago) did *not* feature agriculture but rather hunting and gathering as the predominant form of food supply for humans.

What about the terms "modern" and "extractive"? Why do I use those terms also to describe the form of agriculture that dominates food production on Earth today? Let me first explain the term "extractive", and why it applies to agriculture not only today but stretching back to the days of Jarmo and before.

As noted previously, the foundations of agriculture as seen in Jarmo featured the *annual* planting of seeds in designated plots of land. This process was extractive to the degree that the integrity of the soil that the farmers used for such cultivated agriculture was diminished or compromised by using – or abusing – that soil. Such abuse might take the form, for instance, of either (i) erosion or (ii) fatigue (extracting the soil's nutrients without replacing them). Of course, erosion could be reduced by various prudent farming practices, and fatigue of the soil could be counterbalanced in part by using manure from livestock. Yet some erosion and fatigue still inevitably occurred.

The degree to which this traditional form of agriculture was extractive in character gradually *increased* with the development of more effective means of tilling the soil. Various styles of plow (plough) emerged over time,

---

13 Jason Clay, World Agriculture and the Environment: A Commodity-by-Commodity Guide to Impacts and Practices 14 (2004). Jason Clay, a senior vice president at the World Wildlife Fund, has taught at Harvard and Yale following studies in anthropology, economics, agriculture, and geography at Harvard, the London School of Economics, and Cornell, where he received his PhD. Over a dozen years have passed since Clay noted this figure of $1.3 trillion. As I explain in another context later in this book, I have not attempted to provide inflation-adjustment figures in order to convert such dated dollar amounts (in this case, $1.3 trillion) into current dollar amounts. See note 59 in Chapter 3, *infra*, which also cites a "CPI Inflation Calculator".

including the scratch plow (used in Mesopotamia), the crooked plow (used by the ancient Greeks), the mouldboard plow that was in use at least as far back as the Han Dynasty in China (which started in 206 BCE), and the much lighter Rotherham plow developed in England in the 1730s. Such plows would often have been pulled by draft animals, such as mules, oxen, or horses – or sometimes humans.

Then, in the nineteenth century, two developments suddenly changed agriculture in ways that made it drastically more extractive in character. The first was the creation and manufacture of the steel plow. This implement appeared in the 1830s and triggered what has been termed "the Great Plow-up" leading eventually to the Dust Bowl days.[14] The second development, which occurred in the following few decades, was the introduction of the gasoline-powered tractor. These two developments are so central to the character of modern agriculture that they warrant some more detailed coverage.

Don Worster, the widely-lauded emeritus distinguished professor of environmental history at Kansas University, has offered this description of the first of these developments – that is, the appearance of the steel plow in the 1830s – and how it figured in the rise of modern agriculture on the Great Plains of North America:

> Down to the 19th century the grasslands resisted the farmer's plow. For thousands of years plows had been made of wood, and even when they were given cast-iron edges, they could not penetrate the grasslands. They would break first. Their usable range was limited to exposed soils along the river bottoms or what had once been forest floor.
>
> Not until the nineteenth century did an American inventor named John Deere, followed by other inventors and manufacturers, begin making plows of steel, an alloy of iron and carbon forged with the heat of burning coal. The first steel plow appeared in 1837 near the prairie city of Chicago, Illinois. Such a formidable tool of nearly indestructible steel, pulled in the early days by large yoked teams of oxen or horses, could slice through the toughest sod and expose the deep, fertile soil to the air. Armed with the new plows, farmers could at last, after millennia of avoiding the grasslands, begin to venture out onto them and begin to conquer. They bought John Deere's invention eagerly and [in North America they] began ripping up the midcontinent prairie.
>
> We can trace the waves of conquest decade by decade across the [American] continent: beginning with Iowa and Minnesota in the 1840s

---

14 For a general reference to "the Great Plow-up" in an account of the Dust Bowl, its causes, and its consequences, on the website of the Texas State Historical Association, see www.tshaonline. org/handbook/online/articles/ydd01. See also Elton Robinson, *Abandoned Cotton Acres Recall the Great Plow-Up of 1933* (2011), available on the website of the Farm Press website, at http://deltafarmpress.com/blog/abandoned-cotton-acres-recall-great-plow-1933.

and [18]50s, then across eastern Kansas and Nebraska by the 1860s, then across the mid-latitude grasslands by the 1870s and 1880s, before drought put a stop to the advancing plows. Then in the first three decades of the twentieth century the great plow-up continued westward, all the way to the Rocky Mountains. The original sea of grass had given way to a sea of wheat and corn.[15]

This introduction and energetic use of the steel plow was accompanied and accelerated by a second development, which occurred in the following few decades: the introduction of the gasoline-powered tractor. This development can be seen as having two complementary components. The first component was the invention and improvement of the gasoline-powered internal combustion engine itself, which was the subject of various patents starting around the mid-1800s but which could be made commercially feasible following the discovery of oil by Colonel Edwin Drake just 69 feet below the surface of the ground in Titusville, Pennsylvania in 1859. The second component was the *installation* of such a gasoline-powered internal-combustion engine in the newly-developed tractor, a machine that could be used in lieu of draft animals to pull the plow.[16] The term "tractor" itself was reportedly drawn from the Latin term for "to pull" (*trahere*) and was first used in marketing efforts by Charles W. Hart and Charles H. Parr, who developed a two-cylinder gasoline engine to power the pulling machines for farm use and started selling them in Iowa around 1903.

Accordingly, both of these developments – the introduction of the steel plow and the introduction of the gasoline-powered tractor to pull that plow – made agriculture more *extractive* in character. The steel plow greatly accelerated the attack on the integrity of the soil, especially in the North American prairies that were known for breaking weaker plows. The gasoline-powered internal combustion engine – in farm tractors and of course in all manner of other machinery – began humanity's feverish rush, now about a century and a half old, to extract petroleum and other fossil fuels from the Earth to burn in such engines for power.[17]

---

15 Donald Worster, *The Grasslands in Time: From the Eocene to the Anthropocene* (Keynote address for conference on Comparing Grasslands in China and North America, Chinese Academy of Social Sciences, Beijing, China, September 2011), at 12–13.

16 The first such pulling machines, starting in the 1850s, were powered by steam, but the ones that prompted the most change in farming were powered by the internal combustion engine. Steam power also played an important role earlier, of course, in manufacturing the plows and transporting them by rail to the regions where they were used.

17 The fact that each of these two developments had its origin in North America reflects the mindset that, according to Fred Kirschenmann, prevailed among the Europeans coming to that continent after 1492.

[For those new arrivals,] this land was a catalog of resources waiting to be mined: [b]ison waiting to be harvested; [t]rees waiting to be logged; [g]rasslands waiting to be plowed and

Taking these two developments together, we could say that by the early 1900s, particularly in the USA, the form of "extraction" that had served as the centerpiece of the tradition going back several thousand years had been modified enough to justify using the term "enhanced extraction". I have used that term in preparing a "composite timeline" of certain key technological developments in the tradition of extractive agriculture, in Figure 1.1.[18]

The twentieth century brought the addition of a third form of extraction for agricultural purposes. Since just after the Second World War, great quantities of ammonia (NH_3) – a combination of hydrogen and nitrogen – have been used to create synthetic nitrogen fertilizers to boost crop yields.[19] Indeed, roughly four-fifths of all ammonia produced around the world is devoted to use as agricultural fertilizer. In overall terms, nitrogen fertilizer now constitutes the largest single energy input into industrial agriculture. Production of the ammonia requires a very high-energy source. In the USA, the source is mainly natural gas[20] – a hydrocarbon gas mixture consisting primarily of methane but also including other hydrocarbons (usually ethane) and some impurities. Such natural gas is of course another form of fossil carbon that is extracted from beneath the surface of the Earth in natural gas fields or in "associated" fields (that is, oilfields where natural gas

---

planted into neat rows of corn; [r]ich, black soil and its nutrients waiting to be used; [and] [a]quivers waiting to be tapped and piped into center-pivot irrigation systems to produce huge quantities of grain on marginal land. [In short, early Europeans in North America created a culture] of resource exploitation. Because they approached the land from this perspective, they engaged in the kind of agriculture that became the model for the industrial world.

See Frederick L. Kirschenmann, Cultivating an Ecological Conscience: Essays from a Farmer Philosopher 263 (2010).

18 Several elements of the "composite timeline" in Figure 1.1 (so labeled because it encompasses a variety of developments all tending toward modern extractive agriculture) are overly simplified for ease of presentation. For instance, the development and use of a few chemical or "industrial" biocides began earlier than the latter part of the twentieth century, as the short line at the bottom right corner of the diagram in Figure 1.1 would suggest. An illustration of this can be seen in the use of copper sulfate, dating back to the 1800s, first as a fungicide and then for other agricultural purposes. For an old account of some such uses, see generally George Fiske Johnson, *The Early History of Copper Fungicides*, 9 Agricultural History 67 (1935).

19 The ammonia plants that were used for this purpose were already in production during the Second World War in order to supply nitrogen for use in making explosives. For an interesting account, see the website for Wessels Living History Farm, at www.livinghistoryfarm.org/farminginthe40s/crops_04.html.

20 To appreciate the relationship between prices and supplies of natural gas, ammonia, and nitrogen fertilizer, see Wen-yuan Huang, *Impact of Rising Natural Gas Prices on U.S. Ammonia Supply*, Outlook, Aug. 2007, at 1 (from the USDA's Economic Research Service), www.ers.usda.gov/media/198815/wrs0702_1_.pdf (explaining that "natural gas is the main input used to produce ammonia, which, in turn, is the main input used to produce all nitrogen fertilizers"). The typical modern ammonia-producing plant first converts natural gas (mainly methane) – or, alternatively, liquefied petroleum gas (e.g., propane or butane) or sometimes petroleum naphtha – into gaseous hydrogen (a process referred to as "Steam Reforming"). Then the hydrogen is combined with nitrogen to produce ammonia via the Haber-Bosch process.

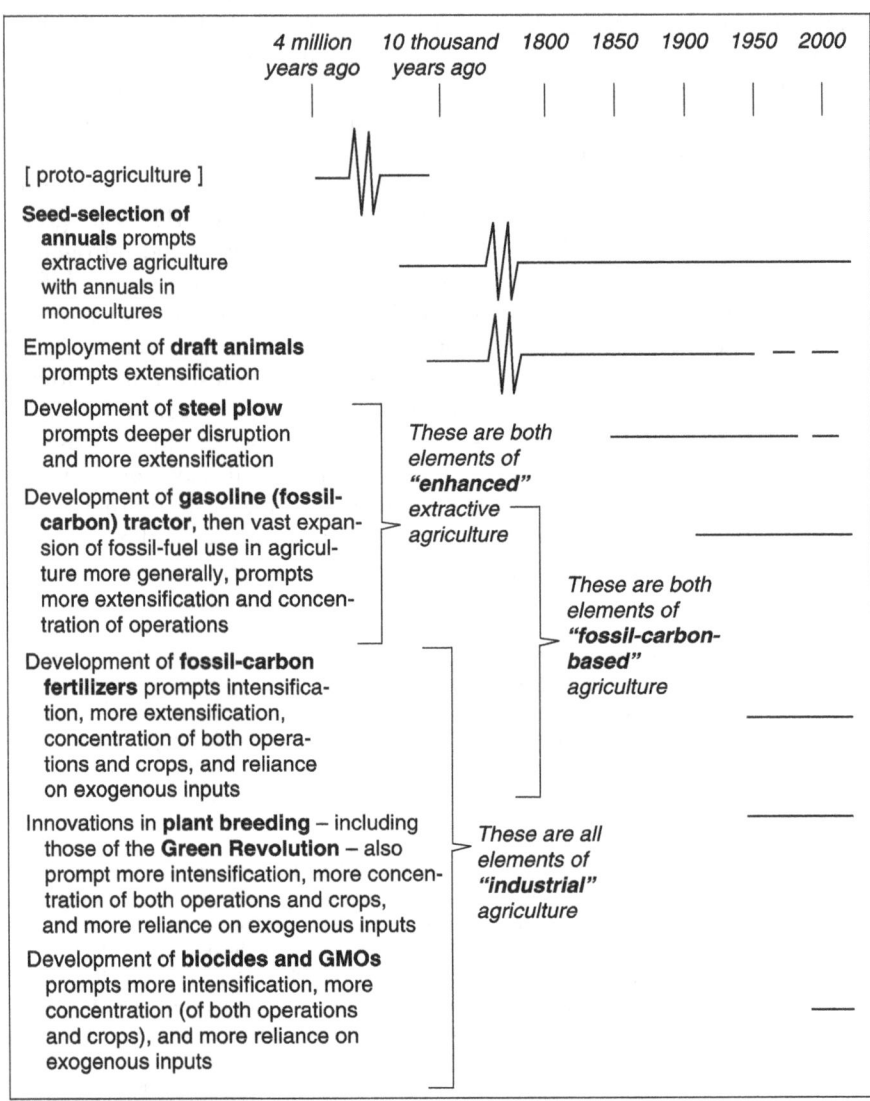

Figure 1.1 A composite timeline of selected key technological developments in the
history of extractive agriculture

is also captured). Hence, as modern agricultural production has come to rely
more and more on ammonia as a form of nitrogen fertilizer, the demand for
– and the extraction of – natural gas has risen accordingly.

By the 1950s, therefore, traditional "enhanced" extractive agriculture
had come to rely heavily on fossil carbon. The earlier development of the

gasoline-powered tractor (and other self-propelled farm implements that soon followed), combined with the rising use of synthetic nitrogen fertilizers, ushered in a new phase that we may refer to as "fossil-carbon-based" agriculture.

As suggested by the "composite timeline" in Figure 1.1, the most recent phase in the tradition of extractive agriculture features a cluster of additional developments that are technological in nature. They fall into two categories. The first category is what is commonly regarded as the Green Revolution, which (as I will explain in more detail later) provided for an enormous intensification of agricultural production – that is, an increase in yields that could be coaxed from plants because of innovative plant breeding and heavy use of newly-developed fossil-carbon-based fertilizers.[21]

The second development, coming just in the past few years, involves a pair of further technological innovations: (i) the emergence of genetically-modified organisms ("GMOs") for agricultural production and (ii) the creation of a much broader range of "biocides". (Both of these technological innovations will be described more in section II of Chapter 3.) The term "biocide" in this context encompasses herbicides (to kill plants), rodenticides (to kill field mice and other rodents), fungicides, insecticides, and other types of "pesticides". The very label "pesticide", and the less commonly used term "biocide", makes clear that all of the life forms being attacked by these poisons are regarded as constituting undesirable forms of *interference* with the field crops being grown in modern agriculture. Nearly all of those field crops in today's agriculture are annuals, and they are grown almost entirely in monocultures. These features make the crops – that is, the maize (corn), the soybeans, the wheat, and so forth – particularly susceptible to such "interference" from other life forms that would tend to compete with them. In practical terms, a single genotype could be a "sitting duck" to an aggressive disease or a swarm of insects looking for a feast, especially if that single genotype has been developed more with a goal of maximizing seed yield than with a goal of maximizing resistance to pests or pathogens. Hence the need for heavy use of such poisons.

In short, a broadly-encompassing label for the type of agriculture that prevails in the world today, especially in the USA and other highly developed economies with agricultural sectors, is *industrial fossil-carbon-based enhanced extractive agriculture*. Yes, it is quite a mouthful. It is:

- *Industrial* in the sense that it shares with other industries a production orientation, based on many exogenous inputs that come from off-farm

---

21 As noted in the last part of Chapter 3, the term "Green Revolution" is not used consistently. In a narrow connotation, it encompasses just the plant-breeding and associated advances emerging from the work of a cadre of researchers (most prominently Norman Borlaug) with financial support from the Rockefeller Foundation and the US Government. In a wider connotation, the Green Revolution encompasses also a range of other developments and innovations that capitalized on advances in fossil-carbon fertilizers and other exogenous inputs.

sources, such as artificial fertilizers and biocides – and because, more fundamentally, it reflects a mindset which assumes that all problems are solvable with human ingenuity alone as opposed to working cooperatively with a landscape's resources and within its limitations.[22]

- *Fossil-carbon-based* in the sense that it relies greatly on fossil carbon in several respects, especially (i) for fuel in farm machinery and (ii) for feedstocks used in the production of synthetic fertilizers and biocides.

---

22 In discussing such a mindset, Wes Jackson – whose work I introduce on page 29 – often refers to "the industrial mind". Although the phrase and its implications are not original with him (it can be traced at least as far back as the twentieth-century American geographer Carl Sauer of the University of California at Berkeley), it is perhaps Jackson who employs the idea of "the industrial mind" most tellingly in the context of agriculture. My colleague Caleb Hall offers these observations about what Wes Jackson means by "the industrial mind":

> Since the time Wes Jackson created the Land Institute and wrote *New Roots for Agriculture*, and probably before then, he has been spreading the environmental gospel of perennial agriculture and denouncing the "industrial mind." But what exactly is the industrial mind? Jackson has such a talent with words that the reader is able to gloss over that term and see how it connects to his message but without actually understanding what the term means. True, the reader understands what Jackson laments about modern agriculture: the deterioration of rural landscapes, the private benefit of agri-business at the expense of farm workers and communities, and the loss of wildness, etc. The reader can also surmise that the industrial mind has played a role, perhaps the predominant role, in creating our modern agricultural predicament . . . . But what exactly is Jackson referring to when he says "industrial mind"?
>
> At its core, we can say that (in Jackson's words) "the industrial mind assumes that our knowledge is adequate to live in the world." By this, Jackson means that the industrial mind assumes that all problems are solvable with human ingenuity alone as opposed to working cooperatively with a landscape's resources and within its limitations. It is this mindset that has fueled the past 250 years of industrial and agricultural development in particular. Beyond that, it is the terrestrial equivalent of dehumanization – that is, seeing all land as substrate rather than soil, as having commodity value rather than character value. The industrial mind is the driving force of "Homo the Homogenizer", the force that spreads the same soy, the same corn, the same malls, and the same trash across the globe.
>
> The industrial mind is not just the idea that all land and resources can be commoditized and exploited, though that is probably the largest part of it. More fundamentally, it includes the foolish belief that we can know enough to solve problems with human ingenuity alone. As long as we presume our knowledge and technology to be adequate we are, according to Jackson (drawing from the work of Wendell Berry), still colonists – exploiting this landscape but never truly discovering it or understanding the "genius of the place".

Caleb Hall, *The Industrial Mind* (2014) (on file with author), drawing from the following sources: *Land Institute Founder: 'Eco' State of Mind Needed*, THE NEWS-GAZETTE, Apr. 5, 2013, www. news-gazette.com/news/local/2013-04-05/land-institute-founder-eco-state-mind-needed.html; Wes Jackson, President of the Land Institute, "Khrushchev in Iowa Speech", address at Coon Rapids, Iowa, Aug. 29, 2009, https://landinstitute.org/library-post/khrushchev-in-iowa-speech-aug-2009; Wes Jackson, Address at the 2004 Prairie Festival, 81 *The Land Report* 14–16 (2005); Emily Hazzard, *A Conversation with Wes Jackson, President of the Land Institute*, THE ATLANTIC, Mar. 23, 2011; Wes Jackson, Address at the Boulevard Brewery (Feb. 12, 2013); Wes Jackson, Interview by Jane Gates (1990), www.youtube.com/watch?v=KEWPeeeWRPs&featu re=youtube.

- *Extractive* because the growing of annuals, as distinct from perennials, creates a constant drain on soil quality (for reasons examined further on pages 89–91) and often a drain as well on soil quantity, through erosion. And I have explained how agriculture's extractive character was greatly *enhanced* with certain technological developments occurring between about 1840 and 1920, especially the emergence of the steel plow and the gasoline-powered tractor.

Instead of using the complicated term "industrial fossil-carbon-based enhanced extractive industrial agriculture", I have opted in most of this book to use the term "modern extractive agriculture". In this usage, the term "modern" serves as a shorthand term to encompass all of the novelties and advances of the past couple of centuries.

It is worth pausing briefly to consider the actual "work-product" of those novelties and advances. That is, what is the overall shape of agriculture today, and what have the developments referred to above – especially those of the last hundred years or so – accomplished in terms of humans' ability to create food?

One way to answer that question is to recall the image I offered near the beginning of this chapter, in which I was riding with Will Bier in a massive corn harvester on my family farm. The crop he was harvesting was a mono-culture – that is, a specific variety of corn covering the entire field. It is an annual crop that must be replanted every year and that requires massive use of highly-dense carbon for fuels and fertilizers. It stands in an area that constituted grasslands for untold numbers of centuries before the steel plow began destroy-ing them in the mid-1800s. Those native grasslands were dominated by peren-nial grasses and forbs, not annuals, and they consisted of a multitude of species, not just a single one of the "big five" crops that now dominate American agriculture: corn, wheat, soybeans, cotton, and rice. For Will Bier, and for my family, the novelties and advances I have summarized above have yielded a form of agriculture that is remarkably productive but remarkably foreign to the type of ecosystem that preceded it in the fields that now comprise our farm.

Another way to answer the question that I posed earlier – "what is the overall shape of agriculture today?" – is to examine the accounts given in Box 1.1 and Box 1.2. Those accounts, along with three maps that follow them (Figures 1.2, 1.3, and 1.4), draw from a variety of public sources to provide a "primer" on key modern agricultural food crops on the global level. The account in Box 1.1 offers straightforward definitions and descriptions of the terms "cereal", "legume", and "pulse", and then Box 1.2 provides some highlights of the history, physiology, breeding, production, and uses of four main crops used for providing food for humans, either by direct human consumption or as use in feeding livestock which are then used for human consumption. These four crops are corn (maize),[23] wheat, rice

---

23 The term "corn" outside North America, Australia, and New Zealand means any cereal crop, its meaning understood to vary geographically to refer to the local staple. In the United States, (Footnote 23 text continued on p. 26)

Box 1.1 Cereals, legumes, and pulses

Maize, wheat, and rice are all cereal grains – hence the inclusion in this box of information on "cereals". Soybean is a legume, with nitrogen-fixing capacity that is important to soil health (and therefore in developing polyculture operations) – hence the inclusion here of the information on "legumes". Including information here on "pulses" reflects the fact that pulses also have nitrogen-fixing capacity and that they are important in human consumption both historically and currently.

I. Cereals

A cereal is a grass cultivated for the edible components of its grain, which is composed of the endosperm, germ, and bran. (The word *cereal* derives from *Ceres*, the name of the Roman goddess of harvest and agriculture.) Cereal grains are grown in greater quantities and provide more food energy worldwide than any other type of crop; they are therefore referred to as "staple crops".

In their natural form (as in *whole grain*), cereal grains are a rich source of vitamins, minerals, carbohydrates, fats, oils, and protein. However, when refined by the removal of the bran and germ, the remaining endosperm is mostly carbohydrate and lacks the majority of the other nutrients. In some economically less developed countries, grain in the form of rice, wheat, millet, or maize constitutes a majority of daily sustenance. In developed nations, cereal consumption is moderate and varied but still substantial.

The first cereal grains were domesticated about 12,000 years ago by ancient farming communities in the Fertile Crescent region. Emmer wheat, einkorn wheat, and barley were three of the so-called "Neolithic founder crops" in the development of agriculture.

The following table shows the global annual production of cereals in 1961, 2010, 2011, and 2012 ranked by 2012 production, as recorded by FAO.

Worldwide production (millions of metric tons)

Grain	2012	2011	2010	1961
Maize	872	888	851	205

A staple food of people in the Americas, Africa, and of livestock worldwide; often called "corn" in North America, Australia, and New Zealand. A large portion of maize crops are grown for purposes other than human consumption.

Grain	2012	2011	2010	1961
Rice	720	725	703	285

The primary cereal of tropical and some temperate regions. Staple food in many parts of Latin America and some other Portuguese-descended cultures, in parts of Africa, and in most of South Asia and the Far East.

Wheat	671	699	650	222

The primary cereal of temperate regions. It has a worldwide consumption but it is a staple food of North America, Europe, Australia, New Zealand, most of the Southern Cone and much of the Greater Middle East. Wheat gluten-based meat substitutes are important in the Far East (albeit less than tofu).

Barley	133	133	124	72

Grown for malting and livestock on land too poor or too cold for wheat.

Sorghum	57	58	60	41

Important staple food in Asia and Africa and popular worldwide for livestock.

Millet	30	27	33	26

A group of similar but distinct cereals that form an important staple food in Asia and Africa.

Oats	21	22	20	50

Formerly the staple food of Scotland and popular worldwide as a winter breakfast food and livestock feed. Processed oatmeal in Latin America is often consumed year-round.

Rye	15	13	12	12

Important in cold climates.

Triticale	14	13	14	35

Hybrid of wheat and rye, grown similarly to rye.

In 2013, global cereal production reached a record 2,521 million tonnes. Maize, wheat, and rice together accounted for 89 percent of all cereal production worldwide in 2012, and 43 percent of all food calories in 2009, while the production of oats and triticale have drastically fallen from their 1960s levels. Other grains, such as buckwheat and quinoa, are important in some places.

While each individual species has its own peculiarities, the cultivation of all cereal crops currently in wide production is similar. Most are annual plants; consequently one planting yields one harvest. Wheat, rye, triticale, oats, and barley are the "cool-season" cereals. These are hardy plants that grow well in moderate weather and cease to grow in hot weather (approximately 30°C, but this varies by species and variety). The "warm-season" cereals are tender and prefer hot weather. Barley and rye are the hardiest cereals, able to overwinter in the subarctic and Siberia. Many cool-season cereals are grown in the tropics. However, some are only grown in cooler highlands, where it may be possible to grow multiple crops in a year.

Some grains are deficient in the essential amino acid lysine. That is why many vegetarian cultures, in order to get a balanced diet, combine their diet of grains with legumes. Many legumes, on the other hand, are deficient in the essential amino acid methionine, which grains contain. Thus, a combination of legumes with grains forms a well-balanced diet for vegetarians. Common examples of such combinations are dal (lentils) with rice by South Indians and Bengalis, dal with wheat in Pakistan and North India, and beans with corn tortillas, tofu with rice, and peanut butter with wheat bread (as sandwiches) in several other cultures, including Americans. The amount of crude protein found in grain is measured as the grain crude protein concentration.

II. Legumes

Well-known legumes include alfalfa, clover, peas, beans, lentils, lupins, mesquite, carob, soybeans, peanuts, tamarind, and the woody climbing vine wisteria. For agricultural purposes, legumes are grown primarily for their food grain seed, for livestock forage and silage, and as soil-enhancing "green manure" (see below).

Many legumes contain symbiotic bacteria called *Rhizobia* within root nodules of their root systems. These bacteria have the special ability of fixing nitrogen from atmospheric, molecular nitrogen (N_2) into ammonia (NH_3). Ammonia is then converted to another form, ammonium (NH_4^+), usable by (some) plants. This arrangement means that the root nodules are sources of nitrogen for legumes, making them relatively rich in plant proteins. All proteins contain nitrogenous amino acids. Nitrogen is therefore a necessary ingredient in the production of proteins. Hence, legumes are among the best sources of plant protein.

In many traditional and organic farming practices, crop rotation involving legumes is common. By alternating between legumes and non-legumes, sometimes planting non-legumes two times in a row and then a legume, the field usually receives a sufficient amount of nitrogenous compounds to produce a good result, even when the crop is non-leguminous. Legumes are sometimes referred to as "green manure".

Farmed legumes can belong to many agricultural classes, including those for use as forage, grain, blooms, and otherwise. The first of these, forage legumes, are of two broad types. Some, such as alfalfa, clover, and vetch, are sown in pasture and grazed by livestock. Other forage legumes, such as *Leucaena* or *Albizia,* are woody shrub or tree species that are either broken down by livestock or regularly cut by humans to provide livestock feed.

Grain legumes are cultivated for their seeds, and are also called pulses. The seeds are used for human and animal consumption or for the production of oils for industrial uses. Grain legumes include beans – soybeans in particular – and lentils, lupins, peas, and peanuts.

III. Pulses

A pulse (Latin "puls", from Ancient Greek πόλτος, *poltos,*), sometimes called a "grain legume", is an annual leguminous plant yielding from one to twelve seeds of variable size, shape, and color within a pod. Pulses are used for food for humans and other animals. Like many leguminous crops, pulses play a key role in crop rotation due to their ability to fix nitrogen. To support the awareness on this matter, the United Nations declared 2016 the UN International Year of Pulses.

India is the world's largest producer and the largest consumer of pulses. Pakistan, Canada, Burma, Australia and the United States, in that order, are significant exporters and are India's most significant suppliers. Canada now accounts for approximately 35 percent of global pulse trade each year.

Box 1.2 Maize, wheat, rice, and soybeans

I. Maize

Maize is a large cereal grain plant domesticated by indigenous peoples in Mesoamerica in prehistoric times. The leafy stalk produces ears which contain the grain (kernels). Maize kernels are often used in cooking as a starch. The word *maize* derives from the Spanish form of the indigenous Taíno word for the plant, *maiz*. It is known by other names around the world – including "corn" in some English-speaking countries.

Maize is the domesticated variant of teosinte. The two plants have dissimilar appearance, maize having a single tall stalk with multiple leaves and teosinte being a short, bushy plant. The difference between the two is largely controlled by differences in just two genes.

Several theories had been proposed about the specific origin of maize in Mesoamerica, but uncertainty remains. Most historians believe maize was domesticated in the Tehuacan Valley of Mexico. Beginning about 2500 BCE, the crop spread through much of the Americas. After European contact with the Americas in the late fifteenth and early sixteenth centuries, explorers and traders carried maize back to Europe and introduced it to other countries. Maize spread to the rest of the world because of its ability to grow in diverse climates. Sugar-rich varieties called sweet corn are usually grown for human consumption as kernels, while field corn varieties are used for animal feed, various corn-based human food uses (including grinding into cornmeal or masa, pressing into corn oil, and fermentation and distillation into alcoholic beverages like bourbon whiskey), and as chemical feedstocks.

Maize is the most widely grown grain crop throughout the Americas, with 332 million metric tons grown annually in the United States alone. Approximately 40 percent of the crop – 130 million tons – is used for corn ethanol.

Maize is widely cultivated throughout the world, and a greater weight of maize is produced each year than any other grain. As is reflected in the shading in Figure 1.2, the United States produces 40 percent of the world's harvest; other top producing countries include China, Brazil, Mexico, Indonesia, India, France and Argentina.

Maize is susceptible to attack by many insects, including army-worms, earwigs, leaf aphid, silkfly, and weevil – but especially to European corn borer and corn rootworms. This problem has led to the development of transgenics expressing the *Bacillus thuringiensis* toxin. "Bt maize" is widely grown in the USA and has been approved for release in Europe. Maize is also susceptible to attack by various types of diseases, including rust, corn smut, leaf blight, and stalk rot.

Maize and cornmeal (ground dried maize) constitute a staple food in many regions of the world. For instance, virtually every dish in Mexican cuisine uses maize, and since its introduction into Africa by the Portuguese in the sixteenth century, maize has become Africa's most important staple food crop. Maize meal is made into a thick porridge in many cultures: from the polenta of Italy, the *angu* of Brazil, the *mămăligă* of Romania, to cornmeal mush in the US (and hominy grits in the South).

Maize is a major source of starch. Cornstarch (maize flour) is a major ingredient in home cooking and in many industrialized food products. Maize is also a major source of cooking oil (corn oil) and of maize gluten. Maize starch can be hydrolyzed and enzymatically treated to produce syrups, particularly high-fructose corn syrup, a sweetener, and also fermented and distilled to produce grain alcohol. Maize is sometimes used as the starch source for beer.

Within the United States, the usage of maize for human consumption constitutes about 1/40th of the amount grown in the country. In the United States and Canada, maize is mostly grown as feed for livestock, as forage, as silage (made by fermentation of chopped green cornstalks), as an ingredient of some commercial animal food products, such as dog food, or as biofuel. "Feed maize" is being used increasingly for heating; specialized corn stoves (similar to wood stoves) are available and use either feed maize or wood pellets to generate heat. Maize cobs are also used as a biomass fuel source. Moreover, as noted above, maize is increasingly used as a feedstock for the production of ethanol fuel. Ethanol is mixed with gasoline to decrease the amount of pollutants emitted when used to fuel motor vehicles. Maize also has other uses, including as fodder, medicines, and a vast range of materials made from its starch. These include plastics, fabrics, adhesives, and many other products.

II. Wheat

Wheat (*Triticum* spp.) is a cereal grain, originally from the Levant region of the Near East but now cultivated worldwide. In fact, this grain is grown on more land area than any other commercial food. Moreover, world trade in wheat is greater than for all other crops combined. Globally, wheat is the leading source of vegetable protein in human food, having a higher protein content than other major cereals. In terms of total production tonnages used for food, it is currently second to rice as the main human food crop and ahead of maize, after allowing for maize's more extensive use in animal feeds.

Wheat was a key factor enabling the emergence of city-based societies at the start of civilization because it was one of the first crops that could be easily cultivated on a large scale, and had the additional advantage of yielding a harvest that provides long-term storage of food. Wheat contributed to the emergence of city-states in the Fertile Crescent, including the Babylonian and Assyrian empires. Wheat grain is a staple food used to make flour for leavened, flat and steamed breads, biscuits, cookies, cakes, breakfast cereal, pasta, noodles, couscous, and for fermentation to make beer, other alcoholic beverages, or biofuel.

Wheat is planted to a limited extent as a forage crop for livestock, although the straw cannot be used as feed. Its straw can be used as a construction material for roofing thatch. The whole grain can be milled to leave just the endosperm for white flour. The by-products of this are bran and germ. The whole grain is a concentrated source of vitamins, minerals, and protein, while the refined grain is mostly starch.

Wheat is the world's most favored staple food. It is a major diet component because of the wheat plant's agronomic adaptability, with the ability to grow from near arctic regions to the Equator, from sea level to the plains of Tibet, approximately 4,000 meters above sea level. In addition to agronomic adaptability, wheat offers ease of grain storage and ease of converting the grain into flour for making foods.

Wheat is the most important source of carbohydrate in a majority of countries. Although wheat protein is easily digested by nearly 99 percent of the human population, as is its starch, several screening studies suggest that approximately 1 percent of some populations may have coeliac disease, a condition that is caused by an adverse immune system reaction to gliadin, a gluten protein found in wheat (and similar grains of the tribe *Triticeae* which includes other species such as barley and rye). For those persons with this reaction, the only known effective treatment is a gluten-free diet.

In the twentieth century, global wheat output expanded about five-fold, but until about 1955 most of this reflected increases in wheat crop area. After 1955, however, there was a ten-fold increase in the rate of wheat yield improvement per year, and this became the major factor allowing global wheat production to increase. Thus technological innovation and scientific crop management with synthetic nitrogen fertilizer, irrigation, and wheat breeding were the main drivers of wheat output growth in the second half of the twentieth century.

Better seed storage and germination ability (and hence a smaller requirement to retain harvested crop for next year's seed) is another twentieth-century technological innovation. In Medieval England, farmers saved one-quarter of their wheat harvest as seed for the next crop, leaving only three-quarters for food and feed consumption. By 1999, the global average seed use of wheat was about 6 percent of output.

Overall world wheat production is shown in Figure 1.3. As partially reflected there, the top wheat producers in recent years have been China, India, the USA, France, Russia, Australia, Canada, Pakistan, Germany, and Turkey.

There are many wheat diseases, mainly caused by fungi, bacteria, and viruses. Plant breeding to develop new disease-resistant varieties, along with sound crop management practices, are important for preventing disease. Fungicides, used to prevent the significant crop losses from fungal disease, can be a significant variable cost in wheat production. Estimates of the amount of wheat production lost owing to plant diseases vary between 10 percent and 25 percent in Missouri. A wide range of organisms infect wheat, of which the most important are viruses and fungi.

In addition, wheat is susceptible to many types of pests. Wheat is used as a food plant, for instance, by the larvae of some *Lepidoptera* (butterfly and moth) species. Early in the growing season, many species of birds and rodents feed upon wheat crops. These animals can cause significant damage to a crop by digging up and eating newly planted seeds or young plants. They can also damage the crop late in the season by eating the grain from the mature spike.

III. Rice

Rice is a major food staple and a mainstay for many rural populations. It is mainly cultivated by small farmers in holdings of less than one hectare. Rice is vital for the nutrition of much of the population in Asia, as well as in Latin America and the Caribbean and in Africa; it is central to the food security of over half the world population. Economically less developed countries account for 95 percent of the total production. As reflected by the shading in Figure 1.4, China and India alone are responsible for nearly half of the world output; heavy concentrations of rice production also appear in Southeast Asia. World production of rice has risen steadily from about 285 million tonnes in 1961 to about 720 million tonnes in 2012.

Rice cultivation on wetland rice fields is thought to be responsible for 1.5 percent of the anthropogenic methane emissions – and, as noted later in this book, methane is over 30 times more potent a greenhouse gas than carbon dioxide. Rice requires slightly more water to produce than other grains, and rice production uses almost a third of Earth's fresh water.

The highest-yielding varieties of rice were developed during the Green Revolution to increase global food production. This project enabled labor markets in Asia to shift away from agriculture and into industrial sectors.

Rice kernels do not contain vitamin A, so people who obtain most of their calories from rice are at risk of vitamin A deficiency. German and Swiss researchers have genetically engineered rice to produce beta-carotene, the precursor to vitamin A, in the rice kernel. The beta-carotene turns the processed (white) rice a gold color; hence the name "golden rice". The beta-carotene is converted to vitamin A in humans who consume the rice.

IV. Soybeans

The soybean (US) or soya bean (UK) (*Glycine max*) is a species of legume native to East Asia, widely grown for its edible bean which has numerous uses.

Although the origin of soybean cultivation remains scientifically debated, recent research indicates that seeding of wild forms started early (before 5000 BCE) in multiple locations through China, Korea and Japan. Soybeans were first introduced to North America in 1765, by Samuel Bowen, a former East India Company sailor who had visited China, but for the next 155 years the crop was grown primarily for forage.

William Morse is considered the "father" of modern soybean agriculture in America. He and Charles Piper took what was an unknown Oriental peasant crop in 1910 and transformed it into a "golden bean" for America, becoming one of America's largest farm crops and its most nutritious. During the Great Depression, the drought-stricken (Dust Bowl) regions of the United States were able to use soy to regenerate their soil because of its nitrogen-fixing properties as described in Box 1.1. Those nitrogen-fixing properties make soybeans rich in plant proteins. Indeed, soybeans can produce at least twice as much protein per acre as any other major vegetable or grain crop besides hemp, five to ten times more protein per acre than land set aside for grazing animals to make milk, and up to 15 times more protein per acre than land set aside for meat production.

Most of the worldwide production of soybeans occurs in five countries: Brazil, the USA, Argentina, China, and India. These countries account for about 90 percent of total worldwide production.

and soybeans. My reason for focusing on those four crops is that they are the most heavily-produced crops in the modern world . . . and in fact the first three of these (corn, wheat, and rice) account for 89 percent of all cereal produced worldwide in 2012 and 43 percent of all food calories in 2009. Modern humans eat grains (and legumes) more than they eat anything else. Modern agriculture supplies these products in vast quantities – over 800 million metric tons per year now of corn, over 700 million of rice, and over 600 million of wheat.

Where in the world are the four major crops produced? Figure 1.2 answers that question for corn (maize), and Figures 1.3 and 1.4 answer that question for rice and wheat. [24]

Canada, Australia, and New Zealand, "corn" primarily means maize; this usage started as a shortening of "Indian corn". "Indian corn" primarily means maize (the staple grain of indigenous Americans), but can refer more specifically to multicolored "flint corn" used for decoration. In Southern Africa, maize is commonly called *mielie* (Afrikaans) or *mealie* (English).

24 For a map of world soybean production for use in livestock feed, see Henning Steinfeld, Pierre Gerber, Tom Wassenaar, Vincent Castel, Mauricio Rosales, and Cees de Haan, LIVESTOCK'S

It is essential to bear in mind that the form of agriculture described in the foregoing account remains fundamentally extractive. As explained above, this makes it "traditional" in character. Despite all the changes that have come in the past two centuries, agriculture has not undergone fundamental change from its earliest days, in which annual crops came to be planted and harvested in monocultures. This central feature of today's agriculture, like the agriculture developed over hundreds of centuries going right back to 10,000 years ago, makes it extractive. In the early twenty-first century, though, agriculture is extractive in more ways, and more intensive ways,

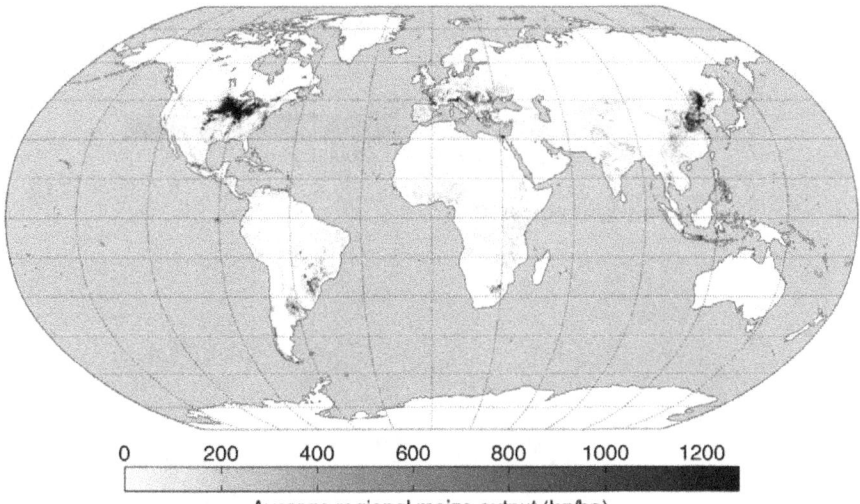

Average regional maize output (kg/ha)

Figure 1.2 Worldwide corn (maize) production[25]

LONG SHADOW: ENVIRONMENTAL ISSUES AND OPTIONS 333 (2006), ftp://ftp.fao.org/docrep/fao/010/a0701e/a0701e09.pdf. That book was prepared under the auspices of the Livestock, Environment, and Development ("LEAD") Initiative, which is supported by the World Bank, the EU, and various other official agencies. The series of maps from which the wheat, maize, and rice production maps (Figure 1.1, Figure 1.2, and Figure 1.3) are drawn does not include a map for soybean production.

25 The world maps of maize, wheat, and rice production in Figures 1.2, 1.3, and 1.4 draw from information compiled by the University of Minnesota Institute on the Environment, with data from C. Monfreda, N. Ramankutty, and J. A. Foley, *Farming the Planet* (geographic distribution of crop areas, yields, physiological types, and net primary production in the year 2000), GLOBAL BIOGEOCHEMICAL CYCLES 22: GB1022 (2008) (Andrew MT user/author of image; image courtesy of Creative Commons Attribution – Share Alike 3.0 unported license), http://creativecommons.org/licenses/by-sa/3.0/deed.en., and are available at: http://commons.wikimedia.org/wiki/File:MaizeYield.png, http://commons.wikimedia.org/wiki/File:WheatYield.png, and http://commons.wikimedia.org/wiki/File:RiceYield.png, respectively.

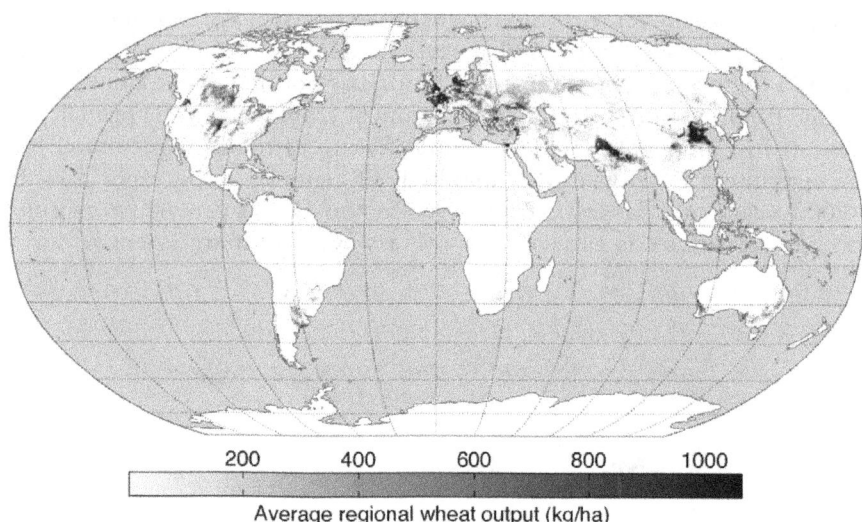

Figure 1.3 Worldwide wheat production

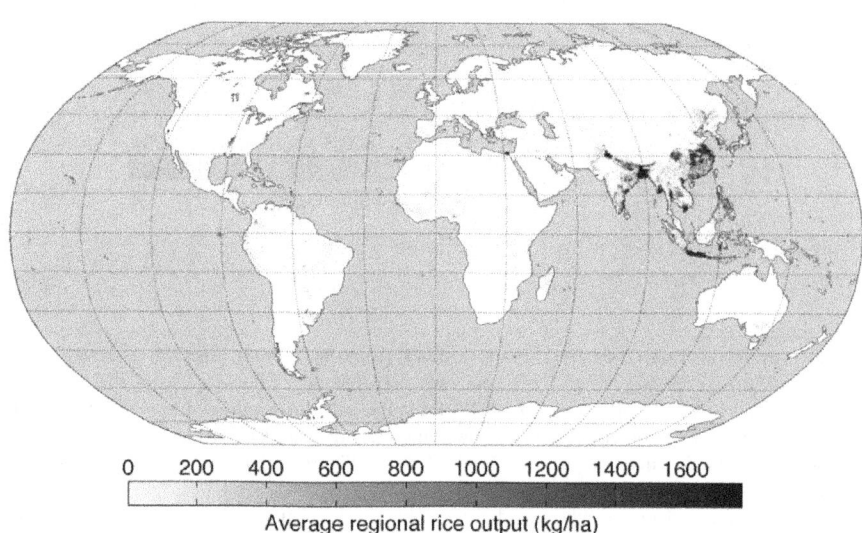

Figure 1.4 Worldwide rice production

than ever before. It relies to a dramatically increased degree on drawing down the planet's "fixed capital" both (i) by extracting nutrient content from the soil[26] – an aspect of our planet that I will describe later in this book as the "thin skin of life stretched over a rock" – and (ii) by extracting fossil carbon deposits from beneath the Earth's surface. Those forms of unsustainable extraction have made it possible to produce the massive yields noted above.

IIC. An even longer view of ecospheric development

None of the foregoing account can be adequately grasped, I believe, without an understanding of deep history – particularly an understanding of how life developed on the Earth over the course of geologic time and of humanity's place in that pageant. Let us turn briefly to that.

About 35 years ago, Wes Jackson of The Land Institute provided a historical perspective that focuses on the central importance of soil. The first chapter in his 1980 book *New Roots for Agriculture* is titled "The Earth in Review: The Rise, Role, and Fall of Soil". Jackson, a 1992 winner of a MacArthur "Genius" Award and a 2000 recipient of the Right Livelihood Award (the so-called "alternative Nobel Prize") portrays soil in that book not with the image I hinted at above – as a "thin skin of life stretched over a rock" – but slightly differently, as "an enveloping membrane or film, a *placenta*".[27] Jackson invites us to see (i) how that placenta supported the development of life over the course of the past 750 million years[28] and (ii) how in just the past few thousand years that placenta has come under attack by the (self-proclaimed) highest life form – *Homo sapiens*. Jackson's technique in telling this story is to compress all of the past 750 million years into a single year.

We start on January 1. By the fifteenth of March we can see several marine invertebrates and we think we can even see lichens on land.

---

26 Indeed, as discussed further on page 119, today's agriculture goes beyond extracting nutrient content from the soil; it largely displaces that nutrient content with exogenous inputs, leaving the soil to serve mainly as a medium by which those exogenous inputs (synthetic fertilizers of one sort and another) can be delivered to the production plants. It is not a huge exaggeration to say that the principal function of the soil in some areas is merely to hold the plants upright while they absorb those exogenous inputs.

27 See Wes Jackson, NEW ROOTS FOR AGRICULTURE 6 (1980, 1985 reprint) [hereinafter Jackson, *New Roots*]. For biographical information about Jackson, who has degrees in biology, botany, and genetics, see the pertinent page on the website of The Land Institute, https://landinstitute.org/.

28 The significance of 750 million years is that, as Jackson explains, that period represents "one-sixth of the total age of the earth". *Id.* at 5. The Earth, he says, "spent five-sixths of its time getting set for the explosive emergence of higher life. Some twenty-five of the major phyla around us today appeared then." *Id.*

Some time after mid-June there are scorpions crawling about and these newcomers are joined by the first bog plants later in the month.

The lung fishes appear in early July. By late August early reptiles inhabit a landscape dominated by swamp forests, and as we approach September we can see the cone-bearing plants become forest trees . . .

Sometime in late October we get our first glimpse of flowering plants. A month later it has become obvious that the dinosaurs are headed for extinction. By December 11 some insignificant little mammals with a larger brain-to-mass ratio than the reptiles have become conspicuous, and by a short week later they are the dominant animal group. The mammals have made it. We are all fascinated as we watch the Miocene uplift that creates a rain shadow east of the Rockies [in what is now the American West], which in turn gives rise to the great North American grasslands. A few days before Christmas we see extensive grasslands in various parts of the planet.

Creatures best described as ape-men appear right after Christmas, and with about thirty hours left in the year, we see a creature which is decidedly human like, even though it shows little promise at first . . . [Then,] with less than three hours of the year's last day (or about 200,000 real years), a creature with a brain almost as large as our own is eking out a livelihood in ecosystems not much different from what we find in many parts of the few wild places left today.

An important system was developing literally under the feet of these diverse life forms. The early dust of the earth was mostly cemented together. It gradually became pulverized by the action of wind and water, plant roots and gravity. The bodies of dead plants and animals were added to this powder. A peculiar type of evolution was under way. This entity teemed with small organisms which secreted chemicals into the powder. Small life forms ingested and egested it, buffered it and burrowed in it. It grew in thickness and began to cover a large area with what we might call "ecological capital." The capital of soil creates "interest" in the form of more soil. This interest then becomes reinvested. Water and wind still carried tons of this capital to the sea to become sedimentary layers, as it always had, but the life forms seemed almost purposefully devoted to retarding this work of gravity . . .

A book written in 1905 by Harvard professor Nathaniel Southgate Shaler entitled *Man and Earth* described the soil and water system as an enveloping membrane or film, a *placenta*, through which the Earth mother sustains life. All life, including humans, Shaler suggested, draws life from the sun, clouds, air and earth through this living film. If the placenta is not kept healthy or intact, life above suffers. If healthy, it is a rich throbbing support system. His message was clear enough: protect the placenta and you protect all Nature's children.

Placenta may not be the best word, for once a birth is complete the placenta is disposed of. And yet Mother Earth is always pregnant with new life and therefore an intact placenta is necessary. Perhaps a better word is matrix. To the biologist a matrix is something within which something else originates or develops. In archaic Latin, *matrix* means a uterus or womb; the word is derived from *mater* or mother . . .

In the early morning hours of December 31, changes took place on the surface of the earth. Later in the day the human population would explode. But before that, the first glaciers came and the placenta was gouged without mercy. The rubble in their wake was altogether unbecoming, but the placenta persevered; in fact, its speed of growth was increased. One glacier would do to a rock in a year what nosing roots would have done to it in a thousand.[29] The adversity of chilling, grinding glaciers had created a richer life-support system . . .

And as for *Homo sapiens*? The human species, ten minutes before year's end, was on all major land masses except Antarctica. It was in the next five minutes – from 15,000 to 8,000 years ago – that something critical happened. Gradually, an invisible claw began tearing at the placenta. It wasn't dramatically ruptured as it had been by the ice; there was just a little scratch which failed to heal in the Middle East, and shortly another like it appeared in middle America. The larger the gash, the larger the concentration of people and their handiwork around it. The placenta itself was being ripped away to build civilization. Within three of those last five minutes, the face of the earth was changed. In some places scarcely anything would grow. Scabs – sterile areas or deserts – increased in size owing to human-directed activity. In the last fifteen seconds of the year, the continent of North America was discovered by the Europeans. The great wildernesses of North America disappeared, and the placenta wasted away faster than it had in any other area of the world.

Nearly half of it disappeared in the year's last eight seconds. In the final three seconds, a new stream of oil began to flow throughout [North America], and out of it, fossil fuel that had been forming for eight months of our telescoped year, was discovered and was about to be used up in six seconds.

It was now being used not only for transportation, but also as feedstocks for chemical fertilizer, in pest control, and in energy for traction in the fields. Clearly a very new thing was happening on earth. Production of living plants was shifting from total dependence on soil to an increasing dependence on fossil fuel. The new reality was clear – agriculture in

---

29 Ruddiman uses a similar image – of the glaciers gouging the surface of the Earth – in explaining how one particular area of the world was affected: "The most productive farms in the American Midwest can thank the ice sheets for their topsoil. Ice repeatedly gouged bedrock and scraped older soils in north-central Canada and pushed the eroded debris south, where streams of glacial melt water carried it into river valleys, and winds blew it across the western prairies." Ruddiman, *supra* note 8, at 193.

America was shrinking the placenta, but the decline was obscured by heavy doses of petroleum-based chemical agriculture.[30]

Wes Jackson's account – compressing 750 million years into a single year in order to illustrate how quickly and recently the face of the Earth has been changed by human activity, and particularly by agriculture and most immediately by human use of fossil carbon – is remarkable in the narrative form he offers. In Figure 1.5, I have provided a diagrammatic presentation of the same narrative. My principal aim in doing so is to illustrate in a more visual form the vast magnitude of geologic time compared with both (i) the short time that human-like creatures (or indeed any mammals at all) have existed on the Earth, and (ii) the extremely brief period that has elapsed since agriculture began. Particularly arresting, in my view, is the fact – made easily visible in the last entry at the bottom of Figure 1.5 – that the extensive below-ground deposits of fossil oil (as distinct from, say, whale oil) that had been forming over the course of about eight months of the compressed "year" became the subject of discovery and exploitation only in the last three seconds of that compressed "year".

After offering his historical perspective on how human activity has changed the Earth – particularly through agriculture – in relatively recent times, Wes Jackson ends his account by shifting from description to prescription. He draws the following lesson regarding the relationship between human beings and the "placenta" of soil that has developed on the Earth during much of the "compressed year" of the Earth's history over the past 750 million (actual) years:

> The intensity of the entire agricultural operation can thus be seen as a frantic last attempt to keep alive a rapidly wasting cancer patient. Unless the health of the placenta is restored, a last convulsion will follow, throughout the countryside and around the world.[31]

Jackson's sweeping account of the "rise, role, and fall of soil" – and of agriculture as "an invisible claw" that has been "tearing at the placenta" that the soil constitutes – offers a useful introductory description to today's agricultural system. It is that system which I describe more fully, and which I criticize rather harshly, in Chapters 2 and 3. My examination there will conclude that the current system of modern extractive agriculture has failed, on grounds that it is economically unsustainable, it is environmentally destructive, it poses unacceptable risks to human health, and it runs historically and socially "against the grain" of human development.

---

30 See Jackson, *New Roots, supra* note 27, at 5–8.
31 *Id.* at 8.

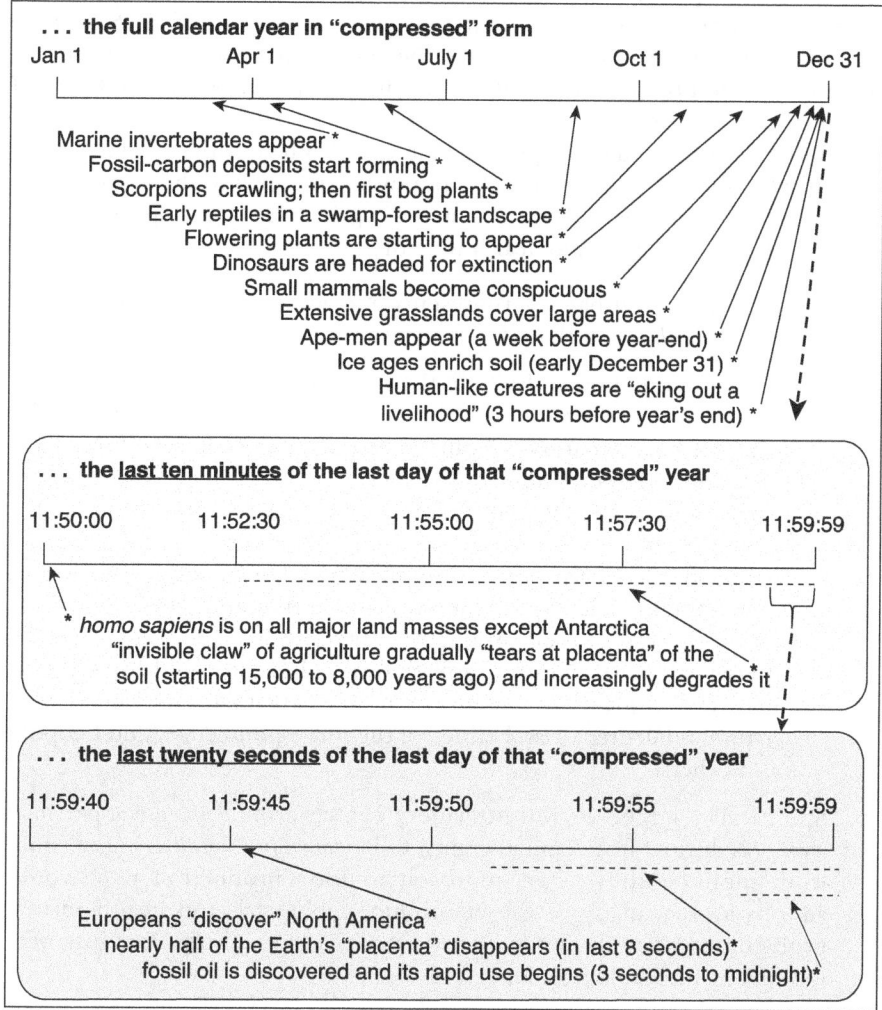

... the full calendar year in "compressed" form

Jan 1   Apr 1   July 1   Oct 1   Dec 31

Marine invertebrates appear *
Fossil-carbon deposits start forming *
Scorpions crawling; then first bog plants *
Early reptiles in a swamp-forest landscape *
Flowering plants are starting to appear *
Dinosaurs are headed for extinction *
Small mammals become conspicuous *
Extensive grasslands cover large areas *
Ape-men appear (a week before year-end) *
Ice ages enrich soil (early December 31) *
Human-like creatures are "eking out a
livelihood" (3 hours before year's end) *

... the last ten minutes of the last day of that "compressed" year

11:50:00   11:52:30   11:55:00   11:57:30   11:59:59

* *homo sapiens* is on all major land masses except Antarctica
"invisible claw" of agriculture gradually "tears at placenta" of the
soil (starting 15,000 to 8,000 years ago) and increasingly degrades it *

... the last twenty seconds of the last day of that "compressed" year

11:59:40   11:59:45   11:59:50   11:59:55   11:59:59

Europeans "discover" North America *
nearly half of the Earth's "placenta" disappears (in last 8 seconds) *
fossil oil is discovered and its rapid use begins (3 seconds to midnight) *

Figure 1.5 Changing the Earth: 750 million years "compressed" into a single year[32]

III. Preview: fundamental changes in agriculture and law

Before embarking on my detailed examination of modern extractive agriculture's shortcomings, let me offer an overview of why this issue matters. The answer in a nutshell, as noted briefly at the beginning of this chapter, has two elements. *First*, it is now possible to develop a high-production

---

32 The details appearing here are drawn largely from Wes Jackson's narrative as quoted above and found at Jackson, *New Roots, supra* note 27, at 5–8.

alternative form of agriculture – one that is *not* industrial, *not* fossil-carbon-based, *not* extractive. I explore that topic in Chapters 4 and 5. *Second*, in order to gain real traction in making the transition to such an alternative form of agriculture, a set of legal and institutional changes can and should be made at the national and the global levels. It is that set of legal and institutional changes that I explore in the last chapters of the book.

I offer the following bullet-point synopsis as a preview to the examination in Chapters 4 and 5 of a "natural-systems" agriculture – and specifically to what I refer to as "agroecological husbandry":

- Agroecological husbandry represents a *dramatic departure from modern extractive agriculture*, in that . . .

    - it focuses on "husbandry"[33] in the sense of a way of life on land, and as part of the natural world, that encompasses conservation and the prudent use of natural resources in the interests of the long-term viability of an ecosystem;[34]
    - it therefore takes the ecosystem as its standard, in the sense that it starts from the assumption that nature's economy – as evident in the native grasslands that constitute the setting for much of the world's agricultural production – should provide the model for a "natural-systems" form of food production; and
    - it gives special attention to soil quality and fertility, and correspondingly to the quality of human communities that farm in concert with nature, reflecting a "land ethic" of the sort espoused by Aldo Leopold and others.

- Specifically, agroecological husbandry centers around *the use of perennial grains* (as distinct from annual grains) to be *grown in polycultures* (as distinct from monocultures) – an approach to food production that would address a great many of the economic, ecological, and health-related problems associated with modern extractive agriculture. For instance, looking at the ecological aspects alone:

    - perennial polycultures can dramatically reduce the required amount of agricultural fertilizer and chemical pesticides, which draw heavily from fossil carbon;

---

33 I explore more fully the meaning and etymology of this term in subsection I of Chapter 4, *infra*.

34 In referring here and elsewhere in this book to "long-term viability", I mean durable, sustained, ongoing, continuous viability. Perhaps even "permanent" and "infinite" could be used in lieu of "long-term", in order to convey the sense that the system would continue *without end* – or at least without end caused by human action – just as such theorists as James Carse envision "infinite games" that do not have a knowable beginning or ending. Such games "are played with the goal of continuing play . . . for the sake of play . . . [and if] the game is approaching resolution [an end] because of the rules of play, the rules must be changed to allow continued play" because the rules are intended "to ensure the game is infinite." James P. Carse, Finite and Infinite Games 4 (1987).

- perennial polycultures can also dramatically reduce the fossil-carbon fuels needed to power farm equipment, and this would further reduce the draw on non-renewable fossil-carbon deposits and reduce greenhouse-gas emissions that contribute to climate change;
- perennial polycultures can arrest the degradation that traditional agriculture causes to soil through erosion, damage to soil structure, and reduction in soil organic matter;
- perennial polycultures can, because of their diversity, better resist attacks by pests and pathogens; and
- perennial polycultures can reduce the loss of water compared with annual grain crops, since the much deeper roots of perennials intercept, retain, and use precipitation better when it comes.

• Perhaps most importantly, agroecological husbandry is not a pipe-dream but instead has *solid prospects for success*, including the sort of commercial success that would obviously be necessary for it to be widely accepted and followed as a practice. In particular:

- efforts to perennialize wheatgrass have already succeeded in creating a domesticated grain-producing perennial (trademarked recently as Kernza®);
- efforts to develop a perennial wheat are proceeding at an accelerating pace, in part under a "dual-purpose" approach that would result not only in grain production but additional forage production;
- a successful perennial rice breeding program has been established in China, and five lines of perennial rice have already been produced;
- although perennial corn research and development has not progressed as far as the grains mentioned above, scientists expect that the recent advances in mapping genetic code and in computer statistical modeling will facilitate corn perennialization as well;
- efforts to perennialize sorghum have already resulted in increased yield and kernel weight as a result of discerning specific ways to divert above-ground biomass into grain;
- perennial versions of certain oilseeds, including sunflowers and rosinweed (*silphium integrifolium*), are also under development – with special promise being found in the fact that the latter of these has a double set of root systems to take advantage of surface water and deeper water reserves; and
- early results are promising also for the design of polycultures, with certain issues of multi-species harvesting having already been addressed successfully and with high expectations that a properly-designed polyculture can secure adequate nitrogen through "biological N_2 fixation" (as distinct from application of synthetic nitrogen produced with fossil carbon) to provide continuous and satisfactory yields of perennial grains.

Drawing on my own background in international law and governance, I provide in Chapters 6 and 7 a detailed explanation of specific legal and institutional changes that can and should be made at the national and the global levels in order to make the transition to such an alternative, "natural-systems" form of agriculture. I summarized those in the Preface to this book. I also acknowledged there that the legal steps and innovations that will be required to facilitate a transition from modern extractive agriculture to a natural-systems agriculture are very substantial. In light of that fact, I will directly address several questions of practicality and political feasibility in the course of Chapters 6 and 7, and I will give additional special attention to these issues in the concluding observations that I offer in Chapter 8.

Part II

The failure of modern extractive agriculture

2 Modern extractive agriculture is economically unsustainable

In the opening paragraphs of Chapter 1, I explained that I am a part owner of a family farm in northeast Missouri. On the largest part of that farm, the actual cropping operations are handled largely by Will Bier, a neighboring farmer. The principal crop produced on the farm in recent years is corn – thousands of bushels of corn – along with some wheat and soybeans. The farm also supports some modest livestock grazing, handled by another local farmer who pays rent for our pasture areas.

Our farm is considered a success, especially for Will Bier and his family. They farm thousands of acres of land in northeast Missouri, and our farm contributes significantly to his "bottom line" – and, I assume, to his quality of life and general well-being. Moreover, the corn and other crops he harvests from the land enter into a global market that helps feed the world – at least that is the message I take special notice of when I pass billboards announcing that "One Missouri Farmer feeds 128 people plus you!" . . . or something to that effect.[1]

How, then, can I (or why would I) claim that this system of agriculture is "economically unsustainable"? Doesn't it provide good income for Will

---

1 The figures vary, of course. According to one source, one American farmer (not just a Missouri farmer) "used to feed about twenty people in the 1940's and, today, one farmer feeds 154 people." Mike Ferguson, *Finding the Next Generation of Missouri Farmers*, http://missouriviewpoints.com/finding-the-next-generation-of-missouri-farmers/ (quoting Kelly Smith of the Missouri Farm Bureau) (hereinafter *Missouri Farmers*).

Bier, for me, for our families? Doesn't it yield food, feed, and fiber for our country and much of the world?

The image of economic success and sustainability presented by modern extractive agriculture – on my family farm, in the USA, in the world at large – is in fact an almost completely false image, with a few short-term winners but many, many long-term losers. In offering this explanation, I will begin by emphasizing what an unusually harsh set of economic and financial realities face most farms and farmers. Then I will explore some of the initiatives that governments in economically developed countries have put in place to take the edge off of that harshness, and how those initiatives have contributed to a dramatic transformation of agriculture in these countries. The most noteworthy aspects of that transformation are (i) a breathtaking drop in the number of small family-operated farms and (ii) a rapid rise in the inefficient use of non-renewable resources at a remarkably high cost – even *before* taking into account the costs of the environmental damage that such use brings. (Modern agriculture's environmental unsustainability is addressed in Chapter 3.) I will conclude my economic assessment by examining the claim that modern agriculture is successful in "feeding the world" – and that therefore the support offered to large agribusiness enterprises helps not only their owners but also vast populations overseas that depend on them. I find that claim deeply suspect.

I. Concentration and its implications[2]

IA. Farming's peculiar perils

From an economic and financial perspective, agriculture is odd. Frederick Kirschenmann, in his book *Cultivating an Ecological Conscience*, offers this summary:

> Farmers are, in fact, caught in a three-way economic squeeze that is wholly dictated to them. They are told what prices they will receive for the products they produce. They are told what prices they will pay for the products and services they have to buy. And they are told how much freight they will pay to ship their grain to the mills and their equipment from the factories. Farmers are the only American businessmen who are forced to buy retail, sell wholesale, and pay the freight both ways. Farmers have no way of influencing prices in relation to the actual costs of production or of passing their increased costs along to consumers.[3]

---

2 This section I reflects extensive research and writing by Caleb Hall, whose work (as emphasized in the Preface and the Acknowledgements, *supra*) has been indispensable in preparing several portions of this book.

3 Frederick L. Kirschenmann, Cultivating an Ecological Conscience: Essays from a Farmer Philosopher 23 (2010) (drawing from a paper presented in 1978). Frederick Kirschenmann has

Those observations focus on farming from a "micro" perspective – that is, from the viewpoint of an individual farmer or farm family. Let us stay with that perspective a bit further. In her important work on agricultural law,[4] Professor Susan Schneider of the University of Arkansas has offered this brief catalogue of some key peculiarities that farming operations present from an individual farmer's perspective:

> The financing of a farming operation raises unique considerations involving 1) the nature of the inputs needed; 2) the special risks associated with production; 3) the typical merger of personal and business assets associated with family farming; and 4) society's basic need for food production . . . . [First, as] noted in the preface to [a leading] agricultural law casebook . . . , "[a]griculture is the only industry where land is the predominant production input.[5] Unlike other resources, land is neither mobile nor fungible."[6] . . . Farmers' business mobility is obviously restricted by their dependence on land as an input. And, the value of their land assets is generally outside of their control . . . . [Moreover,], agricultural economists observe a particular volatility to farmland valuation that can be particularly problematic when financing is at issue . . . . [As one source has observed,] "It has been well documented that farmland values increase more than would be appropriate in response to an increase (decrease) in returns. These periods of overreaction are referred to as 'boom-bust cycles.'"[7]

---

been described as being "one of the most respected critics of the industrial food and farming paradigm", whose work, along with that of Wendell Berry and Wes Jackson (both cited extensively in this book), looks "to the wisdom of Aldo Leopold and Sir Albert Howard for inspiration and guidance". *Editor's Introduction* to Kirschenmann, *supra*, at 2. The same biographical sketch explains that Kirschenmann helped create the Northern Plains Sustainable Agricultural Society, as well as Farm Verified Organic. *Id.* at 3–4. He is an ordained minister with a PhD from the University of Chicago and a distinguished fellow at the Leopold Center for Sustainable Agriculture.

4 Susan A. Schneider, FOOD, FARMING, AND SUSTAINABILITY (2011).

5 Schneider explains that farm real estate accounts for roughly 80 percent of the total value of farm sector assets, confirming the fact that it constitutes a major component of farm wealth – and expense, from the perspective of potential entrants into farming. For this, she cites J. Michael Harris *et al.*, *Agricultural Income and Finance Outlook*, 29, USDA, Economic Research Service, No. AIS-88 (December 2009). Schneider, *supra* note 4, at 206–207.

6 Citing KEITH G. MEYER, DONALD B. PEDERSEN, NORMAN W. THORSON, AND JOHN H. DAVIDSON, AGRICULTURAL LAW: CASES AND MATERIALS xix (1984).

7 For this, Schneider cites *Agricultural Income and Finance Outlook*, *supra* note 5, at 29. Some observers see ample evidence of a boom, or even a bubble, in current US farmland prices today. A Missouri farm publication recently noted that "[t]he number one issue probably . . . is just the cost of getting into farming. The price of farm land continues to escalate . . . and it's not uncommon for an acre of row crop ground to [cost] $8,000 an acre." See *Missouri Farmers*, *supra* note 1. For details on corresponding farmland-price trends in other states, see Michael R. Duffy, *2014 Farmland Value Survey Iowa State University* (January 2015), p. 2, www.extension.iastate.edu/agdm/ wholefarm/pdf/c2-70.pdf (noting that 2013 showed the average cost of Iowa farmland as $8,716 per acre, the highest figure since the survey began in 1947); Kevin C. Dhuyvetter and Mykel

. . . In addition to land, water is a critical natural resource component to the farming operation. In many areas, irrigated land may be essential to farming and the cost of irrigation [is increasingly subject to uncertainties].[8]

In short, Schneider explains, entry costs can be overwhelmingly high. Beyond that, however, she emphasizes that farmers also face a second type and a third type of economic challenge: dependency on factors beyond their control (especially nature and markets), and the merger of family and business interests.

Agriculture's dependency on natural forces, in particular, the weather, combined with the raising of living products[,] brings special risks that are not usually encountered in other industries. Adverse weather, pests, and diseases all threaten production quantity and quality . . . . Moreover, agriculture's traditional role as a "price taker" makes farmers particularly vulnerable to market fluctuations beyond their control. Farmers are generally unable to set the price for the commodity that they produce, but rather must accept the price that the market determines. At times, this price may even be below the farmer's cost of production . . .

[A third peculiarity of the economics of agriculture which is especially applicable to] the family farm system of agriculture is the overlap, even merger, of the family and business interests. The farm family lives and works together on the farm, so the success or failure of the farm has significance even beyond the immediate source of income. The family home is located on the farm, and it is often mortgaged to secure the farm

---

Taylor, *Kansas Land Prices and Cash Rental Rates* (January 2014), p. 2, www.ksre.ksu.edu/bookstore/pubs/MF1100.pdf (demonstrating rising prices for Kansas farmland, with 2013 being the most expensive year). For details on farmland prices in other countries, see *The Wealth Report: a Global Perspective on Prime Property and Wealth* (2011) p. 37, www.privatebank.citibank.com/announce/The_Wealth_Report_2011LowRes.pdf (showing that the prices of certain types of farmland in England, Brazil, Argentina, Canada, Australia, and several other countries have increased). One team of experts offers this troubling assessment of farmland values in several countries: "Over the last decade, the value of agricultural land has risen dramatically, commensurate with other periods of boom and bust." Ben Caldecott, Nicholas Howarth, and Patrick McSharry, *Stranded Assets in Agriculture: Protecting Value from Environment-Related Risks*, (August 2013), p. 108, www.climatechangecapital.com/images/docs/publications/stranded-assets-agriculture.pdf (citing farmland value increases of over 400 percent globally from 2003 to 2013 and spikes of 550 percent in Brazil, 300 percent in Australia, 262 percent in New Zealand, and 75 percent in the USA).

8 Schneider, *supra* note 4, at 207. In addition, equipment costs for row-crop farming – tractors, planters, harvesters ("combines"), etc. – are typically extremely high. By one account dating from a few years ago, a new tractor costs $200,000, a planter $70,000, and a combine $ 300,000. *Id.* See also Harvey Blatt, Americas Food 2 (2008). More contemporary figures, which I obtained by looking at several online sites, suggest that a used (not new) harvester – in this case a 2014 John Deere S780 (a very large rig) with under 450 hours of use – could cost from $375,000 to $410,000. See, e.g., www.tractorhouse.com.

debt. The land may have been passed down from prior generations of farmers, adding historical and personal significance. And, the business success or failure of the farming operation is often obvious for friends, neighbors, and passers-by to see. When failure comes, it comes at a high personal cost for everyone in the family. It may mean not only losing a job and losing a business, it may well mean losing a home and a heritage . . . [9]

Schneider finishes her list of peculiarities of agriculture – especially from a family-farm perspective – by emphasizing some "macro"-level issues:

Given the importance of an adequate food supply, and recognizing the need for adequate capital for production . . . government [agencies in many countries, particularly the USA, have] enacted a variety of measures to support and encourage farm financial investment and to assure the success of farming ventures. Such government-financed farm programs are intended to provide greater income stability and enhanced farm income. The payments received from these programs provide a critical asset [to farmers, but the programs are complex, unreliable, and completely outside a particular farmer's control.] [10]

In sum, the economics of modern agriculture, using the USA as a primary illustration,[11] distinguishes it from most other types of economic activity. The risks are unusually high, the entry and operational costs are also high, and the farmers' control over revenues is woefully low.[12] Partly for these intimidating economic realities – and, ironically, partly because of attempts by government agencies to ameliorate them – the percentages of developed-country populations involved in agriculture have plummeted in recent decades. For instance, the USA has gone from a nation in which 40 percent of the population farmed in 1900, to 12 percent in 1950, and now to only 1.9 percent.[13] The number of farms in the USA actually peaked at 6.8 million in 1935.[14] Now it stands at 2.1 million.[15] Other economically developed

---

9 Schneider, *supra* note 4, at 208–209.

10 *Id.* at 210.

11 For an account of the economics of farming in other developed-economy countries, see generally Caldecott, Howarth, and McSharry, *supra* note 7.

12 For an account of the economic and financial challenges facing many small farmers in the USA, see Bren Smith, *Don't Let Your Children Grow Up To Be Farmers*, THE NEW YORK TIMES, Aug. 10, 2014 (reporting the "dirty secret that the much-celebrated small-scale farmer isn't making a living" and that 91% of all farm households have to "rely on multiple sources of income").

13 See US EPA, *Economic Overview*, June 27, 2012, www.epa.gov/agriculture/ag101/printeconomics. html. See also Blatt, *supra* note 8, at 5.

14 E. Wesley and F. Peterson, A BILLION DOLLARS A DAY 124 (2009).

15 USDA, *Preliminary [Agricultural Census] Report Highlights* (February 2014, p. 1), www.agcensus. usda.gov/Publications/2012/Preliminary_Report/Highlights.pdf. The full report can be found

countries have experienced the same downward trend in the number of farms and the proportions of their populations engaged in farming.[16]

The farms and farmers that remain in the developed countries are predominantly those with large operations.[17] For instance, as of a few years ago, 1 percent of American farmers received 50 percent of the national farm income.[18] As of 1999, 2 percent of American farms accounted for 36 percent of the total amount of US farm land, only 10 percent owned 62 percent of farm land, and the bottom 70 percent owned just 16 percent.[19] This concentration is also exhibited in production values. In 2002, 144,000 farms provided 75 percent of US agricultural production; yet in 2007, that same 75 percent of production value was provided by about 20,000 fewer farms – that is, by just 125,000 farms.[20]

---

at www.agcensus.usda.gov/Publications/2012/Full_Report/Volume_1,_Chapter_1_US/usv1. pdf. Figures of this sort are sometimes difficult to compare directly because of different definitions of farms. Several such definitions are explained in the same report. Mary Jane Angelo notes that "[b]etween 1935 and 2002, the total number of U.S. farms declined by 70 percent while the total acreage of all farms remained the same. This trend was a consequence of larger farms buying out smaller farms." Mary Jane Angelo, *Corn, Carbon, and Conservation: Rethinking U.S. Agricultural Policy in a Changing Global Environment*, 17 George Mason Law Review 593, 620–621 (2010).

16 One authority notes that "the number of farmers around the world is declining in absolute terms. In the United States farmers represent less than 1 percent of the population . . . . Farming populations in France and Germany have fallen by half since 1978. In countries belonging to the Organization for Economic Cooperation and Development (OECD), the number of farms is declining by 1.5 percent per year, and farmers and their families now represent only 8 percent of the population." Jason Clay, World Agriculture and the Environment: A Commodity-by-Commodity Guide to Impacts and Practices 14 (2004). For further details, see Central Intelligence Agency, *The World Factbook 2013–14* (2013), www.cia.gov/library/publications/the-world-factbook/fields/2048.html (showing for various countries the percentages of population occupied in agriculture, and noting these details in particular: United Kingdom, 1.4%; Australia, 3.6%; Germany, 1.6%; Japan, 3.9%; Austria, 5.5%; Belgium, 2%; Canada, 2%; France, 2.9%; EU as a whole, 5.2% – and reporting that the percentages are also low even in such middle-income Latin American countries as Mexico at 13.4%, Argentina at 5%, Brazil at 15.7%, and Chile at 13.2%).

17 As Michael Pollan has explained, the volatility of prices brings financial distress that is often too great for smaller farm operations to withstand; farmers unable to remain in the market with low returns leave, and those larger farms that can survive the downturn accrue more land and produce more in case of another volatile year. Michael Pollan, *The (Agri)Cultural Contradictions of Obesity*, The New York Times Magazine, Oct. 12, 2003.

18 Blatt, *supra* note 8, at 5.

19 USDA, *1999 Agricultural Economics and Land Ownership Survey*, 1999 at Table 68, www.agcensus.usda.gov/Publications/1997/Agricultural_Economics_and_Land_Ownership/.

20 Schneider, *supra* note 4, at 47. Jason Clay also comments on the concentration of ownership that has resulted from the economic strains of small farmers – or, expressed differently, the competitive advantages enjoyed by very large farming operations – and gives special emphasis to livestock operations:

[Although small farmers might have certain advantages in efficiency,] market factors often outweigh local economic or environmental efficiencies in the marketplace. Simply put, it is easier to purchase larger amounts from a smaller number of suppliers. Nowhere is this clearer

IB. Countering the peculiar perils

In summarizing what I called (on page 43) some especially "intimidating economic realities" of farming operations, I referred obliquely to the "attempts by government agencies to ameliorate" those realities. In US experience, these attempts have featured two key elements for the past hundred years – and both have been applied in a rather haphazard and sometimes devastating fashion. The first feature is subsidization of various types for farmers and their crops; the second feature is the exportation of those crops from the USA to overseas destinations.

These two key features of US experience and policy in the agricultural sector warrant some brief attention, because they help illustrate the peculiar economic challenges and vicissitudes of farming, at least in the modern age of increasing globalization of markets. In a book I cited earlier (see page 8) that traces the role of agriculture in world history, Mark Tauger offers a chronology of events over the past hundred years or so. He begins with the period immediately following the First World War:

> After the war America was the main country with the food reserves necessary to feed starving Europe. The U.S. used food as a weapon by withholding aid from Germany until its new leaders accepted the harsh Versailles treaty in 1919. U.S. exports to Europe in [the aftermath of the war] . . . kept U.S. prices high. As Europe recovered from the war and revived domestic production, however, farm prices fell more than 50 percent from June 1920 through 1921, and stayed low throughout the decade. In the US, Europe, and most of the world, farmers in the 1920s faced chronic overproduction and persistent low prices. Farmers,

---

than with livestock. During the past forty years, global per capita meat consumption has increased by 60 percent. To meet this increased demand, livestock production is increasingly industrialized, with several thousand cattle or pigs or 100,000 chickens often raised in a single facility. Over the past fourteen years [that is, from about 1988 to 2002], the average size of animal operation in the United States has increased 1.6-fold for cattle, 2.3-fold for pigs, 2.8-fold for eggs, and 2.5-fold for chickens. In Canada pig operations have increased 2.6 times in size in ten years.

Clay, *supra* note 16, at 15 (citing a 2002 publication by Tilman and others). For more recent figures, see Food and Water Watch, *Factory Farm Nation: How America Turned Its Livestock Farms into Factories*, 2010 at 7, www.factoryfarmmap.org/wp-content/uploads/2010/11/FactoryFarmNation-web.pdf. This source reports that "the number of beef cattle on feedlots larger than 500-head grew by 17.1 percent from . . . 2002 to . . . 2007." It also reports that similar concentrations of operations – and declines in the number of farms and farmers – has occurred in hog production. Specifically, the number of US hog farms fell by 70% between 1992 and 2007, while the actual "number of hogs remained fairly constant as the scale of the remaining [hog] operations exploded" – so that now at least 95% of hogs in the USA are raised on operations with more than 2,000 hogs. *Id.*, citing USDA-sponsored publications. The same source also notes similar increasing concentration in the production of broiler chickens in the USA. *Id.* at 13.

farm politicians, and specialists proposed measures to alleviate the situation, but governments, guided by outdated economic theory and sometimes by farmers' fears of government interference, refused to intervene in the economy until the 1930s crisis.[21]

The crisis that Tauger refers to, of course, is the Great Depression. He offers this explanation of the linkage between the Great Depression and the farm crisis that had already erupted in much of the world:

> While the U.S. stock market crash and bank collapses triggered the Depression, the economic circumstances of the agricultural sector, which employed about two-thirds of the world's economically active population, made a serious economic downturn almost inevitable. By 1929 countries that produced primary products had much larger surpluses, and lower prices, than ever before. During the peak of the crisis, 1929–33, prices declined despite crop failures because aggregate production swamped world markets . . . While traders desperately tried to sell surpluses, governments tried to alleviate farm debt and farm abandonment while maintaining food supplies and low prices for both working and unemployed citizens. In August 1933 the League of Nations belatedly held an International Wheat Conference at which exporting countries agreed to limit exports, but several countries violated them when they saw an opportunity to produce and sell more.[22]

From this global perspective, Tauger then focuses attention again on the USA and explains the efforts in the Hoover Administration to address the farm crisis:

> Farm prices, low and unstable through the 1920s, fell steadily after 1929. Because of America's role in propping up the European economy, declining American prices drove prices down all over Europe and ultimately the world . . .
> President Herbert Hoover refused to repeat wartime market interventions, saying that the market would correct the temporary "slump." The few measures he applied were no help. He established the Federal Farm Board, which purchased eight million tons of surplus grain, yet prices continued to fall. His administration refused to use these reserves for starving southern sharecroppers during the Great Southern Drought of 1930, claiming aid would "demoralize" them and shrink the market, even though sharecroppers were too poor to buy most products . . . . Against the advice of economists and businessmen, Hoover imposed tariffs on thousands of imports. This prompted retaliatory tariffs against U.S. exports and further reduced trade, especially in agriculture.

---

21 Mark B. Tauger, Agriculture in World History 110 (2010).
22 *Id.* at 112.

Meanwhile farmers in the Midwest and Plains states faced plummeting prices, overdue loan payments, and foreclosures. In May 1932, long-term farm activist Milo Reno formed the Farmers' Holiday Association, to organize a farmers' boycott of national markets, and force the government to protect farm prices. In August the farm holiday began in Iowa and Wisconsin with attacks on banks, blocking of roads and trains, and violent conflicts with police, with deaths on both sides. Association members also went to foreclosure auctions and bid pennies, intimidating bankers into accepting these bids. They then returned the farm they bought for a dime back to the former owners. They also worked to have farmers and creditors meet and resolve debts without foreclosures. These, however, were not typical events, and hundreds of thousands of farmers lost their farms to foreclosures.[23]

With that depressing account of the misfortunes of US farmers in the late 1920s and the 1930s, Tauger turns his attention to other countries in the first half of the twentieth century, giving special attention to the attempts by their governments to address the economic distress suffered by their farmers:

Other regimes addressed their agrarian crises in [various] ways. The fascist states harnessed farmers to larger government goals that ultimately included war, but in the process set certain precedents for management of agriculture as part of a total food system. The Mexican revolution renewed peasant villages as ejidos with forced land transfers back to villages from the landed estates that had taken their lands in the past. The Soviet revolution transformed villages into kolkhozy, a type of land reform in reverse. In both of these cases, the revolutionary governments took on the responsibility of supplying the new farms with supplies, technology, and guidance, in the Soviet case also planning production directives. Yet in all of these cases, farmers remained socially and politically subordinated.[24]

In short, as economic challenges of the agriculture sector grew in the twentieth century, a key feature of the attempts by government agencies to address those challenges – not only in the USA but elsewhere – was to implement economic initiatives *within* the countries themselves, by various sorts of price supports, regulation, land reform, and the like. A second approach, though, was to *export* the surpluses that had proved so vexing. This approach could only be taken by a few countries. The USA was one such country, and by mid-century it was exporting extensively – first to supply food to postwar Europe, and then to attract countries away from

---

23 *Id.* at 113.
24 *Id.* at 135.

Soviet influence and to handle US overproduction. The 1954 Agricultural Trade Development and Assistance Act (renamed in the 1960s as the "Food for Peace" Program), provided millions of tons of food aid to scores of countries – sometimes to the detriment of the local farmers in those countries.

That last point warrants attention. At first glance, it might seem counterintuitive that food aid to needy countries might in fact cause injury to them. However, it has been well-documented that the *farmers* in economically less developed countries suffer as a result of the importation of free or low-cost grain and other food supplies from the USA and elsewhere into those countries.[25]

Tauger closes his chronology by noting the effects of the so-called Green Revolution that I referred to earlier and will examine more closely later. According to Tauger, the increases in yields that the Green Revolution produced starting in the 1940s and 1950s were so large as to overwhelm the efforts in the USA to guard against surplus production and the economic problems that they brought to many small farmers:

> New technologies allowed farmers to increase yields and productivity even more than [the government] policies could accommodate. Even with acreage controls, price supports, and export subsidies from the Food for Peace program, millions of U.S. farmers could not cover costs and lost or abandoned their farms in the decades after the war. The number of farmers declined from seven million in 1940 to two million by 2000.[26]

IC. A narrow set of crops

What I have tried to illuminate in the foregoing narrative is this: the "intimidating economic realities" of farming in the modern form of agriculture practiced in economically developed countries have proven so challenging, and have created such distress over the past hundred years or so, that they have prompted an enormous and unprecedented shift in the agriculture sector – a shift toward concentration of ownership and production in the hands of a tiny fraction of the population. Government programs, especially, in the USA, have in fact facilitated this shift – partly

---

25 It warrants emphasis, of course, that food aid almost surely does bring relief to non-farmers in many economically less developed countries. My focus is concentrated here, though, on the farming sector. In many countries, US exports bring displacement and distress to that sector. Moreover, the volumes of the exports are substantial. One observer, commenting about a decade ago, points out that 22% of US farm production was being exported – in an attempt, he says, to stabilize prices in the USA by dumping surpluses onto foreign markets. Alberto Jerardo, *Estimating Export Share of U.S. Agricultural Production*, AMBER WAVES, November 3, 2003 at 47, http://webarchives.cdlib.org/sw1vh5dg3r/http://ers.usda.gov/AmberWaves/November03/.

26 Tauger, *supra* note 21, at 141–142.

by design, partly because they have been so chaotic. And, paradoxically, some of those economic initiatives have brought additional distress to farmers in other countries of the world.

There is another kind of *concentration*, however, that also deserves attention. Like the concentration in farm ownership and agricultural production into the hands of a small minority of the population, this other sort of concentration also has been caused largely by the peculiar economic realities of modern agriculture. It is a concentration of *types of grains* and other agricultural products produced by farmers in the USA and other economically developed countries (and, for that matter, increasingly in economically less developed countries as well). As Jason Clay has observed, 90 percent of the world's food today comes from 30 crop species, even though about 7,000 crop species exist.[27] Moreover, as noted in Chapter 1, corn (maize), wheat, and rice account for 89 percent of all cereal produced worldwide in 2012 and 43 percent of all food calories in 2009. In addition to making agriculture (and our ecosphere generally) more vulnerable to disease and ecological dangers – a matter that we will examine more in Chapter 3 – this concentration in a few crops represents a loss in diversity that has economic implications . . . and causes.

A key economic *cause* for this concentration of farm production in a small number of species – for the USA, the list includes corn, soybeans, wheat, rice, cotton, barley, sorghum, and oats, which together account for a very high proportion of US agricultural production – is that these are the crops most heavily subsidized. Indeed, in the USA, nearly all US government agricultural subsidies go to those eight crops, and 90 percent of the funding

---

27 Clay, *supra* note 16, at 49. This same set of figures – reliance on 30 crop species despite the availability of 7,000 crop species – appears widely in the literature. See, e.g., Chicago Botanic Garden, *Developing Genomic Resources in an Underutilized Crop*, www.chicagobotanic.org/ research/understanding_biodiversity/developing_genomic_resources_underutilized_crop (mentioning the figures of 30 and 7,000 and noting further that "[s]hockingly, more than 50 percent of human food needs are met by only three of those crops: corn, wheat, and rice" – all to emphasize the value that could come from the wider production of breadfruit). Another source substitutes the figure of 20 species for 30 species (as providing 90% of the world's human food) and places this information in an even larger context by explaining that a total of 30,000 plants (out of the total of more than 300,000 plant species) are thought to be edible by humans, although only 7,000 have been cultivated or collected as food – with the result that "tens of thousands of edible plant species remain relatively 'underutilized', with respect to their ability to contribute to the world's increasing food requirements." Pauline Chivenge *et al.*, *The Potential Role of Neglected and Underutilised Crop Species as Future Crops under Water Scarce Conditions in Sub-Saharan Africa*, 12 INTERNATIONAL JOURNAL OF ENVIRONMENTAL RESEARCH AND PUBLIC HEALTH 5685–5711 (2015). For a legal perspective on the narrowness of agricultural biodiversity, and the risks inherent in relying on a limited number of staple crops, see generally Juliana Santilli, AGROBIODIVERSITY AND THE LAW: REGULATING GENETIC RESOURCES, FOOD SECURITY AND CULTURAL DIVERSITY (2011).

is directed at the first five – corn, soybeans, wheat, rice, and cotton.[28] An economic *implication* of this concentration, in turn, is that large agribusiness enterprises produce predominantly this handful of crops. Correspondingly, any small farmers eager to diversify into other crops will suffer, as a relative matter, for doing so. Partly as a consequence of this, US government subsidization pays disproportionately more for larger farm operations.[29] In 2007, $8 billion in federal programs were paid to American farmers, but only 840,000 farms – 38 percent of all US farms – received payments.[30]

II. But what about systemic economic sustainability?[31]

I wish to offer some further observations about subsidies – not just in the USA but in other economically-developed countries as well – but first a reorientation is in order. The discussion thus far has focused on certain economic and financial peculiarities of farming – peculiarities that have increasingly worked (especially in economically developed countries) to the detriment of small farmers, even to the extent of creating a dramatic trend toward concentration in ownership and control over farmland. A dwindling population of farmers, a narrowing of diversity in the crops being produced, a concentration of production, ownership, and financial reward in the hands of large agribusiness enterprises – all these developments characterize modern agriculture of recent decades.

But so what? Why should these developments be matters of concern, instead of grounds for relief or even celebration? After all, the title of this chapter makes the claim that "modern extractive agriculture is economically *unsustainable*". However bleak a picture might emerge from the foregoing discussion in terms of the economic strains suffered by small farmers, or in terms of economic competitiveness more generally in the agriculture sector, the narrative thus far does *not* speak directly to the economic *sustainability* of modern agriculture. Some observers may wish to assert that the developments I have summarized previously are unfortunate or regrettable for some ethical or sentimental reason – for instance, that they run counter to the values

---

28 Blatt, *supra* note 8, at 11. These and other figures at various points in this book are based on agricultural-subsidy data that extend up to roughly two to ten years ago. More recent information would change those figures somewhat, of course, but as I explain later, the general trajectory of US agricultural subsidies – including their concentration in a relatively small number of crops and a relatively small number of companies and individuals – has remained unchanged even under the most recent developments in US agriculture-related legislation. For a summary some key elements of the 2014 farm bill, see Box 6.1 in Chapter 6, *infra*. For commentary regarding the ways in which it departs from, but for the most part continues, agricultural subsidy-and-support programs from earlier years, see text accompanying note 89 in Chapter 6, *infra*.

29 Blatt, *supra* note 8, at 252–253.

30 Schneider, *supra* note 4, at 48. For more recent information regarding US government agriculture subsidies, see Box 6.1 and text accompanying note 89 in Chapter 6, *infra*.

31 This section (II) reflects extensive research and writing by Caleb Hall.

conjured up by agrarian images of the individualistic yeoman farmer – but this does not necessarily mean that agriculture as currently practiced will not endure.

Accordingly, I wish to introduce now, and establish in the remaining portions of this section, the following proposition: The developments summarized in the preceding pages are regrettable – even dangerous – *not* because they contradict some romantic notions of pastoralism but rather for a cluster of entirely different reasons:

- First, that a system of smaller farms, with more diversified land ownerships and crops, has greater potential for widespread poverty reduction than does a system of large corporate farms, and also can in fact yield higher productivity than a system of large corporate farms – and therefore a system that squeezes out those smaller farms is to be resisted.
- Second, that even for large agribusiness enterprises with seemingly high productivity, modern agricultural techniques are profoundly wasteful and inefficient when all the costs of those techniques are accounted for;
- Third, that even with modern agricultural techniques and the robust financial support that governments have afforded them, world per capita food supply has not increased enough (despite the need for it to do so), probably in part because crop yields have peaked – and therefore the "feeding the world" rationalization for modern agriculture does not hold up under scrutiny.

The first of these points – that a system of agriculture that features smaller farms with diversified land ownership and crops is economically superior in certain important ways to concentrated agricultural production – has been widely asserted and supported. For instance, consider the following observations:

- According to one source, "smaller farms [have been found to be] two to ten times more productive than larger ones and ten times more productive per acre than farms of 6,000 acres or more. The smallest farms (1.6 hectares or less) were 100 times more productive per acre than farms of 2,400 hectares or more."[32]
- "The case for setting small farms at the center of agriculture development can be made on two grounds: efficiency and equity . . . . Moreover,

---

32 Clay, *supra* note 16, at 15. The same point is emphasized in John van Zyl, Bill R. Miller, and Andrew N. Parker, *Agrarian Structure in Poland: The Myth of Large Farm Superiority*, The World Bank Policy Research Working Paper 1596, Apr. 1996, at 1 (noting that "the perception that there are real economics of scale present that favor large farms . . . are contrary to international evidence, which indicates that a large-scale, mechanized farm sector is generally inefficient").

small farm households have more favorable expenditure patterns for promoting growth of the local non-farm economy."[33]

- "Some of the ways in which agricultural development can reduce poverty are likely to be enhanced when smallholders raise their production. Compared to larger-scale farmers, small farmers are themselves more likely to be poor, so raising farm incomes directly reduces poverty; while small farmers are more likely to spend additional income locally so that consumption linkages that stimulate the rural nonfarm economy may be greater. Small farms . . . are likely to use more household labor when expanding production, reducing the extent to which they seek additional work off farm, and hence reducing supply to the rural labor market and further driving up rural wages."[34]

- "Many cases support the finding that promotion of agriculture, especially in the developing context, can help the poor to a greater extent than economic growth alone."[35]

- It has been pointed out that smaller, poorer farms often hire poor neighbors – including women – who then sell to others in the local populace, strengthening local food sheds and economies, and also that more labor-intensive work on less land consistently produces more food than energy-intensive concentrated agribusiness operations.[36]

This last point – touting the efficiency of small labor-intensive farms over agribusiness operations relying on other forms of energy – deserves special attention. In order to give it that attention, let us examine the issue of agricultural productivity. In particular, I will consider inputs and outputs in the agricultural sector generally, and then examine inputs and outputs in the context of large, modern, concentrated agribusiness operations in particular.

After all, a fairly straightforward method for assessing whether modern agriculture is sustainable or not would be to measure its inputs versus its outputs. If outputs are higher than inputs, it would stand to reason that there should be ongoing returns that will enable the modern agricultural system to operate sustainably over the long term. As with some of the assessments I have undertaken, I will look to the USA as the model for modern agriculture, since modern agriculture as we know it today is largely

33 Peter Hazell *et al.*, *The Future of Small Farms: Trajectories and Policy Priorities*, 38 WORLD DEVELOPMENT 1349, 1352 (2010).

34 Steve Wiggins *et al.*, *The Future of Small* Farms, 38 WORLD DEVELOPMENT 1341, 1343–1344 (2010).

35 John Donaldson, *Growth is Good for Whom, When, How? Economic Growth and Poverty Reduction in Exceptional Cases*, 36 WORLD DEVELOPMENT 2127, 2130–2135 (2008).

36 See, e.g., Wiggins, *supra* note 34, at 1343–1344. See also Rasmus Heltberg, *Rural Market Imperfections and the Farm Size-Productively Relationship: Evidence from Pakistan*, 26 WORLD DEVELOPMENT 1807, 1807–1826 (1998); Giovanni Cornia, *Farm Size, Land Yields and the Agricultural Production Function: An Analysis for 15 Developing Countries*, 13 WORLD DEVELOPMENT 513, 513–534 (1985); David Montgomery, DIRT: THE EROSION OF CIVILIZATIONS 159 (2007).

a product of the Western world as projected and grafted onto other countries and cultures.[37]

The graphs appearing in Figure 2.1 through Figure 2.6 present an extensive series of data compiled by the US Department of Agriculture ("USDA") on the values of US farm inputs, outputs, and productivity over the period from 1948 to 2009.[38] Each of these three sets of values (inputs, outputs, and productivity) is presented in two forms – first as the *increase or decrease in the rate of change* recorded for certain four-year to ten-year time-sequences during that 60-year period (Figure 2.1 for inputs, Figure 2.2 for outputs, Figure 2.3 for productivity) and then as the *amount of change* (that is, amount of increase or decrease in inputs, outputs, or productivity) from each year to the next during that 60-year period (Figure 2.4 for inputs, Figure 2.5 for outputs, Figure 2.6 for productivity)[39] – all expressed in terms of 2005 dollars.

Looking at the USDA data for farm inputs from 1948 to 2009, a general trend is apparent: overall, although farm inputs (as measured in constant dollars) have steadily increased, they have done so at a progressively slower pace, settling around 1 percent per year (increase) in recent years (see Figure 2.4 in particular); and at the same time, the increase in farm outputs has shown a progressive acceleration, reaching a pace of more than 1 percent per year (increase) in recent years (see Figure 2.5 in particular). Consistent with this pair of trends, the USDA data for farm productivity indicate gradually rising rates at which productivity is increasing – with recent years registering a rate of increasing productivity of more than 1 percent per

---

37 For sources explaining the ways in which modern agriculture has emerged from Western civilization and American methodology and been projected onto other countries via the Green Revolution, see Montgomery, *supra* note 36, at 197–198; Miguel Altieri, *Fatal Harvest: Old and New Dimensions of the Ecological Tragedy of Modern Agriculture*, 30 JOURNAL OF BUSINESS ADMINISTRATION & POLICY ANALYSIS 239 (2002); Harry Cleaver, *The Contradictions of the Green Revolution*, 62 AMERICAN ECONOMIC REVIEW 177, 177–181 (1972).

38 See USDA Economic Research Service, *Agricultural Productivity in the U.S.*, July 5, 2012, www. ers.usda.gov/data-products/agricultural-productivity-in-the-us.aspx. That website notes that the values are expressed in year 2005 US dollars, and it explains that for these purposes the USDA measures total factor productivity, taking account of the use of all inputs to the production process; this way, "[p]roductivity . . . measures changes in the efficiency with which inputs are transformed into outputs", instead of simply the measuring single-factor outcomes (such as corn production per acre or per hour of labor). *Id.*

39 More specifically, the difference between these two forms of presentation can be explained as follows, using the data for farm inputs as an illustration: Figure 2.1 shows how much the *rate* of increase (or decrease) in the amount of farm inputs changed from one time-sequence to the next – using such time-sequences as 1957–1960 (four years) and 1960–1966 (seven years); for instance, in the time-sequence of 1957–1960, the average annual rate of increase in energy-and-nutrient inputs was about 3%, whereas in the time-sequence of 1960–1966, the average annual rate of increase in energy-and-nutrient inputs was a little under 2%. Similarly, the average annual rate of increase in energy-and-nutrient inputs was roughly a negative 2.5% in the time-sequence 2007–2009 – which means that those inputs fell in those years.

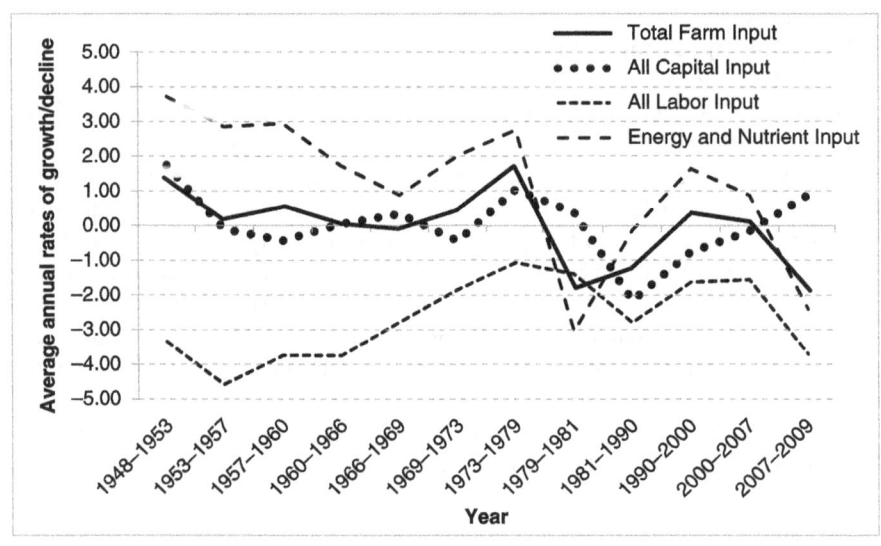

Figure 2.1 US period–average annual farm input % rates of change, 1948–2009

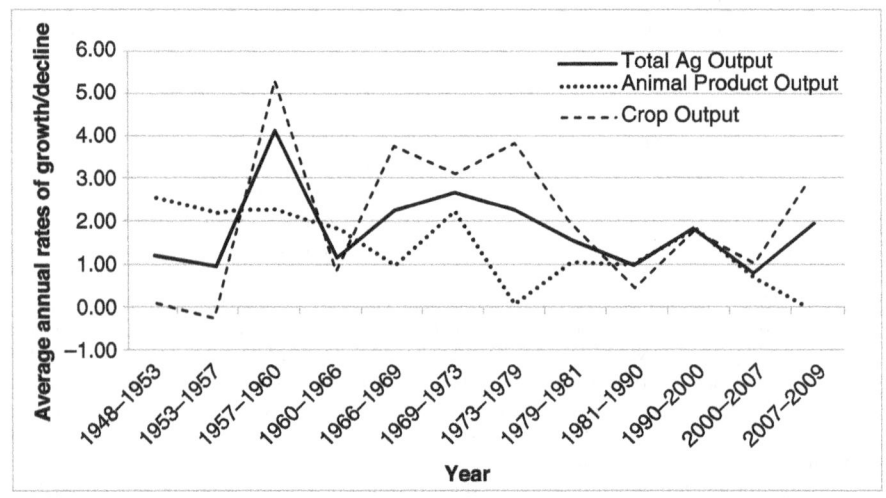

Figure 2.2 US period-average annual farm output % rates of change, 1948–2009

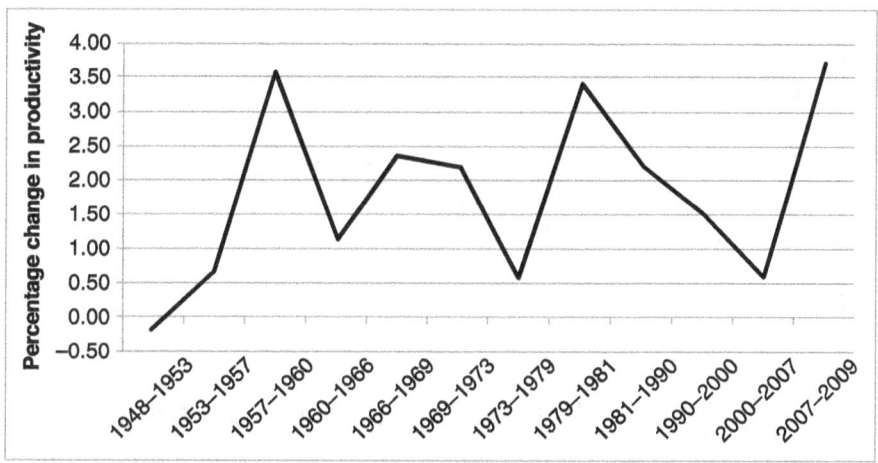

Figure 2.3 US period-average annual farm productivity % rates of change, 1948–2009

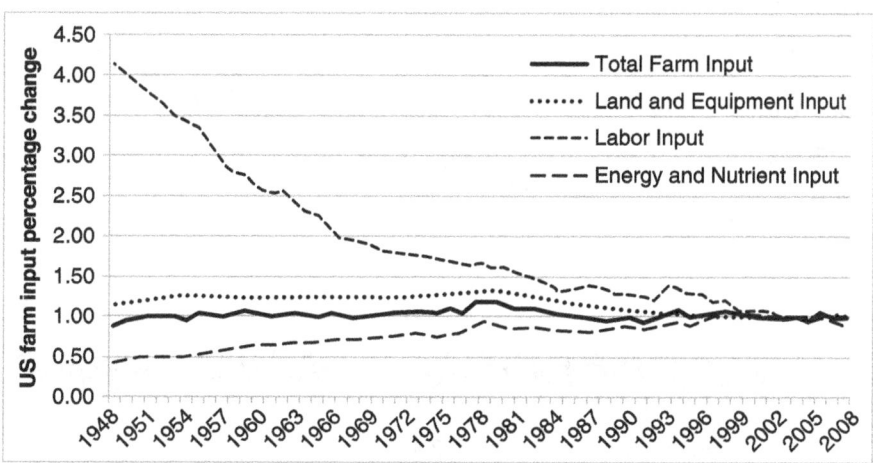

Figure 2.4 US year-by-year % change in farm input, 1948–2009

year (see Figure 2.6 in particular). While the rates of increase have fluctuated from about 0.5 percent to about 3.5 percent per year, they have consistently stayed in positive territory – indicating, that is, steady increase in productivity – for every time-sequence since the early 1950s (see Figure 2.3 in particular).

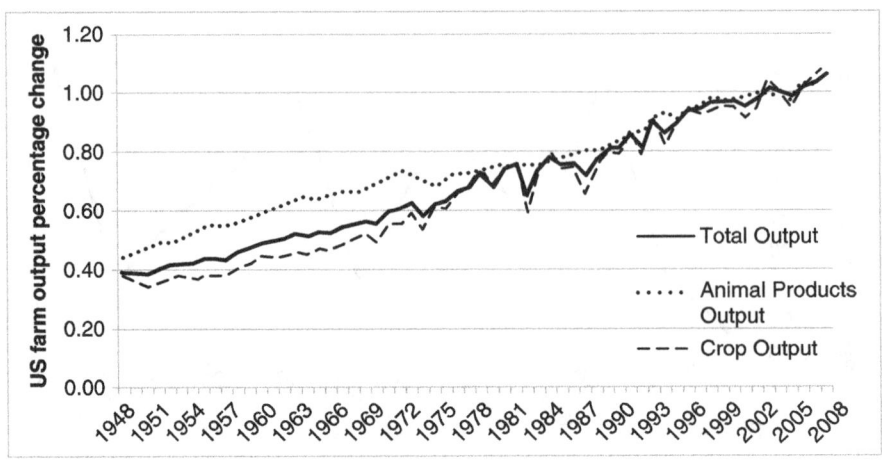

Figure 2.5 US year-by-year % change in farm output, 1948–2009

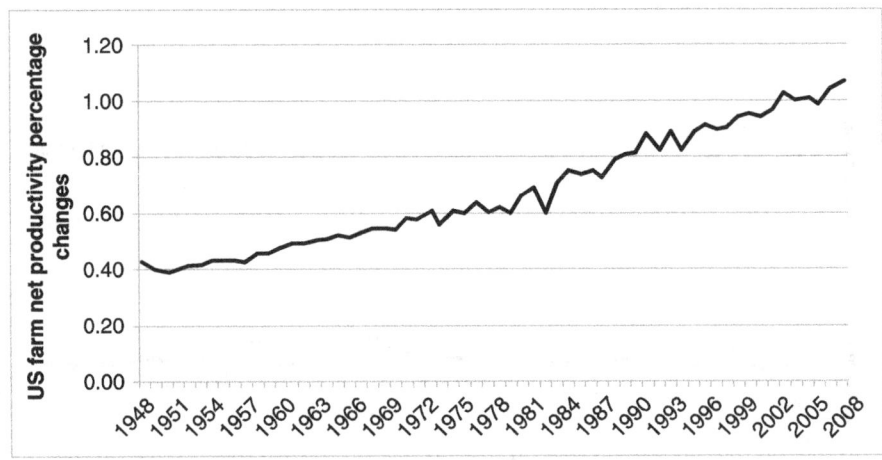

Figure 2.6 US productivity indices, 1948–2009 [40]

It would seem that these data and graphs offer grounds for optimism about the economic sustainability of modern agriculture in the USA – but *only if* they present a complete and reliable picture. Unfortunately, they do not:

---

40 As indicated in note 38, *supra*, the source for the information in Figures 2.1 through 2.6 is USDA Economic Research Service, *Agricultural Productivity in the U.S.*, July 5, 2012, www.ers.usda.gov/data-products/agricultural-productivity-in-the-us.aspx. Graphs were prepared mainly by Caleb Hall.

They fail to reflect properly the full range of factors that we should consider important in assessing modern agriculture on overall economic grounds. Three points in particular stand out in this regard. I will state them briefly in the next three paragraphs and then explain them in more detail.

First, the data and graphs fail to remove from the "output" and "productivity" figures the values of agricultural subsidies that farmers receive. This is a big omission. After all, agricultural subsidies represent an economic transfer to farmers from society as a whole. Through this transfer, society contributes to the value of "outputs", and therefore adds to the apparent "productivity" figures in the USDA reports – assuming that the "outputs" values are reported from the perspective of the farmers themselves. The USDA's explanation confirms that this is in fact the case: "We evaluate industry output from the point of view of the producer; that is, subsidies are added [to] and indirect taxes are subtracted from market values."[41]

Second, those data and graphs also fail to include in the "input" figures the value that farmers receive from fossil-carbon subsidies that are paid not to farmers but to the fossil-carbon industry and that therefore indirectly benefit farmers by keeping down the cost of fuel, synthetic nitrogen, and other petroleum-based inputs such as pesticides. This is also a big omission; the cost incurred in subsidizing the fossil-carbon industry represents a significant economic transfer, although this one affects the "inputs" component of productivity.

Third, the data and graphs also fail to include in the "input" figures the cost of damage done to the ecosystem through such things as soil depletion, species extinction, and poisoning of the air and water on which all species depend. This eco-degradation cost is borne partly by today's society but largely by succeeding generations whose interests today's generation is supposed to (but is failing to) protect.

And it is the fact that *none* of those three categories of value is fully reflected in these graphs that allows the trends shown in those graphs to give the impression – the *false* impression – that modern agriculture is economically sustainable.

Let us examine these propositions more closely, looking first to the value of agricultural subsidies. Again I will use the USA as an illustration, even though the scope of agricultural subsidies in the USA is not in fact as far-reaching as in some other countries or in the European Union.[42]

---

41 See *id.*, and particularly at this page on "documentation and methods" used by the USDA: www. ers.usda.gov/data-products/agricultural-productivity-in-the-us/findings,-documentation,-and-methods.aspx#methods.

42 For explanations about agricultural subsidies in the EU, see European Commission memo, *The common agricultural policy (CAP) and agriculture in Europe – Frequently asked questions*, http://europa. cu/rapid/press-release_MEMO-13-631_en.htm (noting that the CAP, operated by the Council of Agriculture Ministers from the 27 EU countries, spends 70% of its budget on "income support for farmers and assistance for complying with sustainable agricultural practices" including food

(According to one source, for instance, in 2004 the European Union supported 33 percent of its agricultural production and Japan supported 56 percent, while the USA supported only 18 percent).[43] In order to offer a view of US agricultural subsidies over time, I must use data from before the most recent changes brought about by the 2014 farm bill (the Agricultural Act of 2014) came into effect. However, as I have noted in this chapter and will explain more fully in Chapter 6, the general trajectory of US agricultural subsidies has remained unchanged even under the new legislation.[44]

The US agricultural sector has relied heavily over the years on billions of dollars provided in the form of subsidies. Figure 2.7 provides a visual account of these subsidies, focusing on the dollar amounts paid in respect of four crops – corn, soybeans, wheat, and rice – as well as (for comparison purposes) amounts paid in respect of livestock and dairy products.

Subsidies in the USA pay in fact for eight crops – the four crops included in Figure 2.7, plus cotton, barley, sorghum, and oats.[45] However, 90 percent of the funding goes to the first five: corn, soybeans, wheat, rice, and cotton (and the last of these is not a food crop).

Two features of the information presented in figure 2.7 stand out. First, the overall dollar amounts are impressive. Corn subsidies in recent years, for instance, have hovered around $4 billion per year, with wheat and soybean subsidies each hovering around $2 billion per year. Second, there was a huge spike, then a drop, in subsidy payments between 2003 and 2007 for corn. Six factors have been identified to explain this phenomenon: (i) biofuel policies of the USA, Brazil, and the European Union shifted some corn

---

safety standards, animal health and welfare, and direct payments linked to European farmers' compliance with sustainable agricultural practices). It is worth noting in this regard that EU subsidies cover a broader range of food crops than US subsidies cover.

43 Blatt, *supra* note 8, at 11. Another source offers similar comparisons, asserting that by way of the EU's Common Agricultural Policy, over $50 billion equivalent (approximately 44% of the entire budget of the EU in 2011) was distributed among the EU member states in the form of agricultural subsidies. When including EU price supports that were put in place to keep domestic crop prices artificially high, the EU spent over $106 billion equivalent on agriculture subsidies in 2011. *Agricultural Subsidies Remain a Staple in the Industrial World*, Worldwatch Institute, Mar. 11, 2014, www.worldwatch.org/agricultural-subsidies-remain-staple-industrial-world-0). In comparison, the USA spends just over $30 billion in subsidies. Asia spends more on subsidies than the rest of the world combined: China pays the equivalent of $165 billion, Japan $65 billion, Indonesia $28 billion, and South Korea $20 billion. *Id.*

44 For commentary regarding the ways in which the 2014 bill departs from, but for the most part continues, agricultural subsidy-and-support programs from earlier years, see text accompanying note 89 in Chapter 6, *infra*. See also Box 6.1 (also in Chapter 6) for a survey of selected provisions of the US Agricultural Act of 2014.

45 For details, see Blatt, *supra* note 8, at 11. At some points in time, US subsidies have also been extended to some crops of no direct interest to us here, such as peanuts, sunflowers, tobacco, apples, and sugar beets. For some details, see the website of the Environmental Working Group, at http://farm.ewg.org/region.php?fips=00000. For observations about more recent figures for US agricultural subsidies, see note 28, *supra*.

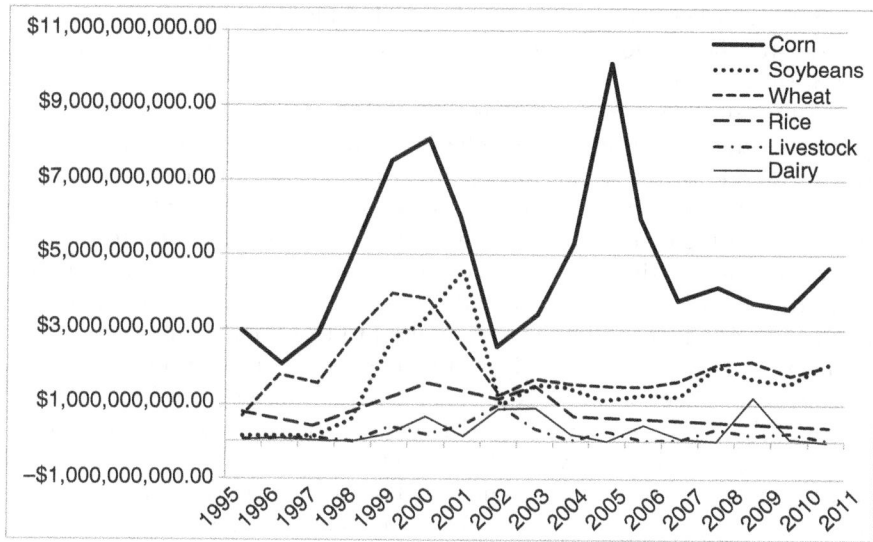

Figure 2.7 US farm subsidy payments, 1995–2011[46]

utilization from food to fuel; (ii) poor weather in Australia and Europe in those years created supply shortfalls; (iii) gradually tightening world food supplies resulted from rapid demand from emerging economies (for example, China, India, and Russia) with slowing crop yield growths; (iv) energy prices used in production moved higher; (v) hoarding and export controls restrained shipments; and (vi) flooding occurred in the Midwestern USA.[47]

Figure 2.7 does not, however, reveal another a crucial feature of US agricultural subsidies: the financial support goes predominantly to the largest agribusiness enterprises. According to one source, between 2003 and 2005, 10 percent of those eligible for farm subsidies – those with the largest land holdings – received 66 percent of the subsidized funding whereas the bottom 80 percent, with the smallest land holdings, received only 16 percent of the subsidies.[48] Whether this distribution is appropriate or not depends in part on what the aims of the subsidies are, or what they should be. If such subsidies should be designed primarily to ensure production of the eight

---

46 The source for the information shown in Figure 2.7 is Environmental Working Group, *EWG Farm Subsidies*, 2011, http://farm.ewg.org/region.php. Subsidy information for livestock and dairy have been included for comparison purposes.

47 Colin Carter *et al.*, Giannini Foundation of Agricultural Economics, University of California, *The Food Price Boom and Bust*, December 16, 2008, http://giannini.ucop.edu/media/are-update/files/articles/v12n2_2.pdf.

48 See Blatt, *supra* note 8, at 11.

crops being funded, then it stands to reason that larger producers would receive more dollars. If, on the other hand, the subsidies should be designed primarily to encourage small farms, the distribution would be skewed in favor of those farms.

In reality, the US system of farm subsidies results in enormous financial benefits for the largest producers. These subsidies have a "self-reinforcing" character: those largest producers benefiting from them might reasonably be expected to use some of the funding to make sure that the subsidies program continues to bring benefits to them. If the subsidies program were to be discontinued, it would dramatically affect the fortunes of the agriculture sector, due to the sector's established natural trend toward capital concentration – and especially if, as is projected, commodity production costs increase in the coming years.[49] Most importantly, if this large category of value – that is, the value of massive subsidies received by farmers and included in the "productivity" figures – were removed, the proposition that modern agriculture is economically sustainable would be suspect.

Having considered the significance of the heavy subsidies provided for agricultural production in some countries, including the USA, it is important that we consider also another kind of subsidy. From a farmer's perspective, it is an *indirect* subsidy because it is paid not to farmers themselves but rather to participants in the fossil-carbon extraction and processing industry. To the extent that the extraction and processing of such fossil carbons[50] as coal, oil, and natural gas receive financial benefits from government subsidies of various sorts – and in the USA, it is a *very* great extent – the agriculture sector also benefits indirectly.

After all, fossil-carbon inputs figure prominently in farm production. One form of fossil-carbon input is diesel to power farm equipment. Another is the wide range of fossil-carbon-based fertilizers, pesticides, and other chemicals crucial to crop production, especially in the economically developed countries. As just one example, consider the importance of natural

---

49 Peter Rosset and Miguel Altieri, *Agroecology Versus Input Substitution: A Fundamental Contradiction of Sustainable Agriculture*, 10 SOCIETY & NATURAL RESOURCES 283, 284–287 (1997) (explaining how smaller farmers are marginalized out of the market by increased capital costs, and how this leads to those remaining being able to reinforce the economic stratification). If projections of commodity production cost increases prove accurate, they would presumably add further financial strain to a system that already has a farm debt that reached about $240 billion in 2008. See Schneider, *supra* note 4, at 210.

50 I use the term "fossil carbon" here, instead of the term "fossil fuel", to reflect the fact that the extraction and processing of oil, natural gas, coal, and other hydrocarbons that exist in fossilized form are not always undertaken for the creation of *fuel*. For instance, when natural gas, such as methane, is used in the creation of synthetic ammonia (widely used as fertilizer in modern agriculture), it is not used as a *fuel* but instead as a component in the chemical process that results in ammonia. For a summary explanation of that chemical process, see note 20 in Chapter 1, *supra*. Some observers explain that there are five principal "carbon pools" on Earth: soil carbon, forest carbon, coal, natural gas, and oil – with the last three of those constituting together the "fossil carbon" pool.

gas to agriculture: roughly four-fifths of all ammonia produced around the world is devoted to use as agricultural fertilizer, and the production of the ammonia is widely dependent on natural gas.[51]

The subsidization of the fossil-carbon industry takes several forms. Some are direct and others are indirect in nature. Principal among the direct subsidies is an array of tax incentives. In the USA, for instance, oil and gas producers are allowed to "expense" a share of intangible exploration and production-drilling costs (thus permitting those costs to be deducted from income for US tax purposes) rather than to "amortize" them over a period of several years. Indirect subsidies to the fossil-carbon industry are also quite numerous. For a brief description and explanation of these various forms of subsidies to the fossil-carbon extraction and processing industry, see Box 6.2 in Chapter 6 of this book. As indicated there, the International Monetary Fund has calculated that the international public paid for the equivalent of $480 billion of fossil fuel profits in 2011 in "pre-tax" subsidies – that is, those that do not take into account the costs of negative externalities and energy consumption itself. Post-tax subsidies have been estimated as reaching up to $1.9 trillion – a figure that amounts to 2.5 percent of the world's GDP or 8 percent of total government revenues worldwide.[52]

In short, farmers in those countries with resources to provide subsidies to certain industries – the USA is certainly one of those countries – benefit both from those subsidies that are designed to flow directly to the agriculture sector and also from other subsidies, principal among them the subsidies provided to the industries involved in the extraction and processing of fossil carbon on which farmers increasingly rely in engaging in modern agriculture.

Before leaving this topic of fossil-carbon subsidies, let us consider more broadly just how it is that fossil carbon has transformed agriculture in the past two centuries. As Tim Crews of The Land Institute has explained, agricultural use of fossil carbon – particularly in the form of fuels – has dramatically changed the source of energy required to grow food. In a 2013 article, Crews uses the term "fossil fuel slaves" (as similar in some ways to human slaves) to highlight the artificiality of modern extractive agriculture. His explanation is so helpful that I will quote from it here extensively.

> In the traditional Mexican farming systems I studied, the fossil fuel share of caloric energy used to grow corn was close to zero. The energy to prepare and plant and weed and harvest the fields came [instead] from the corn and alfalfa that captured energy from sunlight in photosynthesis

---

51 For some details on the relationship between ammonia and natural gas, see notes 19 and 20 in Chapter 1, *supra*.

52 International Monetary Fund, *Energy Subsidy Reform: Lessons and Implications*, (Jan. 28, 2013), www.imf.org/external/np/pp/eng/2013/012813.pdf. This source provides an overview of international fossil fuel subsidy payments as well as profitability and the scope of certain industries involved in fossil fuels.

and went to feed the farmers and their draft animals. David Pimentel of Cornell University estimates that traditional Mexican corn-bean-squash farms like these yielded about 10 calories of food for every calorie of food metabolized by the farmer. [This is typical of such traditional settings.] Most indigenous or traditional agricultures without fossil fuels had ratios between 10 and 40 calories of food out per calorie of food consumed in farming. This ratio defined the amount of energy available to do everything outside farming – create art, play music, worship, fight wars, build things like the Great Wall of China . . .

The fossil fuel share of caloric energy used to grow corn in the US [today, by contrast,] is 99.96%. We are truly *Homo petrolius*. In agriculture . . . we have figured out how to use fossil energy to address virtually every ecological limiting factor . . . such as insect damage, weed competition, temperature, and nutrients, and too much or little water . . .

So in a sense, *modern agriculture relies on the carbon bonds of fossil fuel slaves.* I mean the equivalent work of a human that is accomplished by harnessing the energy of fossil fuels. Some may object to this use of the term slave, as it excludes important aspects of what we need to communicate about slavery, such as human exploitation and suffering. But I use it here because I worry about how interchangeable the two energy sources have been in the past, and could be in the future if we are not mindful. The adoption of fossil fuel slaves began in earnest with James Watt's steam engine patent in 1781. One hundred years later, Andrew Nikiforuk writes in "The Energy of Slaves," the output of the world's coal-fired steam engines, primarily for transportation and manufacturing, totaled 150 million horsepower. These machines collectively exerted the work of more than 3 billion humans working long shifts. The world's population at that time was 1.5 billion. So in 1880 there were at least 2 fossil fuel slaves per human, although not evenly distributed.

Now, if we take the amount of commercial energy consumed in the US today and divide it by the population, and compare this with how much energy a human expends doing physical work, the sobering conclusion is that on average each of us has 80 fossil fuel slaves working the equivalent of 10 hours a day, 365 days a year. That is 25 billion human slave equivalents, 3½ times the world population, just to maintain the lifestyle of US citizens. This conversion is not perfect, because some of the commercial energy we rely on does not come from fossil fuels. But the majority does . . . [Therefore,] this conversion gives us a sense of how deep our dependence goes . . . . [This is also what makes it seem reasonable for us, without thinking twice, to hop] in something that weighs 4,000 pounds, using fossil energy to move it 2 miles to buy a 12-ounce package of cheese, and then driving back.[53]

---

53 Tim Crews, *Will Becoming Local Here Get Us There?*, The Land Institute Land Report, no. 108 (Spring 2014), https//landinstitute.org/wp-content/uploads/2014/11/LR-108 (emphasis added).

In short, modern extractive agriculture, like modern society as a whole, relies heavily on what Crews calls "fossil fuel slaves". As Crews has emphasized to me in person, agriculture of the sort we have grown accustomed to in recent decades would be wholly impossible without this heavy reliance on fossil carbon.[54] This is the reason I have emphasized, in this *economic* analysis of modern extractive agriculture, the importance of fossil-carbon subsidies. Because agriculture cannot exist in its current form without such subsidies, the economic sustainability of modern extractive agriculture cannot be evaluated without including the cost of those subsidies in the calculation of "inputs".

The third category of value that I identified earlier as *not* being reflected in the information presented in Figure 2.1 through Figure 2.6 takes the form of eco-degradation – that is, damage done to the ecosystem through such things as soil depletion, species extinction, and poisoning of the air and water on which all species depend. Having examined the issue of agricultural subsidies and the issue of fossil-carbon subsidies, let us turn to this third category of value. It constitutes a *cost* that is not being borne by the farmers themselves – a fact that allows the input and output trends shown in Figure 2.1 through Figure 2.6 to give the impression (the *false* impression) that modern agriculture is economically sustainable.

In turning to this topic, I am still focusing primarily on economic matters. There are other values – non-economic values – that we assign to the environment, and of course other types of responsibilities that we have to protect it. I will turn to those in Chapter 3. But it is essential, in evaluating whether or not modern agriculture is *economically* sustainable or not, to consider what *economic* costs eco-degradation imposes on us and on those later occupants (human and non-human) of the planet we currently occupy.

Frederick Kirschenmann, whose 2010 book on agricultural history, economics, and ethics I drew from earlier, offers these observations about the necessity of taking such costs into account:

> In a truly free-market economy, every technology should pay its own way. If an insecticide produces human illness or causes environmental damage, then the health care and environmental restoration costs associated with the insecticide should be added to the cost of its use. If livestock manure accumulations from the concentration of livestock in huge feedlots causes high nitrate levels in groundwater, then the cost of such pollution should be added to that method of producing meat.
>
> If we were to institute true cost accounting, the invisible hand of the free market might change the way we determine efficiency and give us a very different kind of agriculture. Estimating calories of energy expended to put a calorie of food on the table is a much better way of

---

54 The modern world's reliance on fossil carbon extends far beyond agriculture, of course. By one account, fossil fuels account for about 77 percent of all energy use today. See note 139 in Chapter 3, *infra*.

determining food production efficiencies. By that formula, modern industrial agriculture does not fare very well. For example, it takes an estimated 100 calories of energy to put a single calorie of potato chips on the table. If all of the costs were included in bringing a hamburger produced on rainforest land to a U.S. consumer's table, the burger would cost $200 . . .

. . . [W]e need economic models that internalize all the costs of doing business. When we defer external costs to the future, depreciate our natural capital, ask farmers to accept substandard wages for their labor, and subsidize exploitive agriculture, we conduct agriculture on the premise that someone besides the end user should pay the bill. To assume that costs don't exist because we have "externalized" them is to believe in economic magic. We need to charge and pay for the true cost of food and fiber, and give farmers a fair return for their labor.[55]

Two points made by Kirschenmann in the quoted passage bear special attention. First, he distinguishes between "true cost accounting" and the sort of cost accounting that overlooks what I have referred to before as "eco-degradation". Second, he distinguishes between production efficiency measured by calories instead of production efficiency measured by dollars currently spent. On that second point, Kirschenmann offers observations later in his same book − *Cultivating an Ecological Conscience* − that emphasize how energy-*inefficient* modern agriculture really is. In doing so, he echoes some of the points I highlighted previously from the analysis offered by Tim Crews of The Land Institute.[56]

Kirschenmann identifies three stages in the move toward this agricultural energy-inefficiency. First, he cites anthropologist Ernest Schusky for the proposition that from an energy-efficiency perspective, hunting and gathering "weren't such bad ways to feed ourselves".[57] When humans began the domestication of plants and animals, Kirschenmann says, another stage was reached: we traded efficiency for the benefits of being able to live in a settled and concentrated society rather than living a more nomadic life. A third stage in the move toward inefficiency, Kirschenmann explains, occurred quite recently:

Around 1930 we embarked on a new era of agriculture that Schusky calls the "neocaloric era" because it is based almost entirely on "old calories," namely fossil fuels.[58] The defining characteristic of our

55 Kirschenmann, *supra* note 3, at 137–138.
56 See text accompanying note 53, *supra*.
57 Kirschenmann, *supra* note 3, at 217, citing Ernest L. Schusky, Culture and Agriculture: An Ecological Introduction to Traditional and Modern Farming Systems (1989).
58 In introducing his book, Ernest Schusky explains why he uses the term "neocaloric": "[This word] is justified, partly because it parallels the term, Neolithic. Just as the Neolithic is not

modern food system is that it replaced human and animal energy with fossil-fuel energy. From an energy-efficiency standpoint, it is the least effective food system ever designed. The industrialized food system consumes more energy than it produces. Schusky cites one egregious example: it takes "about 2200 calories of fossil energy in order to produce a one-calorie can of diet soda," which he suggests is "downright embarrassing to human intelligence."[59]

In making his reference to the "old calories" of fossil fuels, of course, Kirschenmann is acknowledging one of the forms of subsidization that I discussed earlier – the indirect subsidization of modern agriculture by providing subsidies, especially in the USA, for producing energy and by-products from fossil carbon. As noted previously, that form of indirect subsidy gains ever more importance as modern agriculture uses increasing volumes of fuel, fertilizer, and pesticides derived from fossil-carbon sources.[60]

The second point Kirschenmann makes in the passage quoted above – in which he distinguishes between "true cost accounting" and the sort of cost accounting that overlooks what I earlier (on pages 57 and 63) called "eco-degradation" – has been widely discussed. It is often expressed in terms of economic "negative externalities". In this context, a general definition of an "negative externality" is "an unwelcome effect suffered by Party B from an activity undertaken by Party A without taking into account (or paying Party B for) that unwelcome effect."[61] More specifically, the class of "negative externalities" that Kirschenmann refers to (without using that term) – in asserting, for instance, that "the health care and environmental restoration costs associated with the [use of an] insecticide should be added to the cost of its use" – is what I have called "eco-degradation".

Another authority offers these observations, casting the discussion not in terms of negative externalities but specifically in terms of "ecosystem services" that are often, and inappropriately, overlooked in assessing the costs of certain actions:

> Ecologists have coined the term "ecosystem services" to describe processes that nature provides for free and that have real economic value.

---

composed of new stones, neither are there new calories in fossil fuel; indeed, they are most ancient. But humans, in employing fossil energy, embarked on a totally new direction not only in food production but also in their whole way of life". Schusky, *supra* note 57, at xiii.

59 Kirschenmann, *supra* note 3, at 217, citing Schusky, *supra* note 57, at xii–xiii.

60 In this respect, see the earlier references to the importance of ammonia, made from natural gas, at notes 19 and 20 in Chapter 1, *supra*.

61 For other definitions, see various online sources for "externality", including About.com, http://economics.about.com/cs/economicsglossary/g/externality.htm, which defines an externality generally as "an effect of a purchase or use decision by one set of parties on others who did not have a choice and whose interests were not taken into account" and notes that pollution represents a classic example of a negative externality.

For example, trees and other vegetation on hill slopes trap rainfall that would otherwise flow away and erode soils. As the retained water passes into subsurface layers, the soils slowly filter it and transform it into clean, drinkable water that can be retrieved from wells or springs. Some of the water flows into wetlands and is further cleansed there. Nature gives us a large supply of clean water.

But when the trees are cut or the wetlands are filled in, these free services are lost, and society must pick up the cost of doing nature's job. The extra runoff generated by removing trees requires construction of water-impoundment areas, which slowly accumulate silt and require maintenance. The loss of nature's subsurface cleansing of water requires municipal water treatment plants and home water filters, but these remedies rarely return water of the quality nature once provided. So we buy bottled water shipped from other regions or even other continents. All in all, we pay an economic price to replace ecosystem services. Ecologists rightly argue that the costs incurred from losing ecosystem services must be included in complete "economic" analyses of land-use decisions.[62]

As applied to modern extractive agriculture, the point is this: the damage that such agricultural operations inflict on the environment constitutes a cost – a "negative externality" – that must be accounted for if we are to assess properly the economic sustainability of those operations. And as with any negative externality, the responsibility for shouldering its cost should be imposed on the appropriate party. As I emphasize in section I of Chapter 3, modern extractive agriculture is responsible for a broad range of such eco-degradation. Taking the cost of that damage into account will undermine any claim that modern extractive agriculture is economically sustainable. Indeed, the same point can be made not only regarding eco-degradation but also regarding injuries that the products or processes of modern agriculture cause to human health and safety – a topic I will address in section II of Chapter 3.

If, as I have asserted, it is impossible to evaluate a claim that modern extractive agriculture is economically sustainable unless these various types of negative externalities are taken into account, then an obvious question arises: *How*, if at all, can they in fact be taken into account? What sorts of valuation methods can be used in order to account for those externalities, such as eco-degradation and injuries to human health?

In my view, the burden of addressing this question rests *not* with those who question (as I do) the economic sustainability of modern extractive agriculture but instead with those who claim – erroneously, in my view

---

62 William F. Ruddiman, Plows, Plagues, and Petroleum: How Humans Took Control of Climate 191 (2005).

– that modern extractive agriculture is defensible on economic grounds. Consider the matter from a public-works perspective. Modern rules of "environmental assessment" in any developed economy today will require that no new public infrastructure project should be undertaken unless the proponents can demonstrate that the project will produce a positive economic return when all pertinent factors are taken into account. These factors are to include the costs and values of ecological damage or, in some cases, ecological repair or improvement. The burden of identifying and calculating the costs and values rests on the proponents because it is in the very nature of public works projects – highways, airports, power plants, and the like – to change a landscape in ways that will likely disrupt the existing ecology of the site.

Likewise, the massive system of modern agriculture – occupying ever-expanding areas of land on the surface of our shared Earth – should be continued only if its proponents can demonstrate that it produces a positive economic return when all factors are taken into account. These factors should, as in the case of public infrastructure projects, include the costs and values of "externalities" of the kinds mentioned previously. Again, the burden of identifying and calculating those costs and values should rest with those involved in modern extractive agriculture – especially if, as I will explain in Chapters 4 and 5 of this book, there are alternative forms of food production.

Setting aside this "burden of proof" assertion, I would offer the following observations about how to place values on the negative externalities mentioned above. For this, I rely heavily on the work of the work of John Dixon as summarized in a recent book on sustainable development.[63] Under Dixon's explanation, four "approaches to valuation" can be used to capture the values of externalities in this context: "changes in production, hedonic approaches, survey techniques, and surrogate market".[64]

The authors summarizing Dixon's views then explain these various terms and their function in establishing values for non-traded items. For instance, "Hedonic approaches, from the Greek *hedone*, meaning 'pleasure,' deal with those environmental attributes that give pleasure, such as scenic surroundings, an office with a view, and so on." We "can often measure the benefits received from unpriced environmental attributes through what people are willing to pay for them". By contrast, "survey techniques" typically "rely on asking people literally how much they are willing to pay for a cleaner environment or some attribute of the environment".[65]

---

63 See generally Peter P. Rogers *et al.*, An Introduction to Sustainable Development (2008). Chapter 10 of that book, focusing on externalities and environmental sustainability, draws heavily on the work of John Dixon, a former environmental economist at the World Bank. *Id.* at 16, 276.
64 *Id.* at 282.
65 *Id.* at 282–283.

My point is not to explain in detail these and other valuation techniques. Instead, I merely wish to indicate that such techniques do exist, so the fact that economic analyses of modern extractive agriculture typically do *not* employ those techniques in an effort to prove its economic sustainability constitutes a serious shortcoming. What should be undertaken is a full-cost accounting.

Such an accounting would incorporate – again, to rely on the resource cited on page 67 – the totality of "use values and non-use values".

> *Use values* are concrete, such as the value we get from the production of goods; establishing fisheries, forests, and water resources; building dams; and so on. Use values can be broken down into two separate categories, which we call direct and indirect use . . .
>
> A *direct use value* occurs when we directly use something or consume something so that we must produce more of it, such as fish. There can also be non-consumptive direct use, as when we go SCUBA diving in the same marine area that produces the fish. The diver is using the same resource but not consuming it. It is obvious that non-consumptive use is usually more sustainable . . .
>
> *Indirect use values* are usually measures of benefits or services – [such as] ecosystem services, water purification in wetlands, and greenbelts around cities that help reduce pollution in the air and water. There are a lot of ecosystem services that we benefit from directly or indirectly whether we consume them or not, but they are a little harder to measure . . .
>
> Another use value is what is called an "*option value*" or an "option demand." It is an economic or environmental value that is deferred. It is a value where people might say of a recreational area: "I may not want to go there now, but I am willing to pay something to maintain that resource in case I decide to go later." Or it could be that society as a whole may decide that it wants to maintain resources so that future generations can use them . . .
>
> *Non-use values* consist of existence values and bequest values . . . . *Bequest values* . . . are technically the same as option values except that bequest values have a longer time frame . . . *Existence values* . . . are pure benefit – knowing that something exists for the very value of its existence, like whales or mountains. It is the least tangible of these types of values. When we move from direct use to indirect use to option to bequest to existence, and actually try to estimate value, the evaluation techniques become less and less tangible. When evaluating use values we can look at things such as prices and quantities. We can even do this for some indirect use values. However, when we try to measure existence values we always end up relying on survey-based approaches . . . .[66]

---

66 *Id.* at 284–287 (emphasis added).

Any legitimate assessment of the economic sustainability of modern extractive agriculture must reflect these various sorts of values and costs. What I have asserted thus far is that even *without* reflecting them, modern extractive agriculture is *not* sustainable. But assuming my assertion is acknowledged as being correct, are there some other grounds on which modern extractive agriculture can be defended?

III. World food supply as a trump card? [67]

Having offered a range of grounds on which modern extractive agriculture might well be judged *not* to be economically sustainable, I turn to what might be a strong counter-argument. That counter-argument focuses on an issue that is often used as the ultimate justification for today's agricultural practices: world food demand. The counter-argument can be expressed as follows: Even if, on the surface, modern agriculture seems to be unsustainable in an input–output sense – on grounds that it disregards or passes on to others the costs of direct (agricultural) subsidies, indirect (fossil-carbon) subsidies, eco-degradation, and other negative externalities – a more sophisticated view will reveal that modern agriculture is in fact totally justified, and indeed should be celebrated, because it can be relied on to continue providing, year after year, the food supply needed for a growing world population.

This argument fails. For one thing, it overlooks the fact that although agricultural outputs and productivity have gradually risen – again, see Figures 2.3, 2.4, 2.5, and 2.7 for the data showing these trends in the US context – the world's per capita food supply from 1961 to 2011 has risen much more slowly. Data available from the UN Food and Agriculture Organization ("FAO") show that the total daily per capita caloric value available in the aggregate production of food – that is, the world food supply – increased only about 30 percent between 1961 and 2011.[68] In absolute terms, the increase over that 50-year period is from 2,196 kcal/capita/day to 2,870 kcal/capita/day as an average daily per capita caloric value available in the food supply. Not surprisingly, much of that caloric value comes from the top four grain crops – rice, wheat, maize, and soybeans (including soybean oil). These accounted in 2011 for about 19 percent, 19 percent, 5 percent, and 3 percent, respectively, of the total worldwide daily caloric value. By comparison, however, the increases in output and in productivity in US agriculture over that same period – drawing from the data reflected

67 This section (III) reflects extensive research and writing by Caleb Hall.
68 See UN Food and Agriculture Organization, *FAOSTAT*, 2012, faostat3.fao.org/home/index. html (follow "Download data" hyperlink; then follow "Food Balances" hyperlink to "Food Supply – Crops Primary Equivalent" hyperlink). The food supply is measured in numerous ways, including kcal/capita/day.

in Figure 2.5 and Figure 2.6 – are about 104 percent and 109 percent, respectively.[69]

My point in offering these rough-and-ready comparisons[70] is to highlight the mismatch that seems to exist between rhetoric and reality in the assertion that modern extractive agriculture is justified because it is capable of meeting the challenge to "feed the world". In fact, the data show that even though US agricultural productivity – touted as a near-miracle and on that basis adopted in much of the rest of the world[71] – has more than tripled in the past half-century, worldwide average daily per capita caloric supply has increased by only about 30 percent over that same period.[72]

Moreover, this increase remains inadequate to fight malnourishment. That is, as further evidence that modern extractive agriculture is not in fact succeeding in keeping up with the need for an increasing world food supply, consider the trend shown in Figure 2.8, also drawn from FAO data. Figure 2.8 shows that the numbers of malnourished people over the period of 1969–2014 have stayed stubbornly above 800 million people. Other sources confirm the same general point: in the face of increasing global populations, food supply continues to fall short. How far short it will fall will depend in part, of course, on how fast human populations increase. In that respect, it is noteworthy that although a 2004 UN projection called for population to peak at 9.22 billion in 2075, then to drop slightly, the UN has since backed off that projection and raised its numbers. Under the 2012 UN projections, world population would "increase by almost one billion people within the next twelve years, reaching 8.1 billion in 2025, and to further increase to 9.6 billion in 2050 and 10.9 billion by 2100." [73] Even more recently, the UN

---

69 As noted earlier, the US productivity and output data are measured by indices of change relative to the 2005 production year, meaning that 2005=1, as opposed to the method applied in the FAO data, which are measured in kcal/capita/day. The comparison of US and world data therefore must be regarded as rough, not precise.

70 I am aware of the limitations inherent in working with data from several sources and using different methods of measurement – for instance, using US production data focusing on year-on-year changes and using world food supply data focusing on average daily per capita caloric value – a point emphasized in the preceding note. Still, the magnitude of the disparities between rapidly increasing production and slowly increasing supply is great enough to warrant giving credence to such comparisons.

71 In support of using a comparison between US production and global food supply, note that the same methodology is used for similar purposes in Rosset and Altieri, *supra* note 49, at 284.

72 I should emphasize why I focus on *per capita* food supply rather than on *aggregate* food supply. I do so because looking simply at raw food supply ignores how much food is actually available for each person. The naked food supply number also overlooks the realities of luxury crops being grown in developing nations only to be consumed in developed ones, as well as the unequal distribution of resources necessary to secure to food once it is grown, and other socioeconomic issues.

73 See UN Department of Economic & Social Affairs, WORLD POPULATION PROSPECTS: THE 2012 REVISION (KEY FINDINGS AND ADVANCE TABLES) (2013), http://esa.un.org/wpp/documentation/pdf/WPP2012_%20KEY%20FINDINGS.pdf. See also UN Department of Economic & Social

has updated these figures further. The projection for 2050 is now 9.7 billion, up from 9.6 billion. The projection for 2100 is now 11.2 billion, up from the projected 10.9 billion – an increase of 300 million, roughly equal to today's population of the USA.[74]

Accordingly, we must regard with suspicion any attempt to justify the system of modern extractive agriculture on grounds that this system is necessary to, and prepared to, meet the food-supply needs of an increasing global human population – especially when this system continues to intensify the problem by enabling substantial population increases. The pertinent data do not seem to support such an argument.

This is especially troublesome if the *production* of increasing supplies of food per capita is considered necessary to increase people's *access* to food, and therefore to alleviate hunger and poverty. Many observers have emphasized that the overall quantity of food *production* is a less pertinent matter than the *distribution* of food.[75] This is wholly plausible: inadequacies in the systems and

---

Affairs, WORLD POPULATION PROSPECTS: THE 2006 REVISION (2007), www.un.org/esa/ population/publications/wpp2006/English.pdf; Justin Gillis and Celia W. Dugger, *U.N. Sees Rise For the World To 10.1 Billion*, THE NEW YORK TIMES, May 4, 2011, at A1; Jocelyn Kaiser, *10 Billion Plus: Why World Population Projections Were Too Low*, SCIENCE INSIDER (May 4, 2011). See also Population Reference Bureau, 2011 WORLD POPULATION DATA SHEET: THE WORLD AT 7 BILLION (2011), www.prb.org/pdf11/2011population-data-sheet_eng.pdf (the Population Reference Bureau is a US-based private nonprofit organization); US Census Bureau, GLOBAL POPULATION GROWTH (2003), www.census.gov/population/international/files/wp02/wp-02003. pdf. For a discussion of population issues – and specifically the differences between carrying capacity, maximum capacity, and preferred or optimal capacity of global human population – see subsection IIID of Chapter 4, *infra*.

74 UN Department of Economic & Social Affairs, *World population projected to reach 9.7 billion by 2050*, www.un.org/en/development/desa/news/population/2015-report.html (2015), p. 3. Another source explains that 99 percent of projected human population growth over the next four decades "will occur in countries that are classified as less developed", and by 2050 two-thirds of the world's human population will live in urban areas. David E. Bloom, *Demographic Upheaval*, FINANCE & DEVELOPMENT, Mar. 2016, at 6–7.

75 See, e.g., Mark Koba, *A Hungry World: Lots of Food, in Too Few Places* (July 22, 2013), appearing on the CNBC website at www.cnbc.com/id/. See also *Inadequate Food Distribution Systems*, on the website of Mission 2014: Feeding the World (associated with the Massachusetts Institute of Technology) at http://12.000.scripts.mit.edu/mission2014/problems/inadequate-food-distribution-systems (asserting that "[e]nough food is produced worldwide to feed all the people in the world" and that one of the most significant reasons that "nearly 1 billion people are suffering from chronic hunger today" is "poor food distribution"). For a similar point, see John Thackara, HOW TO THRIVE IN THE NEXT ECONOMY: DESIGNING TOMORROW'S WORLD TODAY 69, 78–79 (2015). On the other hand, some observers caution against assuming that global malnourishment results only or predominantly from problems with distribution. For instance, one NGO focusing on nutrition offers this assessment:

Global hunger results from a web of immensely complex factors, including BOTH food scarcity and distribution. Thinking that hunger is mostly a problem of distribution is dangerous in that it leads people to dismiss the issue of scarcity and results in practices that are inappropriate and harmful. Food scarcity at the global level is an issue now with past surpluses being drawn

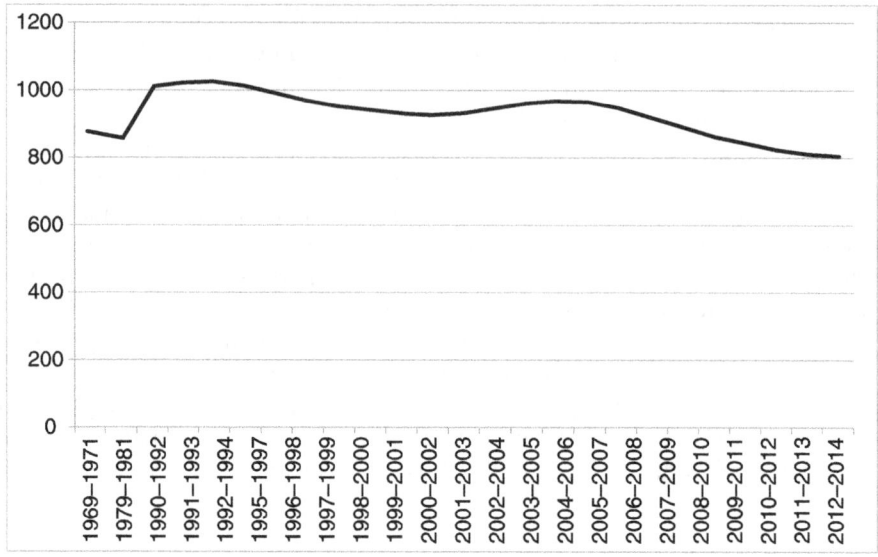

Figure 2.8 Number of malnourished people worldwide (millions), 1969–2014[76]

equities of food distribution, along with the waste of food, are crucial problems contributing to global malnourishment – and those problems, in turn, stem importantly from war, social disruption, economic and political inequality, and a host of other societal problems. Addressing those problems would surely give some relief to the 800 million humans facing malnourishment (see Figure 2.8). That fact, however, still does not add any persuasive power to the claims that the value and necessity of modern extractive agriculture is proven by its ability to meet global human food supply needs. As a practical matter, it has not done so to date, and the fact that global human population is increasing faster than global food production suggests that it will not do so in the future.

The fact that modern agricultural production does not seem to be increasing quickly enough to provide for an ever-increasing global demand

---

down and it is fast becoming a critical issue as our seven billion population expands towards nine billion by 2050. As our population increases, available land, water, energy and other finite resources decrease. So we have more people to feed and fewer resources to feed them.

A Well-Fed World, *Scarcity vs. Distribution*, http://awfw.org/scarcity-vs-distribution/.

76 The sources for the graph in Figure 2.8 are: (i) for information relating to 1969–1971 and 1979–1981, Food and Agriculture Organization, *The State of Food Insecurity in the World: Addressing Food Insecurity in Protracted Crises* (Rome), 2010, at 9, www.fao.org/docrep/013/i1683e/i1683e.pdf; and (ii) for other information, Food and Agriculture Organization, *The State of Food Insecurity in the World 2014* (Nov. 17, 2014), www.fao.org/economic/ess/ess-fs/ess-fadata/en/#.VRoAyeEhEXI (follow "Food security indicators – download the data" hyperlink).

for food might suggest that crop yields have peaked. Some evidence of this can be drawn from Mexico, where wheat yields have stopped increasing, and also in Asia, where rice yields have begun to fall.[77]

Some of these same points about the possible peaking of yields, specifically those yields attributable to the Green Revolution that started in the 1940s, are elaborated by Lester Brown, president of Earth Policy Institute. In his 2011 book *World on the Edge*, Brown offers these observations on the results of – and reasons for – the high-productivity Green Revolution years, and the rather dim prospects for further advances in productivity:

> Prior to 1950, growth of the food supply came almost entirely from expanding cropland area. Then as frontiers disappeared and population growth accelerated after World War II, the focus quickly shifted to raising land productivity. In the most spectacular achievement in world agricultural history, farmers doubled the grain harvest between 1950 and 1973. Stated otherwise, growth in the grain harvest during this 23-year span matched that of the preceding 11,000 years.
>
> This was the golden age of world agriculture. Since then, growth in world food output has been gradually losing momentum as the backlog of unused agricultural technology dwindles, as soil erodes, as the area of cultivable land shrinks, and as irrigation water becomes scarce . . .
>
> [Indeed, there now] are distinct signs of yields leveling off in the higher-yield countries that are using all available technologies. This is illustrated by the plateauing of wheat yields in France (Europe's largest wheat producer), Germany, the United Kingdom, and Egypt (Africa's leading wheat grower) . . . [and also in the plateauing of rice yields in Japan and China] . . . . [In fact, among] the big three grains, corn is the only one where the yield is continuing a steady rise in high-yield countries [such as the USA] . . .
>
> [In short, despite] dramatic past leaps in grain yields, it is becoming more difficult to expand world food output . . .[78]

A further ground for concern regarding the world food supply – and whether modern extractive agriculture can provide it – is that there was as of mid-2013, by some accounts, less than three months' worth of grain-supply reserves.[79] We are still living harvest to harvest.

---

77 See Montgomery, *supra* note 36, at 239. The same source notes that nitrogen inputs have had to increase just to maintain crop yields in the Philippines. *Id.*

78 Lester R. Brown, WORLD ON THE EDGE: HOW TO PREVENT ENVIRONMENTAL AND ECONOMIC COLLAPSE 165–167 (2011).

79 For recent figures on this, see Ned W. Schmidt, *Big U. S. Corn Crop: Feeds World for Two Days* (July 2013), www.financialsense.com/contributors/ned-schmidt/big-u-s-corn-crop-feeds-world-for-two-days. That source explains that although global grain reserves in the 1980s and

In short, despite the advances in agricultural technology and the heavy subsidization that modern extractive agricultural production enjoys, there is precious little cause for confidence. Widespread undernourishment persists, world per capita food supplies are not increasing quickly, and the prospects for further increases in yield are not bright.

IV. A summing-up on economic unsustainability

In the foregoing narrative of this chapter, I have touched on a large number of facts and factors bearing on the economics of modern extractive agriculture as I defined it in Chapter 1.[80] Taken together, these details permit us to formulate some general observations about the sustainability – or rather the unsustainability – of modern extractive agriculture from an economic perspective. Let me now offer this summary of the most important of these economics-related points:

1 *High costs.* Although agriculture has always involved risk – dependence on weather, on markets, and on other factors outside the control of the farmer – the most contemporary "industrial" form of agriculture as it has developed in the last few decades also involves extraordinarily high entry costs and operational costs, presenting to farmers and would-be farmers a financial disincentive or "hurdle" of unprecedented height.

2 *Fewer farms, fewer crops.* This high-cost disincentive helps explain both (i) why there has been such a dramatic decline in the past few decades in the number of people engaging in farming as an occupation – particularly in the developed countries where the most high-cost resource-intensive techniques of agriculture predominate – and also (ii) why agricultural production has come to be concentrated in a remarkably short list of crops, particularly those that government policy subsidizes.

3 *Concentration of power.* As a result, the actual control and operation today of agricultural production is concentrated largely in a few giant agribusiness enterprises, with political power to capture public (state) support that further strengthens the trend away from diversification, despite the fact that smaller farms with diversified ownership and crops probably have greater potential for poverty reduction and efficient productivity if given adequate and equitable support.

4 *Large farm success?* Although small farms and small-scale farmers suffer under modern extractive agriculture as it has evolved in developed countries, particularly the USA – that is, their operations usually do not

---

early 1990s "averaged about 100 days of consumption", such reserves more recently "have averaged only about 70 days". *Id.*

80 For explanations of why I use the terms "extractive" and "modern" (along with "industrial" and other modifiers) to describe the form of agriculture that now dominates the world, see Chapter 1, and especially Figure 1.1, *supra.*

enjoy economic sustainability – *large* farm operators and especially the giant agribusiness enterprises *do* seem at first glance to be succeeding. After all, the direct values of outputs have consistently exceeded the direct values of inputs in the US agriculture sector for the past 70 years.

5 *No, even large operations are not economically sustainable.* Closer examination reveals that even giant agribusiness enterprises do not reflect an agricultural model that is economically sustainable. Without heavy subsidization – that is, without generous public resources from society as a whole in the form of both (i) agricultural subsidies and (ii) fossil-carbon subsidies – the economic viability even of giant agribusiness enterprises would crumble. Indeed, it is that very subsidization that largely explains the productivity of the US agricultural sector over the past 70 years: neither the "direct values of inputs" amount nor the "direct values of outputs" amount referred to in point 4 properly reflects the public subsidies – or other externalities such as environmental degradation, discussed in section I of Chapter 3.

6 *World food-supply needs as justification?* Moreover, what some would see as the ultimate justification of modern extractive agriculture – that it has succeeded in creating an ever-increasing world food supply – is itself deeply suspect. Even aside from the issue of whether we should in fact gear global food production in order to meet exploding human populations (rather than basing population policies on the Earth's appropriate carrying capacity),[81] there is no sign that the system of modern extractive agriculture is meeting the world's needs now, inasmuch as that system is doing a poor job of fighting global malnourishment through increases in the world per capita food supply.

7 *Dangerous fallout from Green Revolution?* Indeed, there are danger signals suggesting that crop yields have peaked, partly because the dramatic gains attributed to the Green Revolution, and the broader yield-enhancing efforts that it contributed to, were like a "flash in the pan". That "flash", moreover, in the process of providing a short-term boost in productivity, has seriously injured farmers and agricultural markets in less developed countries, in two key ways: (i) it has undercut the viability of those farmers and markets by dumping cheap agricultural commodities (exported from the rich countries, especially the USA, in order to absorb their excess production); (ii) it has contributed to a long-term dependency of the developing world on the developed world by substituting Western-developed high-cost crops for traditional local crops. This injury also flies in the face of the fact noted previously at the end of point 3 in this list, that smaller and more ecologically-oriented food production is better in important ways than production through high-cost, fossil-carbon-dependent extractive agriculture.

---

81 For a discussion of the differences between carrying capacity, maximum capacity, and preferred or optimal capacity of human population, see subsection IIID of Chapter 4, *infra*.

8   *Reliance on fossil-carbon "slaves".* The "fossil-carbon-dependent" character of modern extractive agriculture, referred to in point 7, should be of special concern to us. It highlights the central role of *energy* in the system of food production. As Tim Crews has characterized it, today's system would collapse were it not for the fact that fossil carbon has, just within the past century and a half, been brought into service as something like "slaves". To the extent that government subsidies are afforded to the fossil-carbon industry, this undercuts any claim that modern extractive agriculture has economic viability over the long term – if, that is, we recognize that there are limits to the human species' ability to continue relying on such fossil-carbon "slaves".

9   *Fundamental flaws in the system?* This combination of factors should make us question the appropriateness of providing such massive subsidies either (i) to the enterprises in whose hands agricultural production is currently concentrated or (ii) to the industry involved in the production and distribution of fossil-carbon energy. More fundamentally, this combination of factors suggests that the model itself – that is, modern extractive agriculture as described in Chapter 1 – is intrinsically flawed. Expressed differently, modern extractive agriculture is economically unsustainable both (i) from the perspective of small farmers – who, it should be recalled, constituted most of the world's population only a few generations ago – and also (ii) from a global perspective.

10 *Environmental and social externalities increase economic unsustainability.* The foregoing line of reasoning points strongly toward a damning verdict for modern extractive agriculture. But in fact the situation might be even worse than this reasoning suggests: the foregoing analysis does not take fully into account the environmental and social "negative externalities" involved in modern extractive agriculture. As described next in Chapter 3, those externalities represent high costs indeed: modern extractive agriculture causes deep and abiding environmental degradation and poses serious health risks that have not been adequately assessed. If they *were* included in the analysis, the likelihood of finding modern extractive agriculture to be economically sustainable would be even lower.

3 Modern extractive agriculture is ecologically and socially unsustainable

As I explained in the Preface, my overall objective in writing this book is to contribute to the ongoing interdisciplinary efforts aimed at replacing what I have called modern extractive agriculture[1] with a "natural-systems" agriculture concentrating on perennials grown in polycultures instead of annuals grown in monocultures. Chapter 2 explored why a change in global agriculture is warranted from an economic perspective. This chapter examines why a change in global agriculture is warranted on ecological, human-health, and other social grounds. I begin with an ecological assessment.

I. Modern extractive agriculture is ecologically unsustainable

IA. An instructive illustration – ecological destruction and agricultural conversion of grasslands

In my recent book on protecting the world's grasslands, I emphasized how much damage has been inflicted on grassland areas through agricultural conversion – that is, through the conversion of grasslands to agricultural use, including both cropping and livestock production.[2] Now, in the context of this current book, my focus is much broader: I wish to offer a survey of the whole range of environmental ills visited on the Earth through modern extractive agriculture. But for an "opener" to that larger topic, let me offer a summary of what agriculture has done to the world's grasslands, which would provide (but for the degradation inflicted on them by agriculture and other factors) such important benefits as soil conservation, biological diversity, animal habitat, and carbon sequestration.

Numerous studies, groups, and campaigns have drawn attention to the disappearance and degradation of grasslands around the world. One group headquartered in British Columbia offers this assessment:

> Of all the ecosystems on earth, none has been more dramatically affected by humanity than native grasslands. These lands have been widely altered, because they are attractive places for humans to build settlements, grow crops and graze livestock. Although native grasslands at one time covered 40% of the North American Continent, the vast majority has been transformed into agricultural lands, urban settings, and other settlement uses. In places with significant development and agricultural pressures, virtually all native grasslands have disappeared. For example,

---

1 For an explanation of the term "modern extractive agriculture", and how it serves as a shorthand for the more descriptive (but awkward) term "industrial fossil-carbon-based enhanced extractive agriculture", see text accompanying notes 13–26 in Chapter 1, *supra*, as well as the "composite timeline" in Figure 1.1 in that chapter.
2 John W. Head, Global Legal Regimes to Protect the World's Grasslands 41–45 (2012) [hereinafter Grasslands].

98% of the tallgrass prairie east of the Missouri River is gone, and California has lost 99% of its native grasslands.[3]

One of those figures bears repeating: 98 percent of the tallgrass prairie east of the Missouri River is gone.[4] Only slightly less arresting figures apply to the disappearance or degradation of mixed grass and short grass prairies in North America,[5] as well as to the disappearance or degradation of other grasslands all around the world.[6] Here are a few examples:

- Of the 13.8 million acres of native prairie that remains in the eastern Dakotas, 298,000 acres – over 2 percent – were converted to cropland just during the period 2002–2005. This amounts to a 0.55 percent annual conversion rate.[7]
- The conversion rate might be much higher in some regions. For example, research conducted in the Missouri Coteau region of the Dakotas has documented annual loss rates as high as 2 percent in some key areas. To put this into perspective: if that rate were maintained over the long term, half of the remaining native grasslands in those areas would be lost in less than 35 years.[8]
- The loss of grasslands in India, partly through overgrazing, has resulted in a scarcity of habitat for such iconic species as rhinoceros and water buffalo.[9]

---

3 Environmental Law Clinic, University of Victoria Faculty of Law, GRASSLANDS PROTECTION: A PRIMER FOR LOCAL GOVERNMENTS (undated) accessible at: www.rdosmaps.bc.ca/min_bylaws/ planning/rgs/ReportsAndStudies/GrasslandsProtection.pdf, and accessible by link from: www. rdos.bc.ca/index.php?id=316&no_cache=1&sword_list[]=primer.

4 Another source offers a slightly different but equally disturbing measure: "Estimates are that 95[%] to 99% of tall-grass prairie [which] with its spectacular head-high grasses and wildflowers once covered the rich soils of what is now the US Corn and Soybean belt . . . has disappeared, making this one of the globe's most critically endangered ecosystems." John P. McCarty and L. Lareesa Wolfenbarger, *Grassland Birds in Agricultural Ecosystems*, www.unomaha.edu/environmental_ studies/McCartyHomePage/GrasslandBirds.html.

5 The story of America's grasslands – and their disappearance in a mere flash of time – was summarized three decades ago by another observer, who emphasized that after millennia of evolution, those grasslands "were transformed within half a century by the European settlers . . . The existing plants and animals gave way to new species that were either cultivated by human beings or were able to adapt to the conditions created by human beings." As a consequence, "the natural, or 'true,' grasslands in the United States have all but vanished." Lauren Brown, GRASSLANDS 19 (1985) (Audobon Society Nature Guides series).

6 For a survey of problems resulting from the cultivation of grasslands around the world, see J. M. Suite, S. G. Reynolds, and C. Batello, *Grasslands of the World*, PLANT PRODUCTION AND PROTECTION SERIES NO. 34, 2005, www.fao.org/docrep/008/y8344e/y8344e0j.htm.

7 *Sodsaver: Saving America's Prairies*, www.ducks.org/conservation/farm-bill/sodsaver-saving-americas-prairies.

8 *Id.*

9 Dipannita Das, *Need to Protect Our Grasslands* (June 9, 2011), http://articles.timesofindia.indiatimes. com/2011-06-09/pune/29638172_1_grasslands-industrial-land-horticultural-crops.

- The digging and drilling for coal, natural gas, and rare earth elements in the Inner Mongolia Autonomous Region of China – this occupies 12 percent of the total territory of the People's Republic of China – has resulted in widespread grasslands degradation.[10]
- Many grassland areas in Tibet have already been lost.[11]

These changes in the world's grasslands can be credited entirely to one overriding cause: human action. The two main types of human activities triggering this environmental degradation are (i) inappropriate livestock grazing and (ii) agricultural conversion.

In many grasslands areas around the world, native animals that depend on grasslands have had to compete with a number of introduced grazing animals. For example, since the mid-1800s North American grasslands have been used for grazing cattle, horses, and sheep. For most of that time, the grazing practices, combined with the sheer numbers of animals on the grasslands, destroyed or seriously degraded many North American grasslands. Although some grazing practices have been improved, the damage continues, and in a variety of forms:

- Extensive grazing, especially by cattle, damages groundcover that forms protection for ground-nesting birds, mice, and voles, thereby leading to species reduction – not only among *those* species but also among *other* species (for instance, hawks and owls) that feed on them.
- Extensive grazing also can adversely affect indigenous wildlife by reducing the food supply available to them, especially the larger species of grasses.
- Even if grazing is reduced in a particular area, some grasslands have been plowed under in order to grow hay crops for the winter feeding of livestock. (This point also is pertinent to the topic of agricultural conversion, discussed below at pages 81–83).
- Grazing and trampling by livestock have damaged many grassland riparian areas through loss of vegetation, soil erosion, bank erosion, and reduced water quality.
- Invasive plant species that threaten the local vegetation can be introduced to grasslands through hay imported to feed cattle.

While there are alternatives to the types of inappropriate grazing techniques that cause some of these forms of damage – such as seasonal and rotational

---

10 Andrew Jacobs, *Mongolian Protests Gather Momentum*, THE NEW YORK TIMES, May 30, 2011 (describing a campaign by herders against coal-mining vehicles that zigzag across the steppe, chewing up the fragile pastureland and occasionally running down livestock). See also Andrew Jacobs, *Anger Over Protesters' Deaths Leads to Intensified Demonstrations by Mongolians*, THE NEW YORK TIMES, May 31, 2011 (reporting on the "anger over the destruction of the Mongolian grasslands and . . . [the] long-simmering resentment over Beijing's governance" of Inner Mongolia).

11 See Jonathan Watts, *Tibetan Nomads Struggle as Grasslands Disappear from the Roof of the World*, GUARDIAN, Sep. 2, 2010.

grazing, giving grasslands adequate time to recover – in many areas those alternatives are not widely practiced. For instance, the use of vast areas of the American West for livestock grazing has resulted in extensive damage to the grasslands there. One authority offers this account:

> On millions of acres of public lands, grazing has eroded sparse topsoil, scoured and fouled streams, and replaced native grasses with exotic species. Cheatgrass is one; imported to the [American] west more than a century ago, it is now found on 25 million acres of range in the Intermountain West, an area the size of Kentucky. It forms dense, fire-susceptible mats that can wipe out sagebrush, which has led to serious decline of the sage grouse . . . [As a result of this and other damage, many grassland areas of the West] bear little resemblance vegetatively to "potential natural, or climax" plant communities [according to the US Bureau of Land Management].[12]

The same authority notes that, ironically, the damage that grazing causes in the American West does not seem to be justified by a significant return in terms of livestock production:

> While the number of acres involved is vast [at nearly a quarter of a *billion* acres], the amount of meat produced is, compared to total [US] national meat production, quite small . . . . In 1994, the Department of the Interior estimated that stopping all grazing on all federal lands would eliminate only about eight percent of the cows and less than one percent of the sheep in the 11 western states, causing a regional job loss at 18,300 (about one percent of total agricultural employment in the region . . .) with only "slight" effect on beef prices.[13]

Let us turn now from livestock grazing to the other main cause of grasslands degradation around the world: agricultural conversion. Grassland soils, especially those in temperate regions, are often quite fertile. This fertility has tempted settlers to plow the grasslands for agricultural purposes. Doing so can *temporarily* yield large crop production, of course, either for human consumption or for feeding to livestock, but it obviously destroys the grasslands, thereby eliminating all the other benefits that those grasslands can provide.

It is worth reminding ourselves of just *when* the conversion of grasslands to agricultural use began to have a significant impact on the world's grasslands. As portrayed in the account excerpted above from Wes Jackson – compressing

---

12 George Cameron Coggins, Charles F. Wilkinson, John D. Leshy, and Robert L. Fischman, Federal Public Land and Resources Law 769 (6th ed., 2007).
13 *Id.* at 768 (emphasis in original).

750 million years into a single year[14] – the degradation began only quite recently. Up until less than two centuries ago, the world's richest prairies and grasslands were largely intact. One reason for this is population: in the early 1800s the world's human population had only barely reached one billion; it now exceeds seven billion.

Another reality of the early 1800s also explains why the world's richest grasslands, including especially the vast North American prairies, were then still intact: humans had no way to destroy them. This was particularly important in North America, where the rich prairies remained indestructible until they met their match in the mid-1830s with the introduction of the steel plow – a development described above in Don Worster's account of the "waves of conquest decade by decade across the [American] continent" in the form of "the Great Plow-up".[15]

In addition to the direct effect that agricultural conversion has on grasslands – that is, by physically replacing them with fields of crops – some secondary effects also can be seen. Rivers, creeks, and streams in grassland areas have been channeled, dammed, and diverted in order to provide water for agricultural irrigation – that is, water to grow crops.[16] In addition, some ponds and lakes have been drained to make still more land available for further agricultural production. These alterations to bodies of water have in many cases destroyed or degraded the grasslands of which they form an integral part.

The negative impact of the conversion of grasslands to agricultural use takes many forms around the world. A brief survey of the literature reveals such illustrations as these:

• In the Great Plains of North America, "[a] high proportion of the species in the grassland bird community have given indications of population declines over the past decades" as the intensity of agricultural activities has increased.[17]

---

14 See text accompanying note 30 in Chapter 1, *supra*.

15 See text accompanying note 15 in Chapter 1, *supra*. For another account of "the Great Plow-up" in the American Great Plains, see Geoff Cunfer, ON THE GREAT PLAINS: AGRICULTURE AND ENVIRONMENT 16 (2005) (asserting that "[p]lowing is the ecological equivalent of genocide" and that "[p]lowing a grassland is comparable to clear-cutting a forest, but more absolute because the effort to maintain only a single species [in modern farming] continues year after year").

16 For a useful summary of the damage done to grasslands and agricultural areas by irrigation, see Claudio O. Stockle, *Environmental Impact of Irrigation: A Review*, www.swwrc.wsu.edu/newsletter/fall2001/IrrImpact2.pdf.

17 See McCarty and Wolfenbarger, *supra* note 4. Of these grassland birds, one of the most prominent was once the prairie chicken, which is a member of the grouse family. Reports of the near-disappearance of prairie chickens (especially the lesser prairie chicken, but also the more numerous greater prairie chicken) appear regularly in local magazines and newsletters in Missouri and Kansas and neighboring states. See, e.g., *Ghosts of the Grasslands*, RURAL MISSOURI, July 2010, at 8–9 (noting that although it is estimated that before the European encroachment hundreds of

- Likewise, in the Midlands of KwaZulu-Natal province, South Africa, guinea fowl populations and avian diversity overall have "declined with increasingly intensive agriculture and [especially with] the loss of the habitat mosaic [that has resulted from] . . . intensive, modern, monoculture, crop agriculture" in that region.[18]
- In other parts of Africa, a UN Environment Programme publication reports that "species introduced for plantation forestry production are rapidly invading grassland ecosystems where they use vast amounts of water, disturb ecosystem functioning, and reduce biodiversity".[19] (Further references to biodiversity reduction and to foreign species invasion will be made on pages 85–86 and 92–94.)
- In the Cerrado region of South America, "conversion of the land for agriculture has decimated nearly 80 percent of the native vegetation".[20]
- In Europe, changes in agricultural practices and land use pressures mean that grasslands are disappearing at an alarming rate and are currently among Europe's most threatened ecosystems. According to one report, the area of grasslands in Europe declined by 12.8 percent from 1990 to 2003.[21]
- As a consequence of these and other factors, moreover, more than 75 percent of the habitats provided by grasslands areas that do still exist in Europe are in an unfavorable conservation status.[22]

IB. Environmental damage done by agriculture more generally

Now let us broaden the field of view. The observations offered above relate only to one aspect of the ecological damage caused by modern extractive agriculture, focusing on how both crop production and livestock production have seriously degraded and destroyed grassland areas around the world.

---

thousands of prairie chickens lived in the territory now occupied by Missouri, "fewer than 100 native birds [of that species] still exist in Missouri").

18 C. S. Ratcliffe and T.M. Crowe, *The Effects of Agriculture and the Availability of Edge Habitat on Populations of Helmeted Guineafowl Numida meleagris and on the Diversity and Composition of Associated Bird Assemblages in KwaZulu-Natal Province, South Africa*, 10 BIODIVERSITY AND CONSERVATION 2109 (2001).

19 *Forest Cover and Quality in Southern Africa*, in AFRICAN ENVIRONMENT OUTLOOK, www.unep.org/dewa/africa/publications/aeo-1/139.htm.

20 The Nature Conservancy, *Grasslands: Cover Story*, available on the Nature Conservancy's website at: www.nature.org/earth/grasslands/coverstory.html.

21 See João Pedro Silva, Justin Toland, Wendy Jones, Jon Eldridge, Edward Thorpe, and Eamon O'Hara, LIFE AND EUROPEAN GRASSLANDS (2008), http://ec.europa.eu/environment/life/publications/lifepublications/lifefocus/documents/grassland.pdf. For a technical report from 2001 providing additional data about changes in the condition of European grasslands over recent years, see *Protection of Grasslands*, available on the website of the European Environment Agency at:www.eea.europa.eu/data-and-maps/indicators/protection-of-grasslands#toc-0.

22 LIFE AND EUROPEAN GRASSLANDS, *supra* note 21 (reporting on data provided by Member States under Article 17 of the EU Habitats Directive of 21 May 1992).

Important and troubling as these effects may be, they represent only a small proportion of the overall ecological damage – or, to employ the more commonly used term, environmental damage – that agriculture has caused.

Numerous other sources provide information regarding the types and magnitude of that environmental damage,[23] so I will offer here only a synopsis, beginning with observations made just over a decade and a half ago and then continuing with a series of more specific details.[24]

In a law journal article published in 2000, J. B. Ruhl drew critical attention to modern extractive agriculture's "dramatic impact on our planet's landscape and environmental systems".[25] Ruhl's main point in pressing this point was to underscore – and bemoan – the lack of agricultural law and regulation to address environmental issues, a matter that I will examine in some detail later, especially in Chapter 6. It is instructive first, however, to read Ruhl's "inventory of environmental harms that farms cause":

> Consider the typical farming process: first, remove all existing vegetation from the land and level it; second, deploy a single-species regime of crop or livestock; third, cultivate the crop or livestock with water and chemicals; finally, remove the crop or livestock and associated waste products from the land and start over. A number of environmental harms flow directly and necessarily from that basic reality of farming: (1) habitat loss and degradation; (2) soil erosion; (3) water resources depletion; (4) soil salinization; (5) chemical releases; (6) animal waste disposal; (7) water pollution; and (8) air pollution. In each of these categories, farms are a significant source of environmental harm.[26]

Having offered this list of ways in which agriculture – in its modern extractive manifestation – causes environmental damage, Ruhl then elaborated on each one, often in deeply depressing detail. In the following pages of this section, I will summarize these and other details, drawing partly from Ruhl's article, but mainly from a broad range of other writings. There are many such writings to draw from: the various environmental harms that J. B. Ruhl

---

23 One especially good set of explanations appears in Mary Jane Angelo, Jason J. Czarnezki, and William S. Eubanks II, FOOD, AGRICULTURE, AND ENVIRONMENTAL LAW (2013).

24 Several of the points emphasized here are succinctly set forth also in chapter III of Susan Schneider's work on agricultural law: Susan A. Schneider, FOOD, FARMING, AND SUSTAINABILITY (2011).

25 J. B. Ruhl, *Farms, Their Environmental Harms, and Environmental Law*, 27 ECOLOGY LAW QUARTERLY 263, 265 (2000) (citing various other observers who also emphasized agriculture as having brought about vast human alteration to the global environment).

26 *Id.* at 274. In a footnote, Ruhl acknowledged that to some extent these eight categories overlap – for instance, irrigation practices can contribute both to water-resource depletion and to soil salinization – but he points out that "the literature on the impacts of farming on the environment tends to break the problem down into these discrete topics, each of which is susceptible to measurement and study". *Id.* at 274 n.36.

blamed on modern farming in the year 2000 have been identified, measured, and castigated in countless other publications by experts in science, law, policy, history, and development. Some of those publications concentrate only on a narrow range of such harms, but many of them offer, as did Ruhl, a fairly broad inventory of those environmental effects.[27]

IB1. Habitat loss and degradation

J. B. Ruhl's article begins its cataloguing of the consequences of modern extractive agriculture by focusing on habitat loss and degradation:

> The consequences of modern agriculture on wildlife habitat are undeniable, from habitat elimination to more direct effects on water and wildlife species . . . . Despite the ability of perennial, vegetationally diverse agro-ecosystems with complex structure to provide important habitats for many birds and other animals typically found in undisturbed habitats, farms pose an enormous net negative to wildlife.
>
> [In a sense, however,] farming no longer poses a [new] significant direct threat of habitat loss. Most direct loss of habitat resulting from conversion of land areas to farming has already occurred . . . . [A]fter all, at one time virtually all of the 930 million acres currently in farming uses [in the USA] were undisturbed habitat. The fact that these habitat losses were experienced in the past does not [reduce, of course,] the seriousness of their continuing impacts to wildlife in the present.[28]

What Ruhl says about habitat loss rings true to me from a personal perspective. Growing up on our family farm in the 1950s and 1960s, I felt surrounded by an opera of life – mourning doves calling at dawn, toads escaping my bicycle tires, field mice rustling and running in leaf-piles, pigeons prancing and bowing with worried chortling in the feedlot and the hayloft, worms (and half-worms) wriggling in an overturned shovel of garden soil, bees stinging, moles burrowing, squirrels chattering, flies circling, startled rabbits scattering wildly, red-wing blackbirds perching on fence posts, and crickets and tree frogs making their pulsing buzz at dusk. Nearly all this gentle,

---

27 See generally Jason Clay, WORLD AGRICULTURE AND THE ENVIRONMENT: A COMMODITY-BY-COMMODITY GUIDE TO IMPACTS AND PRACTICES 14 (2004) (summarizing agriculture's impacts on soil quality, biodiversity, habitats, pollution and effluents, water depletion, climate change, energy consumption, and more, and then detailing the environmental impacts of each of 21 specific crop production or livestock practices); R. M. Harrison *et al.*, ENVIRONMENTAL IMPACTS OF MODERN AGRICULTURE (2012) (examining the environmental impact made by agriculture, with special attention to soil quality, greenhouse gas, water-borne pathogens, surface water chemistry, groundwater, agricultural pesticides, and aspects of agricultural production of biofuels); Andrew Kimbrell, THE FATAL HARVEST READER: THE TRAGEDY OF INDUSTRIAL AGRICULTURE (2002).

28 Ruhl, *supra* note 25, at 275.

insistent, chaotically choreographed opera is now finished; the farm seems largely silent and sterile. Granted, the fields produce higher crop yields and profits now than they did in my growing-up years, but the music and drama have stopped. I think it is because the former players – the divas and tenors and musicians in the pit – have disappeared or been driven away as their habitats have gradually disappeared.

It was this sort of silence and sterility that Rachel Carson wrote of in her 1962 classic, *Silent Spring*, and that numerous other authors have mourned in other works. It is a sort of death – a communal multi-species death – due in large part to habitat loss and degradation. In the opening pages of this chapter, I identified one set of victims by noting that a high proportion of the species in the grassland bird community in the North American Great Plains "have given indications of population declines over the past decades" as the intensity of agricultural activities has increased.[29] There are of course countless other victims of habitat loss and degradation due to modern extractive agriculture. One source reports that "[g]lobally, over 4,000 assessed plant and animal species are threatened by agricultural intensification, and the number is still rising. Over 1,000 (87%) of a total of 1,226 threatened bird species are impacted by agriculture."[30]

IB2. Soil erosion

J. B. Ruhl continued his "inventory of environmental harms that farms cause" with an examination of soil erosion and sedimentation. He pointed out that in the North American context "topsoil is replenished at a rate of less than one inch in 200 years",[31] a rate that is dwarfed by that at which topsoil has been eroded away by wind and water in recent years.[32] While acknowledging that total soil erosion on all cropland in the USA decreased by 42 percent between 1982 and 1997 because of improved soil management technology and practices – these include "low-till" or "no-till" cropping operations – Ruhl emphasized that the 1997 figures for US soil erosion still amounted, according to US government figures, to nearly two billion tons. Moreover, he pointed out that "even these improved rates are 12 times higher than soil formation rates".[33] Numerous other observers have echoed Ruhl's concerns in this regard.[34] Some have drawn special attention to the

---

29 See *supra* note 17 and accompanying text.
30 See United Nations Environment Programme, *The Environmental Food Crisis* 65 (2009), www.grida.no/files/publications/FoodCrisis_lores.pdf, at 65.
31 Ruhl, *supra* note 25, at 278 (citing a 1992 article in a science journal on soil and water conservation).
32 *Id.* at 277–278.
33 *Id.* at 279.
34 See, , R. P. C. Morgan, Soil Erosion and Conservation (2005); Scott Faber *et al.*, *Plowed Under: How Crop Subsidies Contribute to Massive Habitat Losses*, Environmental Working Group

fact that soil erosion has contributed importantly to the collapse of numerous civilizations.[35]

In order to get a clearer view of soil erosion and how it relates to agriculture, let us consider how soil is formed in the first place. Lester Brown, president of the Earth Policy Institute offers an excellent explanation that emphasizes the difference between (i) the "geological time scale" on which soil formation operates and (ii) the human time scale on which soil erosion occurs.

> The thin layer of topsoil that covers much of the earth's land surface and is typically measured in inches is the foundation of civilization. Geomorphologist David Montgomery, in *Dirt: The Erosion of Civilizations*, describes soil as "the skin of the earth – the frontier between geology and biology." After the earth was created, soil formed slowly over geological time from the weathering of rocks. It was this soil that supported early plant life on land. As plant life spread, the plants protected the soil from wind and water erosion, permitting it to accumulate and to support even more vegetation. This relationship facilitated an accumulation of topsoil that could support a rich diversity of plant and animal life.
>
> As long as soil erosion on cropland does not exceed new soil formation, all is well. But once it does, it leads to falling soil fertility and eventually to land abandonment. Sadly, soil formed on a geological time scale is being removed on a human time scale.
>
> Journalist Stephen Leahy writes in *Earth Island Journal* that soil erosion is "the silent global crisis." He notes that "it is akin to tire wear on your car – a gradual, unobserved process that has potentially catastrophic consequences if ignored for too long."
>
> Losing productive topsoil means losing both organic matter in the soil and vegetation on the land, thus releasing carbon into the atmosphere. Rattan Lal, a soil scientist at Ohio State University, notes that the 2,500 billion tons of carbon stored in soils dwarfs the 760 billion tons in the atmosphere. The bottom line is that land degradation is helping drive climate change.
>
> Soil erosion is not new . . . . What is new is that it has gradually accelerated ever since agriculture began. At some point, probably during the nineteenth century, the loss of topsoil from erosion surpassed the new soil that is formed through natural processes . . . Today, roughly a

---

(Feb. 2012), www.defenders.org/publications/plowed-under-how-crop-subsidies-contribute-to-massive-habitat-loss_0.pdf; Clive Potter, *Beyond Soil Conservation*, 38 ENVIRONMENT, Issue 7, p. 25 (Sept. 1996); Eric Roose *et al.*, SOIL EROSION AND CARBON DYNAMICS (Advances in Soil Science) (2005).

35 See generally David Montgomery, DIRT: THE EROSION OF CIVILIZATIONS (2007); Daniel Hillel, OUT OF THE EARTH: CIVILIZATION AND THE LIFE OF THE SOIL (1991).

third of the world's cropland is losing topsoil at an excessive rate, thereby reducing the land's inherent productivity.[36]

In another passage, Brown identifies the two principal causes of soil erosion: overgrazing and overplowing. After referring to "the Great Plow-up" that occurred in the USA in the run-up to the Dust Bowl days, and in the Soviet Union in the 1950s,[37] Brown offers this grim report on today's soil-degradation situation in two particularly hard-hit regions of the world:

> Today, two giant dust bowls are forming. One is in the Asian heartland in northern and western China, western Mongolia, and central Asia. The other is in central Africa in the Sahel – the savannah-like ecosystem that stretches across Africa, separating the Sahara Desert from the tropical rainforests to the south. Both [of these dust bowls] are massive in scale, dwarfing anything the world has seen before. They are caused, in varying degrees, by overgrazing, overplowing, and deforestation.
>
> China may face the biggest challenge of all [because of its livestock populations.] . . . The United States, a country with comparable grazing capacity, has 94 million cattle, a slightly larger herd than China's 92 million. But when it comes to sheep and goats, the United States has a combined population of only 9 million, whereas China has 281 million. Concentrated in China's western and northern provinces, these animals are stripping the land of its protective vegetation. The wind then does the rest, removing the soil and converting rangeland into desert.[38]

Similar dust-bowl problems are developing in India, which Brown says is struggling to support 17 percent of the world's people and 18 percent of its cattle, with only 2 percent of the world's land area. He cites a report that "24 percent of India's land area is slowly turning into desert."[39] Likewise, various regions in Africa are suffering from massive amounts of soil erosion, with wind-borne soil leaving Chad in quantities ten times as much as in 1947 when measurements began and with Nigeria losing nearly 900,000 acres of rangeland and cropland each year to desertification.[40]

Sounding the same themes, Jason Clay points out how global a problem soil erosion is. Soil losses average 30 to 40 metric tons per hectare per year, he explains in many parts of Asia, Africa, and South America, whereas erosion rates in undisturbed forests range from only 0.004 to 0.05 metric

---

36 Lester R. Brown, World on the Edge: How to Prevent Environmental and Economic Collapse 35–36 (2011).

37 During that period, the Soviets "plowed an area of grassland roughly equal to the wheat area of Australia and Canada combined", and a dust bowl followed. *Id.* at 37–38.

38 *Id.* at 38–39.

39 *Id.* at 39–40.

40 *Id.* at 41. Brown offers similarly alarming statistics for Afghanistan, Iran, and Iraq. *Id.* at 42–43.

tons per hectare per year.[41] Consistent with these figures are some other arresting statistics: roughly 30 million acres of farmland go out of production each year because of soil loss,[42] and 75 billion tons of soil are lost to erosion worldwide each year.[43]

IB3. Soil degradation and depletion

Whereas soil erosion causes a decline in the *quantity* of soil (in a particular location), soil degradation and depletion causes a decline in the *quality* of soil – or, expressed more colloquially, poisoning it and wearing it out. A broad literature has developed over soil degradation, focusing attention not just on the USA but on agricultural production worldwide.[44] For a personal and historical perspective on soil degradation caused by farming, I look again to Frederick Kirschenmann:

> Some mistakenly romanticize farming of the past and believe the land was better cared for then. However, my grandfather's style of farming was to break sod and grow wheat for about six years until the soil's fertility was depleted. Then he broke new ground. Livestock manure was spread on the closest fields simply to get it out of the barns without much thought to restoring the health of the soil.[45]

Soil degradation started millennia ago, of course, and in that respect Kirschenmann's grandfather probably behaved no differently from scores of earlier generations of grandfathers. But a new form of degradation occurred starting in the middle of the twentieth century:

> By the 1940s, this prairie-busting style of farming was running out of new land to "mine."[46] Following the war, petroleum companies

---

41 Clay, *supra* note 27, at 47, citing David Pimentel *et al.*, *Environmental and Economic Costs of Soil Erosion and Conservation Benefits*, 267 SCIENCE 1117–1123 (Feb. 24, 1995) [hereinafter Pimentel-1995].

42 Pimentel *et al.*, *supra* note 41, at 1117; Pierre Crosson, Commentary, *Soil Erosion Estimates and Costs*, 269 SCIENCE 5223 (1995).

43 H. Eswaran, R. Lal and P. F. Reich, *Land Degradation: An Overview*, appearing in E. M. Bridges, I. D. Hannam, L. R. Oldeman, F. W. T. Pening de Vries, S. J. Scherr, and S. Sompatpanit (eds.), RESPONSES TO LAND DEGRADATION, Proceedings of the Second International Conference on Land Degradation and Desertification, Khon Kaen, Thailand (2001), www.nres.usda.gov/wps/portal/nres/detail/soils/use/?cid=nrcs142p2_054028.

44 See, e.g., Morgan, *supra* note 34, Faber *et al.*, *supra* note 34, Potter, *supra* note 34, and Roose *et al.*, *supra* note 34. See also B. Söderström, E. Baath and B. Lundgren, *Decrease in Soil Microbial Activity and Biomasses Owing to Nitrogen Amendments*, 29 CANADIAN JOURNAL OF MICROBIOLOGY 1500–1506 (1983).

45 Frederick L. Kirschenmann, CULTIVATING AN ECOLOGICAL CONSCIENCE: ESSAYS FROM A FARMER PHILOSOPHER 41 (2010).

46 The idea of "mining" the soil echoes the perspective offered by Sir Albert Howard in 1947, when he published his second classic volume, *The Soil and Health*. He warned there that, as

sought new markets and found one in farmers looking to boost slumping yields. They seized upon the nineteenth-century theories of German scientist Justus von Liebig, who maintained that plants needed only nitrogen, phosphorus, and potash to stimulate growth. Nitrogen can be made from petroleum processes; the other two elements are mined and processed.

Farms saw these new inputs as their salvation. County extension agents and top farmers assured others that the synthetic products wouldn't hurt the land, and many farmers saw positive proof in their own fields. In the early years, the mechanical technology to apply fertilizers wasn't too advanced, and the fertilizer wasn't well granulated. Often the drive belt would slip or the fertilizer applicator would plug up, resulting in part of the field not being treated. Visually, it was obvious which part of a field wasn't fertilized, convincing my father he could never again farm without fertilizer.

Shifting from mining new soil to making existing acres more productive by purchasing outside inputs was an easy transition. Use of synthetic fertilizer encouraged farmers to raise specialized crops and abandon crop rotation. Farmers began to raise continuous crops of whatever produced the best returns in the marketplace. To attain ever-higher yields, they needed to buy more fertilizer. Because monoculture also invites weed, insect, and fungus problems, new markets for chemical companies were created.[47]

Should the narrative Kirschenmann offers in the above passage be considered one of soil degradation or (as some agrichemical enterprises might claim) soil *enhancement*? Surely it must be soil degradation, because of the damage that the chemicals – fertilizers, pesticides, etc. – inflict on the countless forms of life in the soil. Microbes, worms, rodents, insects, and other creatures are for the most part eliminated, stripped away by the chemical additives – and of course with the disturbance of the ground by plowing, disking, planting, cultivating, and the like.[48]

Kirschenmann summarizes it, "the industrialization of agriculture was taking us in the wrong direction." Kirschenmann, *supra* note 45, at 214. Industrial agriculture, which focuses on "quantity at all costs" by adding artificial fertilizers to the soil (the "NPK mentality", with N standing for nitrogen, P standing for phosphorus, and K standing for potassium), paid almost no attention to the health of the soil. The lack of attention to managing soil for health, Howard argued, led to "mining the land," which he considered a "form of banditry." *Id.*

47 Kirschenmann, *supra* note 45, at 41–42.
48 For further details on the "architecture of life" that constitutes rich and fertile soil, see text accompanying note 4 in Chapter 4, *infra*. Don Worster is cited there for his reference to "the humus, the organic residue of roots, carrion, feces, bone, and leaves mixed through the mineral components" that make up such soil, as well as to "an incomprehensibly large number of bacteria working away, decomposing the dead, fixing nitrogen, forming nitrates to feed the living".

This disturbance of the soil's architecture of life, including the killing of most of the creatures in it, occurs not only in those cropland fields to which the chemicals are directly applied, of course, but in many "downstream" ecosystems in various parts of the planet. I turn to another one now: aquatic ecosystems.

IB4. Dead zones and other aquatic poisoning

In a passage that builds on his points about habitat loss and degradation, J. B. Ruhl identified in his 2000 article some of the pernicious effects of the chemical "ingredients" used in modern extractive agriculture:

> [Although much of the loss of habitat occurs on the farms themselves,] the truly pernicious effects of farming on habitat today occur offsite. For example, gaseous and dissolved nitrogen oxide and ammonia emitted from agricultural ecosystems are transported to and deposited in downwind and downstream terrestrial and aquatic ecosystems. This deposition causes inadvertent fertilization, which can lead to acidification, eutrophication, shifts in species diversity, and effects on predator and parasite systems. Transport of pesticides beyond farm boundaries also causes severe damage to wildlife and habitat functions. Similarly, because evaporation and concentration effects cause irrigation return-flows to carry greater concentrations of salt and minerals than found in irrigation water sources, fish and wildlife populations downstream often suffer. Also, high erosion rates associated with cultivated agriculture can lead to sedimentation in reservoirs and lakes, which reduces the lifetime of these water systems as aquatic habitat.[49]

Numerous publications document the connection between fertilizer run-off from farms and dead zones in various parts of the world.[50] Among their findings are that over the past century, the development of new agricultural practices has drastically disrupted the nitrogen cycle, leading to extensive eutrophication of fresh waters and coastal zones, often resulting in dead zones.[51] Indeed, the dead zones that result from such run-off have now been

---

49 Ruhl, *supra* note 25, at 276–277.
50 See, e.g., *Agriculture: A Threat to Ocean Life?*, 47 ENVIRONMENT, issue 4, pp. 8–9 (May 2005); J. Michael Beman, Kevin R. Arrigo, and Pamela A. Matson, *Agricultural Runoffs Fuels Large Phytoplankton Blooms in Vulnerable Areas of the Ocean*, 434 NATURE 211–214 (Mar. 10, 2005); Donald E. Canfield, Alexander N. Glazer and Paul G. Falkowski, *The Evolution and Future of Earth's Nitrogen Cycle*, SCIENCE, 192–196 (Oct. 8, 2010); Robert J. Diaz and Rutger Rosenberg, *Spreading Dead Zones and Consequences for Marine Ecosystems*, SCIENCE 926–929 (Aug 15, 2008); NOAA National Coastal Data Development Center, GULF OF MEXICO HYPOXIA WATCH, www.ncddc.noaa.gov/hypoxia/.
51 Eutrophication (or more precisely hypertrophication) is the ecosystem response to the addition of artificial or natural substances, such as nitrates and phosphates, through fertilizers or sewage,

reported from more than 400 river systems, affecting a total area of more than 245,000 square kilometers.

In another discussion of dead zones, Paul Gilding places special emphasis on the role of one key ingredient that is central to the carrying out of modern extractive agriculture: nitrogen.

> [O]ur efforts to increase agricultural productivity have led to us dramatically exceeding the earth's capacity to absorb our emissions of nitrogen.
>
> Nutrients in the form of nitrogen are added to the land as fertilizer to boost crop production. However, when they are washed into the oceans, they have the opposite effect, as they encourage algal blooms and deplete oxygen levels to the point where nothing else can survive. So while in this case significant economic benefit comes from higher food productivity, significant economic loss comes from loss of drinking water, loss of fisheries, and dead rivers. It was estimated that the total economic losses from freshwater eutrophication in the United States was $2.2 billion in 2009 alone.[52]

The costs of dead zones and other forms of aquatic poisoning extend beyond economic costs. As Frederick Kirschenmann has pointed out, there are serious food-supply implications as well.

> [T]he increased use of toxic inputs to increase yields may result in a net *decrease* of available food. For example, 60 percent of the world's population today [i.e., as of the mid-1990s] depends on fish and seafood for 40 percent of its annual protein. There is ample evidence to confirm that land-based industrial agriculture's environmental impact on aquatic and marine habitats is seriously diminishing the food supply from this critical source. Increasing the yields of land-based agriculture (assuming it can be done) at the expense of aquatic agriculture is, therefore, hardly the way to feed more people.[53]

IB5. Reduced biodiversity

One way in which farming practices often reduce biodiversity is by destroying or degrading wildlife habitat, as highlighted in J. B. Ruhl's "inventory of environmental harms that farms cause". Beyond that, however, modern

---

to an aquatic system. One example is the "bloom" or great increase of phytoplankton in a water body as a response to increased levels of nutrients. This can often lead to hypoxia, the depletion of oxygen in the water, which induces reductions in specific fish and other animal populations. In some cases, one species may experience an increase in population that negatively affects other species.

52 Paul Gilding, THE GREAT DISRUPTION 43 (2011).

53 Kirschenmann, *supra* note 45, at 56–57.

extractive agriculture reduces biodiversity in other ways as well. Numerous observers have emphasized, for instance, the danger of homogenization of agriculture – that is, the trend toward uniformity in food crops.[54] Large-scale agriculture concentrates on such uniformity, thus reducing and threatening the gene pools of humanity's food crops. Since diversity typically serves as a source of resistance to disease, this homogenization (narrowing of diversity) of such crops increases susceptibility to the spread of pestilence. As is summarized on the website of the Convention on Biological Diversity, agriculture is "a major driver of biodiversity loss. The Earth's biodiversity is being lost at an alarming rate, putting in jeopardy the sustainability of agriculture and ecosystem services and their ability to adapt to changing conditions, threatening food and livelihoods security."[55]

The reference in the preceding paragraph to the food-security risks posed by reduced biodiversity bears some elaboration. Just as Frederick Kirschenmann has warned of the implications of dead zones on food supplies, he has also warned of the long-term effects of reduced species biodiversity more generally. In the following passages, he links this issue to the claims of the Green Revolution, human population pressures, and even the value of termites:

> [In questioning the long-term viability or value of the "Green Revolution," we must realize that] the imbalance of humans relative to the millions of other species with whom we coevolved now disrupts the

---

54 For explanations of this homogenization and the risks its poses, see, e.g., Cary Fowler and Patrick R. Mooney, SHATTERING: FOOD, POLITICS, AND THE LOSS OF GENETIC DIVERSITY (1990), and Colin K. Khoury *et al.*, *Increasing Homogeneity in Global Food Supplies and the Implications for Food Security*, 111 PROCEEDINGS OF THE NATIONAL ACADEMY OF SCIENCES 4001–4006 (2014). See also various reports and press releases from the Food and Agriculture Organization ("FAO"), including FAO, *Loss of Domestic Animal Breeds Alarming* (press release) (Mar. 31, 2004), www.fao.org/newsroom/en/news/2004/39892/; FAO, *The State of the World's Animal Genetic Resources for Food and Agriculture* (2007), www.fao.org/docrep/010/a1250e/a1250e00.htm; and FAO, *The Second Report on the State of the World's Plant Genetic Resources for Food and Agriculture* (2010), www.fao.org/docrep/013/i1500e/i1500e00.htm. Although concerns over diversity and agriculture have gained momentum in recent years as agricultural production has become ever more homogenized, such concerns have been voiced for several decades. For a broad discussion of biological diversity that was published near the time of the adoption of the Convention on Biological Diversity, see Edward O. Wilson, THE DIVERSITY OF LIFE (1992, republished 2010). Wilson touches on some of the same themes in his most recent book. See Edward O. Wilson, HALF-EARTH: OUR PLANET'S FIGHT FOR LIFE (2016). Even in the early 1970s the genetic vulnerability of agricultural crops to disease was recognized and studied in response to a 1970 epidemic disease that swept over the corn crop of the USA. See National Academic of Sciences, GENETIC VULNERABILITY OF MAJOR CROPS (1972).

55 Convention on Biological Diversity, *About Agricultural Biodiversity: What's the Problem?*, www.cbd.int/agro/whatstheproblem.shtml. Another page in the website provides a series of case studies around the world detailing impacts on agricultural biodiversity, benefits of such biodiversity, the role of rare species, and related matters. See Convention on Biological Diversity, *Agricultural Biodiversity Case Studies*, www.cbd.int/agro/casestudies.shtml.

biotic community and the delicate ecological relationships that have evolved over billions of years. This disruption and deterioration now threaten the food supply of the human species. The reason we need to consider an alternative [form of agriculture] is that our current model of industrial agriculture is contributing, dramatically, to this ecological disruption. So while the green revolution may have enjoyed success in increasing the yields of a few crop varieties for the short term, it now threatens the ability of future generations to feed themselves.

. . . [This raises a crucial question about the biological and ecological relationships involved in agriculture:] how do we regain and maintain the evolutionary stability of the various ecological neighborhoods in which we humans live [. . . or, expressed differently, the complex blend of] elements that make human life possible on this planet[?] . . . [In short, what] kind of agriculture can best mirror and maintain that evolutionary stability?

Niles Eldredge gives us some examples of our utter dependence on these complex biological relationships. Insects, which humans generally hold in low regard . . . , are so important, says Eldredge, that "humanity probably could not last for more than a few months" without them . . . . [Specifically, termites,] because of their symbiotic relationship with spirochete bacteria, are one of the few creatures on the planet that can digest cellulose. Consequently, we humans are absolutely dependent on termites for a huge portion of the recycling of the world's biotic material . . .

What is often posed as the food/population problem can more accurately be described as a population/ecology problem. In other words, the real problem with the unprecedented increase in human population is that it has led to the disruption and deterioration of the natural functioning of Earth's biotic community and that is what threatens our future . . .[56]

IB6. Concerns over pesticides

Much of the concern expressed over the interrelated issues of reduced biodiversity, dead zones, habitat degradation, and soil degradation – all

---

56 Kirschenmann, *supra* note 45, at 142–143. Kirschenmann's references to "the ability of future generations to feed themselves" (at the end of the first paragraph quoted above) and to the "disruption and deterioration of the natural functioning of Earth's biotic community" as the development that "threatens our future" (at the end of the last paragraph quoted above) evoke questions that are both legal and ethical in character. For instance, do yet-unborn future generations of humans have rights to life and sustenance? If so, how can the interests of those future humans be represented and protected? Can the legal concept (used in the USA) of "virtual representation for unborn remaindermen" be extended to represent those interests? I address some of these issues in later chapters, and particularly in section III of Chapter 4 and in Chapter 7.

summarized previously – revolves around the use of pesticides in modern extractive agriculture. Even aside from the drumbeat of concern that has intensified in recent years about the direct effects of pesticide use on *human* health – a matter I will address below in section II of this chapter – the literature is thick with studies and assessments on how pesticides[57] (that is, insecticides, herbicides, fungicides, and other agents classified by the target organism)[58] can have detrimental environmental effects in other ways.

After all, modern extractive agriculture relies very heavily on pesticides. In his 2000 journal article, J. B. Ruhl referred to farms as "massive users of chemicals" and offered this summary regarding agricultural pesticide use in the USA:

> The pesticide industry involves about 30 major manufacturing companies, 100 smaller companies marketing the active ingredients of pesticides, 3,300 product formulators who take the raw pesticide ingredients and produce finished pesticide products, and over 29,000 pesticide distributors. About 600 distinctive groups of active ingredients are found in the 45,000 pesticide products that are marketed in the United States. About 1.2 billion pounds of pesticides, valued at over $6.5 billion, are sold each year in the United States, over 70% of which are used in farming.[59]

Such pesticide use has continued, of course, over the past decade and a half since Ruhl's article was written – especially as some crops have been developed that are resistant to the killing effects of glyphosate (marketed as

---

57 In general terms, a pesticide is a substance meant to prevent, destroy, or mitigate the damage from any pest. Sometimes the term "biocide" is used in place of "pesticide". The most common use of pesticides is as plant protection products, which in general protect crops from damaging influences such as weeds, diseases, or insects. This use of pesticides is so common that the term *pesticide* is often treated as synonymous with *plant protection product*, although the former is in fact a broader term, as pesticides are also used for non-agricultural purposes. According to the Stockholm Convention on Persistent Organic Pollutants, nine of the twelve most dangerous and persistent organic chemicals are pesticides. For a list of some prominent agricultural pesticides currently in use, including some of the commercial names under which they are marketed, see *infra* note 91.

58 Pesticides can be classified by target organism, by chemical composition (e.g., organic or inorganic), and by other means.

59 Ruhl, *supra* note 25, at 282 n. 97. Assuming the figure of $6.5 billion that Ruhl gives for the value of pesticides used in the USA dates from around the time his article was published (2000), the value in current, inflation-adjusted terms would probably exceed $9 billion. In the meantime, pesticide use and value have both increased. Similar inflation adjustments would need to be made, of course, in order to appreciate the precise significance of numerous values appearing in other passages of this book, especially values dating from more than a few years ago. With only a few exceptions, I have not attempted to provide such inflation-adjustment figures. For guidance in making such inflation adjustments, see CPI Inflation Calculator, http://data.bls.gov/cgi-bin/cpicalc.pl?cost1=6.50&year1=2000&year2=2015.

"Roundup")[60] – and this has prompted ever more research into the impact that pesticides have on water, air, soil, and wildlife. Some of the research suggests that the impacts include the destruction of beneficial species, increases in pest resistance, reduction in pollination, crop losses, ground and surface water contamination, and more.[61]

IB7. Degraded water quality and air quality

Agricultural operations can contribute to water quality deterioration through the release of various materials into water. One of these – fertilizers – was discussed above in subsection IB4 of this chapter. Others include sediments, pesticides, animal manures, and other sources of inorganic and organic matter.[62]

These agricultural releases produce environmental damage of various sorts. As with the other forms of environmental degradation referred to above, a

---

60 Here is a general description of glyphosate, based on reports and analyses from the World Health Organization, the European Union, US government agencies, and academic sources, as compiled in the Wikipedia entry for "glyphosate": Glyphosate (N-(phosphonomethyl)glycine) is a broad-spectrum systemic herbicide used to kill weeds, especially annual broadleaf weeds and grasses known to compete with commercial crops grown around the globe. It was discovered to be a herbicide by Monsanto chemist John E. Franz in 1970. Monsanto brought it to market in the 1970s under the trade name "Roundup". Glyphosate was quickly adopted by farmers, and in 2007 glyphosate was the most used herbicide in the US agricultural sector, with 180 to 185 million pounds applied. Glyphosate's mode of action is to inhibit an enzyme involved in the synthesis of the aromatic amino acids tyrosine, tryptophan and phenylalanine. It is absorbed through foliage and translocated to growing points. Some crops have been genetically engineered to be resistant to glyphosate. With these crops – such as "Roundup Ready" soybeans, also created by Monsanto Company – farmers can use glyphosate as a post-emergence herbicide against both broadleaf and cereal weeds. However, the development of similar resistance in some weed species is emerging as a costly problem, given glyphosate's heavy use in agriculture. See, e.g., Michael Wines, *Invader Storms Rural America, Shrugging Off Herbicides*, THE NEW YORK TIMES, Aug. 12, 2014 (explaining how the weed palmer amaranth – called "careless weed" – has developed a resistance to glyphosate and thereby "devastated Southern cotton farms and is poised to wreak havoc in the Midwest"). While glyphosate has been approved by regulatory bodies worldwide and is widely used, concerns about its effects on humans and the environment persist.

61 See, e.g., J. Patrick Madden and Scott G. Chaplowe, FOR ALL GENERATIONS: MAKING WORLD AGRICULTURE MORE SUSTAINABLE (1997) (addressing, among other things, pest control in agriculture); *State of the Nation's Pesticides*, 48 ENVIRONMENT, issue 4, pp. 6–7 (May 2006); David Pimentel, *Environmental and Economic Cost of the Application of Pesticides Primarily in the United States,* 7 ENVIRONMENT, DEVELOPMENT AND SUSTAINABILITY 229–252 (2005) [hereinafter Pimentel-2005]; David Pimentel *et al.*, HANDBOOK OF PESTICIDE MANAGEMENT IN AGRICULTURE (2nd ed. 1990) [hereinafter Pimentel-1990]. See also the website of the National Pesticide Information Center. http://npic.orst.edu/ (identifying environmental effects of individual pesticides) and the website of the USDA's Marketing Service "Pesticide Data Program".

62 In addition to these various types of materials released into water as a direct result of agricultural operations are several other types of materials released as an indirect result of such operations. For instance, as more food production supports more humans, the release of human waste into bodies of water is an increasingly significant contributor to water quality deterioration.

broad literature has developed documenting these forms of damage, which include salinization, transmission of infectious diseases, threats to biodiversity (also noted above), bioaccumulation – by which some aquatic species accumulate pesticides more readily than terrestrial organisms – and other degradation.[63] Susan Schneider draws attention to the "massive fish kills" that, according to the US Environmental Protection Agency, have occurred in a number of locations in the USA because of wastes that are released from animal feedlots and that then make their way into surface water and groundwater.[64] Moreover, the sheer volume of water involved in modern extractive agriculture – by one estimate, agriculture consumes about 70 percent of fresh water worldwide[65] – is cause for concern.

Of similar cause for concern is the effect that modern extractive agriculture has on air quality. In addition to the profound effects they have had on global climate change through the release of greenhouse gases (discussed more fully on pages 98–104), farming operations also can damage local and regional air quality in a variety of other ways. These include the increase of

---

63 See, e.g., E. D. Ongley, *Control of Water Pollution from Agriculture* (FAO Irrigation and Drainage Paper 55, 1996); David Pimentel *et al.*, *Water Resources: Agricultural and Environmental Issues*, 54 BioScience 911–915 (Oct. 2004) [hereinafter Pimentel-2004]; Guye H. Willis and Leslie L. McDowell, *Pesticides in Agricultural Runoff and Their Effects on Downstream Water Quality*, 1 Environmental Toxicology and Chemistry, 2672–2679 (1982); Peter H. Gleick, ed., Water in Crisis: A Guide to the World's Fresh Water Resources (1993) (publishing essays on a range of issues, including the use of water for food production).

64 Schneider, *supra* note 24, at 135. Schneider goes on to highlight the enormous quantities of animal waste produced by concentrated animal feeding operations ("CAFOs"). She cites in this regard a 2008 US Government Accountability Office report noting that "some large farms can produce more raw waste than the human population of a large U.S. city". The example given in that report focuses on "a very large hog farm, with as many as 800,000 hogs", which "generates . . . more than one and a half times the sanitary waste produced by the . . . residents of Philadelphia" in a year. *Id.* at 136. See also *Putting Meat on the Table: Industrial Farm Animal Production*, A Report of the Pew Commission on Industrial Farm Animal Production in America (2009), www.ncIFAP.org/. Michael Pollan has highlighted an irony in the rise of large-scale livestock feedlots and the pollution they create. In a letter to "the president-elect" just before the 2008 US presidential election, Pollan offered this observation, drawing also from the work of Wendell Berry:

> [I]f taking the animals off farms made a certain kind of economic sense, it made no ecological sense whatever: their waste, formerly regarded as a precious source of fertility on the farm, became a pollutant – factory farms are now one of America's biggest sources of pollution. As Wendell Berry has tartly observed, to take animals off farms and put them on feedlots is to take an elegant solution – animals replenishing the fertility that crops deplete – and neatly divide it into two problems: a fertility problem on the farm and a pollution problem on the feedlot. The former problem is remedied with fossil-fuel fertilizer; the latter is remedied not at all.

Michael Pollan, *Farmer in Chief*, The New York Times: The Food Issue (Oct. 12, 2008), http://michaelpollan.com/articles-archive/farmer-in-chief/.

65 Pimentel-2004, *supra* note 63, at 911.

particulate matter, the release of odors, the creation of acid rain, and the spread of toxins.[66]

In that last regard (spread of toxins), concerns about air quality overlap with the concerns about pesticide use that were summarized above. "Pesticide drift" occurs when pesticides suspended in the air as particles are carried by wind to other areas. "Pesticide volatilization" refers to the evaporation or sublimation of a herbicide. Both pesticide drift and pesticide volatilization pose threats to wildlife and other components of an ecosystem. In addition, large-scale livestock feedlots can emit high levels of hydrogen sulfide and ammonia nitrogen.[67] As in the case of pesticide use, degraded air quality resulting from agriculture has been tied to direct effects on human health – a matter that I will turn to in section II of this chapter.

IB8. Contribution to climate change

One of the most recent topics of critical review regarding modern extractive agriculture revolves around global climate change. This current interest reflects a growing understanding of the main ways in which agricultural and livestock production release greenhouse gases ("GHGs") into the atmosphere, and how this release of greenhouse gases bears on climate change.

In order to examine this effectively, let us start with a long-term view of the effects that humans have had on Earth's climate. William Ruddiman offers this explanation and proposition:

> For most of the time that human beings and our recognizable ancestors lived on Earth, we did not affect climate. Few in number, and moving constantly in search of food and water, our Stone Age predecessors left no permanent "foot-prints" on the landscape for several million years. Throughout this immensely long span of time, climate changed for natural reasons, primarily related to small cyclical changes in Earth's orbit around the Sun. Nature was in control of climate.
>
> But the discovery of agriculture nearly 12,000 years ago changed everything. For the first time, humans could live settled lives near their crops, rather than roaming from area to area. And gradually, the improved

---

66 One of these, acid rain, has risen in prominence. As one NGO's website announces, "acid rain is a continuing and growing problem; forests and animals all over the world (including the U.S. East Coast) are indeed facing catastrophe." This catastrophe, however, is different from the acid rain that received great attention some years ago due to sulfur dioxide emissions from coal-fired power plants: "the No. 1 source of today's acid rain pollution is no longer sulfur dioxide, as it was 20 years ago. It's nitrogen oxide emissions from factory farms." Tom Laskawy, *Acid Rain is Back, and Thanks to Farming, Worse than Ever* (June 24, 2010), appearing on the website of the Seattle-based environmental-issues group Grist, at http://grist.org/article/food-acid-rain-is-back-and-thanks-to-farming-worse-than-ever/.

67 For a brief assessment of these forms of air pollution from agricultural operations, see Ruhl, *supra* note 25, at 292.

nutrition available from more dependable crops and livestock began to produce much more rapid increases in population than had been possible in the earlier hunting-and-gathering mode of existence. As a result, the growing human settlements began to leave a permanent footprint of ever-increasing size on the land . . .

Until very recently, scientists thought that humans first began altering climate some 100 to 200 years ago, as a direct result of changes brought about by the gassy effusions of the Industrial Revolution. But here I propose a very different view: the start of the switch-over from control of climate by nature to control by humans occurred several thousand years ago, and it happened as a result of seemingly "pastoral" innovations linked to farming. Before we built cities, before we invented writing, and before we founded the major religions, we were already altering climate. We were farming.[68]

Having advanced this theory – that in fact anthropogenic (human-caused) climate change actually started in a highly significant way thousands of years ago with the advent of agriculture – William Ruddiman presses a point about the "unnatural" character of farming:

The agricultural life, often referred to as "pastoral" or "rural," seems a natural world to modern-day city dwellers and suburbanites. A weekend drive in the country is a time for relaxing and "going back to nature." Yet *farming is not nature, but rather the largest alteration of Earth's surface from its natural state that humans have yet achieved.* Cities and factories and even suburban mega-malls are still trivial dots on maps compared to the extent of farmland devoted to pastures and crops (more than a third of Earth's land surface). So when those people in Mesopotamia 11,500 years ago created agriculture, they set humanity on a path that would transform nature.[69]

That transformation of nature, Ruddiman emphasizes, involved climate change through the release of greenhouse gases – or, as Ruddiman expresses it, by "taking control" of methane and carbon dioxide. The net impact of

---

68 William F. Ruddiman, PLOWS, PLAGUES, AND PETROLEUM: HOW HUMANS TOOK CONTROL OF CLIMATE 4 (2005).

69 *Id.* at 63. Ruddiman's perspective on farming as constituting a fundamental alteration of the Earth's surface resembles a point made by Bill Vitek and others at a recent conference on ecospheric studies: the Earth can be visualized differently at various phases of its existence. First, and for countless millennia, it was "Earth as ecosphere", with "ecosphere" encompassing all living and nonliving components of the surface, the immediate sub-surface, and the atmosphere of the Earth. Then, for the last several thousand years until quite recently, it has been "Earth as farm" (corresponding to the point Ruddiman makes in the passage quoted above). In the past several decades, since the advent of industrial agriculture, it has been "Earth as machine". Discussions at Conference on Ecospheric Studies, The Land Institute, June 14–16, 2015.

human activities, he says, has been "a long slow rise in greenhouse-gas concentrations prior to the industrial era, and then much more rapid increases during the last 200 years of industrialization."[70]

If Ruddiman's analysis is correct, it tends to "greatly complicate studies of climatic changes over the last several millennia", because it suggests that in fact "no part of the last several thousand years of climate change is actually free from potentially significant human impacts".[71] However, he emphasizes that the past two centuries have seen by far the most important extent of human-caused climate change, and he underscores the role that agricultural innovations – and the growing populations agriculture has been called on to feed – have played in that trend:

> During the 1800s and 1900s, human population increased from 1 billion to 6 billion, an explosion unprecedented in human history . . . . As a result, our already sizable impact on Earth's surface increased at a much faster rate. Early in this millennium, we live in a world that has largely been transformed by humans.
>
> [Even if] the explosive population increase still under way will end near AD 2050 as global population levels out at some 9–10 billion people,[72] [human impact on the Earth's surface will continue to expand. Why? Because] as affluence and technology continue to spread, increased pressures on the environment will occur for that reason alone. If a billion or more people in China and India begin to live the way Americans and Europeans do now, their additional use of Earth's resources will be enormous. Even without population increases, humanity will continue to alter the environment in new ways.[73]

Ruddiman closes his book on a grim note. He speculates that because of the massive unsustainable use in the past 200 years of the Earth's resources, "around the year 2075, I suspect it will be clear to all that some of the 'free' gifts from Nature were not an unlimited resource" and that people living then "will look back on the era between the late 1800s and the early part of the twenty-first century as a brief bubble of good fortune, a time when a

---

70 Ruddiman, *supra* note 68, at 95. In an especially interesting portion of his book, Ruddiman asserts that the "control" that humans exerted for thousands of years over greenhouse grass emissions actually lapsed – or at least changed – briefly, with temporary reductions in human populations. Several times in the past 2,000 years, major pandemics killed tens of millions of people, with the effect of "reversing the gradual clearance of forests for agriculture, and contributing to short-term climatic cooling, including the Little Ice Age interval between 1300 and 1900". *Id.* at 117.

71 *Id.* at 146.

72 For recent UN population projections contradicting Ruddiman's assumption that human population will "level out" in 2050, see text accompanying notes 73–74 in Chapter 2, *supra*.

73 Ruddiman, *supra* note 68, at 190.

lucky few human generations consumed most of these gifts, largely unaware of what they were doing".[74]

I have drawn from these observations by Ruddiman because I believe they provide important context for an understanding of what effect agriculture has *today* on global climate change. It is an enormous effect – probably drastically more substantial than is generally understood. Here are some details, drawing from materials prepared by the World Resources Institute ("WRI"), and particularly from a "World GHG Emissions Flow Chart" prepared by WRI a few years ago[75]:

- 6.0 percent of greenhouse gas ("GHG") emissions come from "agricultural soils", 1.4 percent from "agricultural energy use", 5.1 percent from "livestock & manure", 1.5 percent from "rice cultivation"[76] and 0.9 percent from "other agriculture" – for an overall total of 13.5 percent of world GHG emissions attributable to agricultural activities, including livestock-related activities.[77]
- Except for the "agricultural energy use" category, all agriculture and livestock-related emissions shown on the WRI chart are of two specific GHGs: nitrous oxide (N_2O) and methane (CH_4).
- In fact, nearly all nitrous oxide emissions in the world come from the "agricultural soils" category mentioned above. Why? Because nitrous oxide is a by-product, along with dinitrogen, of a microbial process working on the nitrogen-based fertilizers that are used heavily in modern agriculture. In addition, when farmers turn the soil to prepare the land for crops, nitrous oxide is often released.
- Significantly, nitrous oxide is said to be about 300 times more potent, molecule for molecule, than carbon dioxide (the much more prevalent

---

74 *Id.* at 192.

75 See World Resources Institute, *World GHG Emissions Flowchart*, http://pdf.wri.org/world_greenhouse_gas_emissions_flowchart.pdf. For another view of GHG "flows", see www.ecofys.com/files/files/asn-ecofys-2013-world-ghg-emissions-flow-chart-2010.pdf.

76 With respect to the contribution that rice cultivation makes to GHG emissions, and therefore to climate change, one authority makes this interesting observation: "Plans for reducing emissions in agriculture must consider the consequences for less advantaged populations. The growing of rice in flooded fields releases methane from waterlogged soils, but rice feeds a third of the world's population and is the staple diet of many poor people in Asia." Kirstin Dow and Thomas E. Downing, THE ATLAS OF CLIMATE CHANGE: MAPPING THE WORLD'S GREATEST CHALLENGE 50 (2006) [hereinafter CLIMATE CHANGE ATLAS].

77 This figure of 13.5% does not include GHG emissions coming from the conversion of land use from forest to agricultural production. This is a very large source of GHGs (18.3% of all GHG emissions), and although it does not relate directly to the *manner* in which modern agricultural production occurs, the conversion of land from forest cover to agriculture would be less necessary – perhaps not necessary at all – if the global need for agricultural products (for use as food, feed, and fiber) could be satisfied either by increasing supply or decreasing demand. See *World GHG Emissions Flowchart*, *supra* note 75.

GHG) in terms of causing climate change, a matter which prompted its inclusion as one of the GHGs regulated by climate-change treaties negotiated in the 1990s.[78]

- As for methane, a very high proportion (around 50 percent or more) of all methane emissions in the world are attributable to livestock.[79] Methane is said to be well over 30 times more potent than carbon dioxide in terms of its impact on global climate change.[80]

- In short, then, agriculture and livestock production, viewed as a single composite source, accounts for a significant overall contribution to GHGs, and practically all of those emissions are of extremely high-impact GHGs – namely, methane and nitrous oxide.[81]

---

78 For details on this, see the website of the Nitrous Oxide Focus Group (University of East Anglia), at www.nitrousoxide.org/. See also *Agriculture and Climate Change*, on the website of the International Industry Fertilizer Association at www.fertilizer.org/ifa/HomePage/SUSTAINABILITY/Climate-change/Agriculture-Climate-Change.html.

79 With respect to the contribution that livestock-sourced methane makes to greenhouse gas emissions, and therefore to climate change, one authority offers this observation: "Livestock are . . . a source of methane, but those on the small holdings of poor farmers and pastoralists produce much less gas than the well-fed cattle in large-scale commercial enterprises." Climate Change Atlas, *supra* note 76, at 50. On the other hand, some studies show that the methane released by cattle in high-grain feedlot situations is relatively lower in quantity on a gross-energy (GE) basis than in other settings, including grass-fed settings. K. A. Johnson and D. E. Johnson, *Methane Emissions from Cattle*, 73 Journal of Animal Science 2483–2492 (1995) (citing data suggesting that for most US beef cattle herds, "methane losses vary from approximately 5.8 to 6.5 of GE for all categories and classes except for the unique high grain feedlot situation in which typical methane loss may drop to approximately 3%"). More recent reports offer similar observations; see Judith L. Capper, *Is the Grass Always Greener? Comparing the Environmental Impact of Conventional, Natural and Grass-Fed Beef Production Systems*, 2 Animals 127–143 (2012) (citing a publication sponsored by the Tennessee Beef Cattle Improvement association.) Not surprisingly, the figures on livestock-based methane emissions vary. Some sources report that livestock operations are responsible for about 18% of greenhouse gas emissions globally and over 7% of greenhouse gas emissions in the USA – and that of these livestock-based GHG emissions, methane is most important, with livestock currently contributing nearly 30% of anthropogenic methane emissions (only about one-eighth of this is from manure, with the remainder coming mainly from gases produced during digestion and then released); some projections suggest that this 30% figure will rise to 60% by 2030 unless the trends of livestock production are changed.

80 A few years ago the Intergovernmental Panel on Climate Change revised upward its estimate for the global warming potential of methane – from about 25 times more potent to 34 times more potent than carbon dioxide on a 100-year time scale. See Gayathri Vaidyanathan, *Methane's Warming Potential Rises in Latest Report*, Energy & Environment, Oct. 1, 2013). Perhaps more significantly, methane is about 86 times more potent than carbon dioxide on a 20-year time scale. Gayathri Vaidyanathan, *How Bad of a Greenhouse Gas is Methane?*, Scientific American Climatewire, Dec. 22, 2015.

81 For some further discussion of agriculture's role in GHG emissions, see *Agriculture Is Single Most Important Contributor to Climate Change*, http://bifurcatedcarrots.eu/2009/01/agriculture-is-single-most-important-contributer-to-climate-change/.

Given these facts, along with some others having to do with soil carbon sequestration[82] and with climate change feedback effects relating to agriculture,[83] it is hardly surprising that a great deal of critical attention has been directed toward modern extractive agriculture as a major culprit in global climate change.[84]

The issue of climate change more broadly, of course, has gained such intense attention recently because it is one of the key developments that

---

82 As will be discussed further in Chapter 5, current forms of agricultural production reduce in many ways the ability of the Earth's land cover to sequester carbon, or to serve as a carbon "sink". The logic, in general terms, is as follows: much of the land used for agricultural crop production has been converted to that use from either grasslands areas or forest cover; both grasslands and forests can serve to capture carbon released into the atmosphere from a range of sources, and indeed recent studies confirm that the overall potential of carbon sequestration by grasslands compares favorably with the potential for carbon sequestration by rain forests; hence the use of vast areas of the Earth's surface for raising annual crops (as distinct, that is, from perennial crops) not only *releases GHGs* – thereby contributing to global climate change – but also *reduces the carbon sequestration potential* that could (up to a "steady state" level of sequestration) help reduce or mitigate such climate change. See GRASSLANDS, *supra* note 2, at 62–63; R. Lal, J. L. Kimble, R. F. Follett, and C. V. Cole, THE POTENTIAL OF U.S. CROPLAND TO SEQUESTER CARBON AND MITIGATE THE GREENHOUSE EFFECT (1998); C. Neely, S. Bunting, and A. Wilkes, eds., *Review of Evidence on Drylands Pastoral Systems and Climate Change: Implications and Opportunities for Mitigation and Adaptation*, FAO Land and Water Discussion Paper 8, at 13–14 (2009). For a link to that publication, and more general information on the subject, see *Fighting Climate Change with Grasslands*, available on the FAO website at: www.fao.org/news/story/en/item/38916/icode/.

83 There is a further, and more ironic, angle to the relationship between agricultural conversion and climate change. Recent studies have verified the hypothesis that increases in temperatures – which are themselves caused in part by agricultural activities and agricultural conversion – will then trigger reductions in crop yields in corn, which is a principal crop grown in the grasslands areas that have been destroyed through agricultural conversion. For an explanation of this "feedback" effect, see *One Degree Over*, ECONOMIST, Mar. 19, 2011, at 91 (reporting on research showing that "a 1°C rise in average temperature will reduce yields across two-thirds of the maize-growing region of Africa, even in the absence of drought", and that *with* drought, "that effect spreads over the entire area").

84 See, e.g., L. K. Mann, *Changes in Soil Carbon Storage After Cultivation*, 142 SOIL SCIENCE, pp. 279–288 (Nov. 1986); G. P. Robertson, E. A. Paul, and R. R. Harwood, *Greenhouse Gases in Intensive Agriculture: Contributions of Individual Gases to the Radiative Forcing of the Atmosphere*, 289 SCIENCE 1922 (Sep. 2000); Keith Paustian *et al.*, *Agriculture's Role in Greenhouse Gas Mitigation* (Pew Center on Global Climate Change 2006), www.c2es.org/publications/agriculture-role-greenhouse-gas-mitigation (concluding that "globally about one-third of the total human-induced warming effect due to GHGs comes from agriculture and land-use change" and that "U.S. agricultural emissions account for approximately 8 percent of total U.S. GHG emissions when weighted by their relative contribution to global warming"); Juha Siikamaki, *Climate Change and U.S. Agriculture: Examining the Connections*, ENVIRONMENT 50, issue 4, 36–49 (Jul. 2008); Tristram O. West and Gregg Marland, *Net Carbon Flux From Agriculture: Carbon Emissions, Carbon Sequestration, Crop Yield, and Land-use Change*, 63 BIOGEOCHEMISTRY, issue 1, pp. 73–83 (Apr. 2003). In addition to the attention given in the academic and technical literature to the role of agriculture in climate change, significant steps have been taken also by international financial institutions to the matter. For a survey of pertinent projects and initiatives undertaken by the World Bank in recent years, see GRASSLANDS, *supra* note 2, at 144–148.

threaten the resilience of the Earth's ecosystems[85] Expressed differently, climate change is one of the ways in which the Earth's "planetary boundaries" have already been exceeded.[86]

IC. *A summing-up on environmental unsustainability*

The preceding discussions present a rather damning picture of the environmental effects of modern extractive agriculture. Personally, I find these details to be deeply troubling: I regard myself as an environmentalist, and I contribute money and time to environmental-protection efforts. At the same time, as I have already noted, I grew up on a farm and I now own farmland myself. Although my family and I direct most of our attention elsewhere, the fact remains that we are engaged to some significant degree – financially, culturally, personally – in modern extractive agriculture. Accordingly, I am hardly comfortable playing the role of "anti-farmer", nor should the observations I have made thus far (and there are still more to come in this chapter) be so construed. Perhaps my position, particularly regarding the relationship between agriculture and the environment, is similar to the one expressed by J. B. Ruhl in his article of about 15 years ago:

> To acknowledge that farms pollute and degrade the environment should neither indict farming as a way of life nor denigrate the ideals farmers hold. Farming in America is a deeply-rooted cultural institution with many noble qualities and important economic and social benefits, but . . . [it causes serious environmental damage. Acknowledging this fact should not be] regarded as an attack on the people or the institutions involved [in farming] . . . . The plain truth is that farms pollute ground water, surface water, air, and soils; they destroy open space and wildlife habitat; they erode soils and contribute to sedimentation of lakes and rivers; they deplete water resources; and they often simply smell bad.

---

85 For an excellent explanation and evaluation of the notion of "resilience", and of how this notion has different application in science disciplines versus social science disciplines, see Lennart Olsson *et al.*, *Why Resilience Is Unappealing to Social Science: Theoretical and Empirical Investigations of the Scientific Use of Resilience*, 1 SCIENCE ADVANCES 1400217 (2015), http://advances.sciencemag.org/content/1/4/e1400217. See also, Robert L. Glicksman, *Ecosystem Resilience to Disruptions Linked to Global Climate Change: An Adaptive Approach to Federal Land Management*, 87 NEBRASKA LAW REVIEW 833, 838 (2009) (noting the "scientific paradigm shift from an equilibrium to a disequilibrium model and the relevance of that shift to the importance of striving to achieve resilience as a resource management technique").

86 For further details of this perspective on climate change and the other "planetary boundaries" – which are identified by some as biosphere integrity, land-system change, freshwater use, biochemical flows, ocean acidification, atmospheric and aerosol loading, stratospheric ozone depletion, and "novel entities" – see Will Steffen *et al.*, *Planetary Boundaries: Guiding Human Development on a Changing Planet*, 347 SCIENCE 1259855 (2015).

These effects are and always have been consequences of farming in general.[87]

A crucial challenge that I see – it is the challenge that has prompted me to study the subject and write this book – is to bring fundamental change, if at all possible, to this ages-old reality of agriculture. Although J. B. Ruhl is surely correct in saying that farming operations "are and always have been" responsible for serious environmental degradation, and although the intensity and reach of that degradation has increased dramatically in recent decades, perhaps there is some radical change that can be made to arrest the degradation and even reverse it. My attention will turn to that – first from a scientific perspective and then from a legal and institutional perspective – in subsequent chapters of this book. First, however, there are some further troubling aspects of the current system of agriculture that need to be considered. They relate to human health and safety.

II. Modern extractive agriculture poses undue risks to human health

IIA. Introductory comments

One of the biggest areas of concern emerging in recent years about modern extractive agriculture is its effects on human health. The rising consumer interest in organic foods, along with rising public demands for more revealing labeling on food products – especially to identify genetically modified or genetically engineered inputs – reveal a growing public distrust of the modern "industrial" food production system.[88]

However, there are two sides to the story. At the same time that some critics allege serious agriculture-related and food-related health risks, others

---

87 Ruhl, *supra* note 25, at 266. In quoting this same passage from Ruhl's article, Susan Schneider adds historical commentary:

> Farming has caused widespread environmental degradation for centuries. For example, the January 1849 *Scientific American* included a report of the practice, common in England at the time, of steeping wheat in an arsenic solution before sowing it to prevent loss of the crop to worms and birds. Although successful in achieving its intended agricultural purpose, the magazine condemned the practice for the adverse effect it had on partridges and pheasants, concluding "we can afford to feed both men and birds."

Schneider, *supra* note 24, at 120 n.2 (citing a 1999 *Scientific American* article and a 1985 *Journal of Soil & Water Conservation* article).

88 For discussions of these issues, see Hiroki Uematsu and Ashok Mishra, *Organic Farmers or Conventional Farmers: Where's the Money?*, 78 ECOLOGICAL ECONOMICS 55, 55 (2012); Ulf Hjelmer, *Consumers' Purchase of Organic Food Products. A Matter of Convenience and Reflexive Practices*, 56 APPETITE 336, 336–337 (2011); Anne Bellows et al., *Gender and Food, a Study of Attitudes in the USA Towards Organic, Local, U.S. Grown, and GM-Free Foods*, 55 APPETITE 540, 540 (2010).

respond by *praising* the modern system of food production and delivery for providing the world's human population with food that is safer than ever before. Complicating the discourse further is a fundamental difference between cultures and countries as to the role of government and the public sector generally in addressing health risks – with some urging a strongly proactive "precautionary" approach and others emphasizing the benefits of a more reactive approach that aims to handle specific problems efficiently if and when they arise.

It is against that backdrop that a close examination is warranted regarding public health issues relating to today's food system. For two reasons, however, my treatment of those issues in this section II differs importantly from the discussions of economic and environmental effects of modern extractive agriculture, as presented in Chapter 2 and in section I of this chapter, respectively. First, there is a big difference between (i) the human-health effects that flow from today's food system *in its entirety* – that is, encompassing the production, processing, and delivery of food, particularly in (allegedly) advanced societies – and (ii) the human-health effects that flow from *agriculture*, which is just the "front-end" element of the food system as a whole. My aim in this book is to focus just on agriculture and the set of ecological realities in which it can exist. Hence I am reluctant to venture too far into issues relating mainly to the human-health implications of the "follow-on" elements of the food system as a whole – that is, those relating mainly to what use is made of the agricultural products once they have left the field.

Second, I see some artificiality in distinguishing between (i) environmental risk and injury of the sort that I examined above in section I of this chapter and (ii) risk and injury to the health of *humans* in particular. If we accept the proposition, as I do, that humans are part of, not separate from, all of nature (that is, the ecosphere) – and indeed that humans' interests should in some cases be subordinated to the interests of other species – then agriculture's effects on *human* health would logically constitute just a subset of agriculture's effects on the ecosphere more generally. However, a good reason for giving at least a limited treatment here to modern extractive agriculture's implications for human health is, simply put, that humans are special in at least one way that is relevant here: they (we) have a heightened responsibility for the well-being of other species. We cannot fulfill this responsibility unless we are healthy ourselves.

My colleague Caleb Hall recently prepared a detailed survey of issues that relate to food and agriculture in today's world. That survey, published in early 2016,[89] gives special attention to (i) the use of fertilizers, pesticides, and

---

89 Caleb Hall, *Agriculture, Food, and Human Health: A Survey of Issues, Concerns and Implications*, CITA WORKING PAPER #5 (2016), available on the website of the Center for International Trade and Agriculture, http://law.ku.edu/cita.

other agricultural chemicals, (ii) the very recent introduction and use of advanced genetic technology in certain crops, (iii) food-borne illnesses that are attributable in part to agricultural processes, and (iv) the obesity epidemic that has descended with full force recently on the USA and that threatens some other countries as well. I would urge readers interested in the details of that survey and analysis to see that article. I have drawn from that article to present in the following few paragraphs a "nutshell" account of how I see certain shortcomings in modern extractive agriculture as posing hazards to human health.

IIB. Agricultural chemicals: too much use and too little caution

Of the four main topics enumerated above, none is more directly related to agriculture – and to the notion of "agroecological husbandry" that I will turn to in Chapter 4 – than that of how agricultural chemicals have come to play a central role in modern extractive agriculture. Such agricultural chemicals fall into two main categories: fertilizers and pesticides. As for fertilizers, I explained above in Chapter 1 that since the Second World War, great quantities of ammonia (NH$_3$) – a combination of hydrogen and nitrogen – have been used to create synthetic nitrogen fertilizers to boost crop yields.[90] As for pesticides, the results of scientific development have, if anything, been even more remarkable. The six key types of pesticides are (i) insecticides, (ii) mineral oils, (iii) herbicides, (iv) fungicides and bactericides, (v) plant growth regulators, and (vi) rodenticides.[91]

---

90 See text accompanying notes 19–20 in Chapter 1, *supra*. As pointed out there, the ammonia plants that were used at first for this purpose were already in production during the Second World War in order to supply nitrogen for use in making explosives. Now, roughly four-fifths of all ammonia produced around the world is devoted to use as agricultural fertilizer. Indeed, nitrogen fertilizer now constitutes the largest single energy input into industrial agriculture.

91 Within (i) the *insecticide* group there are chlorinated hydrocarbons (e.g., DDT, mirex, aldrin) used for arthropod pests generally, organo-phosphates (e.g., Dursban, Nuvan Top, Vapona, Kontrol) used for arthropods generally as well as in shampoos and pet care products, carbamates (e.g., Temik, Award, Logic, Larvin) used for arthropods generally but especially for Hymenoptera, and pyrethroids (e.g., Biomist, Scourge, Anvil) used widely in home arthropod control. There are also certain insecticides used to treat seeds such as organo-phosphates, carbamates, and pyrethroids. Within (ii) the *mineral oils* group, various products (e.g., Saf-T Side, Orchex, Trilogy) are used for arthropod pests generally, especially in organic farming or applications. Within (iii) the *herbicide* group there are triazines (e.g., Miracle Gro, Scotts Bonus S Weed and Feed, Monsanto Lariat) used to kill broadleaf plants, phenoxy hormone products (e.g., Weed-B-Gone, Weedmaster) used to kill broadleaf plants, amides (e.g., Frontier, Outlook, Stampede) used to kill broadleaf plants and yellow nutsedge, carbamates (e.g., Chlorpropham, Sulfallate, Phenmedipham) used as photosynthesis inhibitors, dinitroanilines (e.g., Prowl H$_2$O) used to kill grasses and small seed broadleaf plants, urea derivatives used primarily in pre-emergence application against broadleaf plants, bipiridyls (e.g., paraquat, diquat) used as an enzyme inhibitor, and uracil (e.g., bromacil, isocil) used as a photosynthetic inhibitor. Another highly prominent herbicide is glyphosate, marketed beginning in the 1970s as "Roundup". Glyphosate (described in more detail in note 60, *supra*) is used for comprehensive elimination of plants – with the

I summarize in the following list several specific points that strike me as most important regarding this array of agricultural chemicals.

- Aside from hybrid crops, the development of agricultural chemicals – both fertilizers and pesticides – was one of the principal areas of innovation emerging out of the Green Revolution and the broader set of twentieth-century agricultural advances of which the Green Revolution was a prominent part. Introducing those chemicals into farm operations helped increase crop yields dramatically, but they also caused serious damage and injury – such as the near-extinction of the bald eagle due to DDT use.

- Although regulations have been put in place in many countries, including the USA, to guard against the worst human-health effects flowing from these agricultural chemicals, occupational hazards associated with their use still remain. The 16,000 pesticides currently marketed in the USA cause thousands of farm-worker poisonings each year, according to the US Center for Disease Control. Some other countries, most notably Mexico, have reported even more troubling figures on occupational hazards stemming from the use of agricultural chemicals.

- Some of these agricultural chemicals are known to be carcinogenic, but foods in whose production they are used can remain on the shelves – that is, in the chain of commerce and in use by consumers – on grounds that the pesticide residues carried in those foods are below specified levels. However, the rationale that permits their continued sale is suspect in some cases because of the nature of the chemicals themselves: they are endocrine disruptors, acting like imposters to trick an organism's body in ways that sabotage its normal functions. Moreover, some of the chemicals are persistent organic pollutants ("POPs") with long life cycles[92] – and some can travel around the globe through condensation

---

exception of some grain crops (such as "Roundup-Ready" corn) that have been developed to survive exposure to glyphosate. Within (iv) the *fungicide and bactericide* group there are inorganics (including sulfur, sodium azide, potassium azide, marketed under various brand names), dithiocarbamates (e.g., Polyram, Manzate, Zineb), benzimidazoles (e.g., Spectrum), triazoles and diazoles (e.g., Stratego YLD), and diazines and morpholines (marketed under various brand names), each product being used to kill fungi and bacteria on different crop species. There are also those fungicides used to treat seeds such as dithiocarbamates, benzimidazoles, and triazoles diazoles. Pesticides in (v) the *plant growth regulators* group (e.g., Stinger, Paramount, Clarity) are used to make weeds grow uncontrollably until they die. Within (vi) the *rodenticide* group there are anti-coagulants (e.g., Contrac, Terad₃ Blox, Rentokil), cyanide, hypercalcaemics (e.g., Agrid₃, CyKill), and narcotics, all of which are used against a variety of mammalian pests. See Food & Agriculture Organization, *FAOSTAT*, United Nations (2013), http://faostat.fao.org/site/424/default.aspx#ancor.

92 For instance, "pesticides are less likely to leach [into the soil] when their hydrolysis half-life is less than six months and their soil half-life is less than three weeks." U.S. Environmental Protection Agency, *Half-Life*, Ag 101 (Jun. 27, 2012), www.epa.gov/oecaagct/ag101/pesthalflife. html. However, although sources disagree as to the half-life of DDT in water (some saying as

and volatization (the "grasshopper effect"), and this can spread their effect widely. Even beyond these characteristics, some of these chemicals can be accumulated in human body fat, making their overall long-term toxicity difficult to ascertain.

- Ironically, the most recent increase in use of agricultural chemicals has not been accompanied by a concurrent increase in crop productivity.
- The use of agricultural chemicals in less-developed economic and regulatory systems warrants special attention.[93] For instance, use of pesticides between 1990 and 2010 has dropped in the USA but risen in Latin American countries.
- Claims of causal connections between the use of agricultural chemicals and the incidence of human illness – for instance, cancer – are still subject to uncertainty. The response to this uncertainty in some countries, especially in Europe, is to follow a robust version of the Precautionary Principle.[94] But in much of the rest of the world, a more lax approach is taken because of either (or both) (i) policy choices that highlight individual freedom and responsibility over a communitarian or paternalistic approach or (and) (ii) insufficiencies of government resources to provide robust protection to the public.
- An illustration of this difference in approach – what we might roughly label as "proactive paternalism" versus "reactive individualism" – can be seen in the case of atrazine. Many US streams and rivers, along with many public drinking water systems, contain atrazine. Atrazine use is banned in the EU.

---

much as 150 years), DDT's half-life in soil is anywhere from 2 to 15 years. See National Pesticide Information Center, *DDT (General Fact Sheet)* (1999), http://npic.orst.edu/factsheets/ddtgen.pdf; Extension Toxicology Network, DDT (dichlorodiphenyltrichloroethane), Cornell University, http://pmep.cce.cornell.edu/profiles/extoxnet/carbaryl-dicrotophos/ddt-ext.html (stating that DDT's half-life is 56 days in lake water and 28 days in river water). For more soil half-lives for numerous pesticides, and classifications of herbicide families, see Troy Bauder *et al.*, *Best Management Practices for Agricultural Pesticide Use to Protect Water Quality*, Colorado State University Extension (Feb. 2010), www.ext.colostate.edu/pubs/crops/xcm177.pdf.

93 For details on agricultural-chemical regulation in economically less-developed countries, see Dun Tri Phun *et al.*, *Pesticide Regulations and Farm Worker Safety: The Need to Improve Pesticide Regulations in Viet Nam*, 90 BULLETIN OF THE WORLD HEALTH ORGANIZATION 468, 468 (2012) (emphasizing that Vietnam is particularly ineffective in regulating the use of agricultural chemicals, with one survey finding 2,800 pesticide retailers operating without a license, reflecting the fact that Vietnam has not developed any legal measures to address concurrently increasing usage of pesticides since 1990). One authority has explained that the lack of domestic regulation leaves many economically less developed countries with only the FAO's "International Code of Conduct on the Distribution and Use of Pesticides" as a guide. See Kees Jansen, *The Unspeakable Ban: The Translation of Global Pesticide Governance into Honduran National Regulation*, 36 WORLD DEVELOPMENT 575, 575 (2008).

94 See *infra* note 96.

The survey that Caleb Hall has recently published[95] provides extensive details and substantiation regarding these and numerous other points regarding agricultural chemicals and their role in modern extractive agriculture. On these and other grounds, I arrive at the following assessment, expressed in a "nutshell" form:

- Notwithstanding (i) the fact that most agricultural chemicals (that is, fertilizers and biocides) have long life cycles and (ii) the fact that we do not yet fully understand the long-term effects of their concentrated accumulation over time, modern extractive agriculture continues to use them widely.
- True, some controls and regulations are in place (varying from one country to another, with little international coordination), but those controls and regulations are rather spotty and porous. Even where we do understand, after bad experience, how some chemicals, such as DDT, injure human health, the response to that knowledge has been weak: for instance, the USA has in a sense outsourced the danger of DDT by exporting it to be used elsewhere in the world. This practice seems ill-advised. Even though we do not know much about the long-term effects that agricultural chemicals (for instance, atrazine) will have on humans from ingesting even small amounts of them through water, we *do* know that farm workers exposed directly to some agricultural chemicals can be seriously injured.
- Given all this, and given the fact that the chemicals seem to be having less productivity-boosting effect now anyway, there is a serious mismatch between the amount of *use* of the chemicals and the amount of *caution* that surrounds that use. As for caution, it is short-sighted and a dereliction of duty for public servants (in legislative and regulatory-agency positions) in many countries to take too much of a hands-off "reactive" approach to agricultural-chemicals regulation. Instead of declining to take serious anticipatory protective action until *after* a specific injury has been clearly shown, public officials in all countries should take (and multilateral initiatives should be put in place to press for) the more public-oriented approach of the Precautionary Principle as practiced in Europe[96] – as through requiring heavy testing and then imposing adequate safety requirements and appropriate mitigation guarantees.
- *In a sentence: The key problem relating to agricultural chemicals is that they are still used too much and with too little caution.*

---

95 See generally Hall, *supra* note 89.
96 I will be giving further attention to the Precautionary Principle in Chapter 6; see text accompanying note 26 in Chapter 6, *infra*. For a discussion of the adoption of the Precautionary Principle specifically in Europe, see note 15 in Chapter 7, *infra*. For a survey of its reflection in international instruments, see note 16 in Chapter 7, *infra*.

IIC. Genetic technology: privatization, short-sightedness, and anthropocentrism

A second topic relating to the human-health aspects of modern extractive agriculture concerns recent advances in genetic engineering, genetic modification, and biotechnology[97] in general – all as they relate to modern extractive agriculture. As with the topic of agricultural chemicals, discussed above, I offer the following "nutshell" assessment, drawing from the separately-published survey that Caleb Hall has undertaken:

• Recent advances in genetic modification ("GM"), genetic engineering ("GE"), and biotechnology generally show great promise, and they should be encouraged, not condemned out of hand. Indeed, many such advances are not inconsistent with natural-systems agriculture innovations, including the perennial-polyculture research program described below in Chapters 4 and 5 as part of a new system of agroecological husbandry.
• However, the current trajectory of GM/GE/biotechnology work is dangerous and ill-managed in several respects. For instance, current law allows an inappropriate degree of privatization of new agriculture technology. Instead of placing responsibility for, and control over, the development of biotechnology in private hands with aims of maximizing profits, such responsibility and control – for biotechnology and GE research specifically but also for agricultural research more generally[98] – should be placed and kept in the public domain.
• Doing so would reflect certain crucial values. One of these values is an appreciation of and respect for the complexity of nature. Another is the encouragement of innovation through proper incentives.[99] And another

---

97 Susan Schneider offers this definitional guidance:

> Karl Ereky, a Hungarian engineer, coined the term "biotechnology" in 1919 to refer to the science and the methods that permit products to be produced from raw materials with the aid of living organisms. According to the Convention of Biological Diversity, biotechnology is "any technological application that uses biological systems, living organisms, or derivatives thereof, to make or modify products or processes for specific use" (Article 2). According to the FAO's statement on biotechnology, "interpreted in a narrow sense . . . [biotechnology] covers a range of different technologies such as gene manipulation and gene transfer, DNA typing and cloning of plants and animals."

Schneider, *supra* note 24, at 536.

98 For observations about the need for agricultural research in general to take place in public institutions, see text accompanying notes 84–87 in Chapter 7, *infra*.

99 For a discussion of the role of incentives in privatization of new agricultural technology, and whether the responsibility for and the results of research aimed at developing it should be placed and kept in the public domain, see Gregory Graff, Amir Heiman, Cherisa Yarkin, and David Zilberman, *Privatization and Innovation in Agricultural Biotechnology*, 6 AGRICULTURAL AND RESOURCE ECONOMICS UPDATE 5–7 (2003), available on the website of the Giannini Foundation of Agricultural Economics at the University of California, at: http://giannini.ucop.edu/media/

is the importance of broadly-shared benefits rather than a concentration of benefits in the hands of only a few persons or entities.[100]
- Therefore, GE research into technological manipulation of agricultural plant species – research that is possible now as never before[101] – should proceed:

---

are-update/files/articles/v6n3_2.pdf (concluding that "[e]stablishing proprietary rights over knowledge enhances the incentives for commercial development, but it may constrain future innovations in both the public and private sectors"). See also John King, Andrew Toole, and Keith Fuglie, *The Complementary Roles of the Public and Private Sectors in U.S. Agricultural Research and Development*, Economic Brief Number 19 (September 2012), available on the website of the USDA's Economic Research Service, at: www.ers.usda.gov/media/913804/eb19.pdf (observing that "the private sector underinvests in R&D from a societal perspective" and that "private firms also have incentives to limit the availability of new technology and knowledge through trade secrets, patent protection, and other efforts to appropriate a greater share of R&D benefits . . . [and this] encourages investment but can hamper social benefits to R&D" in several ways). See also *Agricultural Biotechnology* on the website of the Government of the Netherlands Department for International Development, at www.odi.org/sites/odi.org.uk/files/odi-assets/publications-opinion-files/3163.pdf (noting that "[p]atent systems need to be implemented so as to ensure a balance between commercial incentives and [public] access" to biotechnological innovations in crops and seeds, especially in the global South).

100 My colleague Caleb Hall has offered this summation on the need for a cautious skepticism to guide the embrace of biotechnology in agriculture, especially if it unduly concentrates benefits:

Historically, whenever the agricultural community has "embraced" a new technology it has done so to the point of nearly devotional delusion – with the result that no one seriously questions (i) whether the use of the new technology should be tempered for the sake of the environment or (ii) whether the new technology disproportionately benefits a small cluster of extraordinarily powerful established business interests, to the detriment of farming communities.

Accordingly, he urges that while patents and copyrights generally deserve protection, it must be recognized that in the context of biotechnology such protections can impede research into other potentially novel crops. One manifestation of this is the application of patents to prevent farmers from saving their own seed so as to breed crops more suited for their environments. "Because of patent rights, a farming practice as old as agriculture itself has been stopped because of the profit motives of a few biotech companies." Correspondence with Caleb Hall, March 2016. See also David Catechi, *Two Wrongs Don't Make a Patent Right*, 56 HASTINGS LAW JOURNAL, 769, 778–779 (2005); Nathan Busch, *Genetically Modified Plants are not "Inventions" and are, Therefore, not Patentable*, 10 DRAKE JOURNAL OF AGRICULTURAL LAW 387, 387 (2005). For observations of this issue from the perspective of a "pro-poor strategy", see Robert Tripp, ed., BIOTECHNOLOGY AND AGRICULTURAL DEVELOPMENT xxiii (Foreword by Ray Offenheiser and Kimberly Pfeifer, of Oxfam America) (2009) (noting that a "pro-poor strategy . . . would champion the notion that governments should support appropriate institutional mechanisms to maintain a public domain for agricultural technology").

101 Although farmers have been artificially modifying the genetics of their plants for several thousand years, it is just within the past 30 years that new technological methods of genetic engineering have emerged to introduce plants with phenotypic traits that would have been impossible through cross breeding, or at least have taken countless generations. As noted elsewhere, these new techniques have been employed to make crops – most notably corn and soybeans – produce their own pesticide, to make them resistant to a pesticide, to make them produce more of a desired compound, to increase their temperature resistance, and to increase

i with great caution and strict regulatory oversight, intent on avoiding unintended consequences,

ii not with a search for a "silver bullet" or series of such "silver bullets" – for example, addressing just one food-crop disease at a time – but instead in a long-term effort to improve food-crops steadily;

iii in such a manner as to encourage farmers to nurture crops that are best situated for their own ecosystems, rather than developing a small cluster of monoculture crops with aims to grow them in many disparate ecological settings;

iv with an eye to ensuring a broad sharing of the benefits such improvements will bring to farmers and to society at large – and, to this end, GE research should be stripped of its private-sector and profit-motive features to the extent that those features interfere with that sharing of benefits;[102]

v with a perspective that looks beyond just our own species and instead recognizes that "fiddling" with plants without considering an extremely broad range of implications for the biosphere as a whole is shortsighted and potentially dangerous;[103] and

vi with a long-term perspective that looks to the benefit of future human generations and overall long-term genetic diversity on the planet.

- The fact that the foregoing standards and perspectives are *not* being applied now can be seen in the fact that agriculture-related biotechnology research is not currently subjected to serious and effective public regulation – especially at the multilateral level – and that, in particular, inadequate scrutiny is being given to transgenic manipulation, which

---

their shelf life. For various reasons, wheat has not thus far been the subject of the sorts of extensive genetic modification that has transformed the production of corn and soybeans. For an assertion that this situation should change, see Jayson Lusk and Henry I. Miller, *We Need G.M.O. Wheat*, THE NEW YORK TIMES, Feb. 3, 2014, p. A19. A few years ago, reports circulated about the "escape" of GMO wheat – which was the subject of earlier research (1994–2004) by the Monsanto company but never proposed for approval by US governmental authorities. See Carey Gillam and Julie Ingwersen, *U.S. Discovery of Rogue GMO Wheat Raises Concerns Over Controls*, REUTERS, May 31, 2013 (noting that this incident "joins a score of episodes in which biotech crops have eluded efforts to segregate them from conventional varieties" but that this was the "first time that a test strain of wheat, which has no genetically modified varieties on the market, has escaped the protocols set up by U.S. regulators to control it"). "Rogue" GMO wheat was found in 2014 in Montana. Carey Gillam, REUTERS, Sep. 26, 2014. For details about GMO wheat development in various countries, see the website of GM Freeze, www.gmfreeze.org/gmwheatnothanks/gm-wheat-news.

102 The removal of those features will almost surely require revisions to intellectual property laws in order to ensure that GM crops cannot qualify for government-protected monopoly rights.

103 For specific grounds for concern over such "fiddling", see the various entries on the website of The Nature Institute (http://natureinstitute.org), under its project titled "Unintended Effects of Genetic Manipulation".

poses peculiar risks. Moreover, recent increases in the consumption of meat have negative consequences not only for many of the individuals themselves but also by adding urgent pressure to increase grain production for use as livestock feed in concentrated operations. This pressure is in turn hastening the development of super-crops through transgenic manipulation. By contrast, imposing incentives through various fiscal and regulatory means to reduce high meat consumption[104] – and in favor of locally-sourced meat – would, in addition to the health benefits this could provide, also reduce the pressure for such hasty efforts toward purported super-crops through transgenic manipulation.

• *In sum: While careful and broad-based GM/GE/biotechnology research and innovation holds promise for agriculture and should be encouraged and supported, a cluster of key problems relating to how such research is conducted now revolve around the fact that there is too little attention to the complexity of nature and to the public interest. As a consequence, such research currently places too much reliance on private-sector control over agriculture-related biotechnology research. This, in turn, inserts profit-motive biases into the process. The result is a view that is both short-sighted (just looking for one silver bullet, then another, without adequate consideration of long-term complexities and consequences) and anthropocentric (just considering the effects of agricultural-biotechnology innovations on humans).*

IID. Food-borne illnesses: haste, waste, and resistance

A third topic relating to the human-health aspects of modern extractive agriculture concerns food-borne illnesses as they relate to modern extractive agriculture. As I have noted earlier, most of my attention in this book focuses on the food-crop aspects of agriculture and not on the role of livestock in farming or the food system generally. I make some exceptions to this approach, however, in some specific cases. For instance, one reason it makes sense to consider livestock operations as well as cropping operations for purposes of examining the role that agriculture plays in global climate change is that methane, which is released in vast quantities in cattle operations, is an especially

---

104 For a report on a group of scientists' "contentious step of suggesting methane emissions be cut by pushing up the price of meat through a tax or emissions trading scheme", see John Metcalfe, *To Fight Climate Change, the Entire World Will Have to Eat Less Meat* (Mar. 31, 2014), appearing on the website of the Atlantic Monthly CityLab, at www.citylab.com/work/2014/03/fight-climate-change-eat-less-meat/8753/. For a comprehensive analysis of the global role of livestock – including discussions of health and environmental issues and the role of various types of incentives – see generally Henning Steinfeld, Pierre Gerber, Tom Wassenaar, Vincent Castel, Mauricio Rosales, and Cees de Haan, Livestock's Long Shadow: Environmental Issues and Options (2006), ftp://ftp.fao.org/docrep/fao/010/a0701e/a0701e09.pdf. That book was prepared under the auspices of the Livestock, Environment, and Development ("LEAD") Initiative, which is supported by the World Bank, the EU, and various other official agencies.

potent greenhouse gas.[105] Likewise, the worldwide destruction and degradation of grasslands is attributable mainly to a combination of (i) their conversion to cropping operations and (ii) their use for extensive grazing of livestock for meat production. This combination has been nearly fatal, for instance, to the prairies that formerly covered the Great Plains of North America.

Another exception to my concentration on cropping operations, not livestock operations, applies here: just as the production of food-crops plays a central role in the world's overall food system, so does the production of livestock. Moreover, those two types of operations are closely related to each other both traditionally and today. The family farm I grew up on was typical in involving a variety of crops – wheat, corn, soybeans, fescue, and others – and a variety of livestock, mainly Angus cattle but also a smattering of hogs, sheep, and poultry. Such integrated operations have nearly disappeared in the USA now, but a great proportion of crop production – particularly corn production – is devoted to the growing of feed for livestock in large feedlot operations, as a supplement to or substitution for pasturing on grass.

In short, livestock operations are central to the overall food system. Because of that, Caleb Hall's survey of human-health concerns gives some attention to food-borne diseases associated both (i) with livestock operations and (ii) with food-crop operations. As with the topic of agricultural chemicals, discussed above, I offer the following "nutshell" assessment, drawing from his survey:

- To the rather limited extent that modern extractive agriculture contributes to this category of problems, it is attributable mainly to haste in handling food products – especially meat but also vegetables. Transmission of pathogens, including E. Coli, Salmonella, Campylobacter, and Listeria, increases with the proliferation of concentrated animal feedlot operations ("CAFOs") and other mass processing facilities – all of which are geared for hasty preparation of products for sale, without proper safeguards to prevent those products from getting contaminated by animal waste, especially feces.

- More public oversight of such facilities and of food handling more generally – via stronger laws and regulations and better enforcement – can help reduce food-borne illnesses, and so can some other approaches that have not yet been adequately explored, such as:

  i   reducing the role of meat in high-meat-consuming societies by incentivizing heavy meat-eaters to eat less meat (as mentioned briefly in subsection IIC) – thereby taking some demand pressure away from the process – or at least to make sure that the meat being eaten comes mainly from local suppliers; and

---

105 For the role of livestock in methane production and release, see notes 79–80, *supra*, and accompanying text.

  ii providing better public education, since most such illnesses can be cooked out of the food products.

- An especially serious failing relating to these aspects of human health, however, is the over-prescription and improper use of antibiotics – particularly for animals but also for humans. There is a culture of over-medication and sloppy administration in both, and this has led to the rapid emergence of strains of pathogens that have developed resistance to antibiotics.

- In these respects, today's system of high-volume livestock operations reflects many of the same underlying problems that plague modern extractive agriculture more generally: In a reflection of what Wes Jackson calls "the industrial mind",[106] extractive agriculture has deconstructed the ages-old integration of livestock and livestock operations – a deconstruction that, as Wendell Berry has observed, replaces an elegant solution (animals replenishing fertility that crops deplete) with two problems (declining fertility plus livestock-based pollution)[107] – and it disregards such principles as the "law of return"[108] and an "agriculture of restraint"[109] that have until recently characterized food production within the confines of the natural world, not as an enterprise that can be undertaken in ignorance or defiance of that world.

IIE. *Obesity: yet-uncriminalized stacking of the subsidies deck*

A fourth and final topic relating to the human-health aspects of modern extractive agriculture concerns obesity. Like the discussion of food-borne illnesses immediately above, a discussion of obesity lies squarely within the scope of the modern food system but is only indirectly related to agriculture. Given this fact, I offer only this brief assessment, again expressed in a "nutshell" form drawing from the separately-published survey that Caleb Hall produced:

- Although a lack of personal discipline might be partly to blame for the obesity crisis facing the USA (and soon to face several other countries), the deck is stacked against several segments of the population that are especially vulnerable to obesity. Because of the structure of US agriculture subsidies, the American public is essentially financing the obesity epidemic.[110]

---

106 See note 22 in Chapter 1, *supra*.
107 See *supra* note 64.
108 See *infra* note 117.
109 See text accompanying note 118, *infra*.
110 For details relating to this proposition, see the general examination of agricultural subsidies in subsection IC and section II of Chapter 2, *supra*.

- Although one approach to this issue would be to restrict and/or criminalize sales of unhealthful "food" items, perhaps a better approach would be (i) to subsidize the production and sale of more healthful foods and dietary choices – fruits and vegetables, for instance – and (ii) to modify or restrict the subsidies currently provided less healthful items, and also (iii) to make similar reductions in subsidies that indirectly work in that same way, particularly subsidies for fossil-carbon extraction and processing. More generally still – but outside the immediate ambit of agriculture law and policy – an intense public attack should be mounted against the causes of poverty and ignorance that lead to poor diets.[111]

- *In a sentence: The key problem relating to agriculture's role in obesity is that the structure of agricultural subsidies, especially when combined with fossil-carbon-related subsidies, is deeply flawed in a way that encourages production and sale of unhealthful products and discourages the production and sale of healthful food.*

III. Modern extractive agriculture goes historically and socially "against the grain" of human development

The points of criticism that I have discussed in Chapter 2 and in earlier sections of this chapter have focused on how modern extractive agriculture is economically unsustainable as well as environmentally unsustainable, and on the ways in which modern extractive agriculture poses unduly large risks to human health. In this final section of Chapter 3, I wish to take a somewhat different approach. I aim to explain some key ways in which the most recent manifestation of modern extractive agriculture – which has been transformed for reasons discussed below into what is now widely called "industrial agriculture" – represents a radical departure from the course that human development has taken up through about the end of the eighteenth century. At least in the West, social development in general, and especially agriculture, has in the past few decades been "industrialized".

IIIA. Before industrialized agriculture: going "with the grain"

As pointed out in section I of this chapter, a transition from hunter-gatherer life to agricultural life started around 11,000 or 12,000 years ago. As several authors have observed, it was not a foregone conclusion that such a transition would be made. William Ruddiman gives this explanation:

> [T]he switch from the hunting-fishing-gathering life to agriculture was not inevitable. For one thing, hunter-gatherers derive food from many

---

111 As for the second of these – ignorance leading to poor diets – perhaps a useful example of an "intense public attack" that has met with some success is the anti-smoking campaign that has taken many forms over the past several decades in the USA.

sources, and the wide variety of easily available plants and animals in regions like the Fertile Crescent naturally promoted nutritional balance. By comparison, overreliance on food sources from just one or two crops can cause malnutrition because of loss of sufficient protein or fat. From this perspective, it did not necessarily make sense for people to rely increasingly on just a few crops after 12,000 years ago. . . .

Even though the origin of agriculture was not inevitable, several factors caused people living in the Fertile Crescent to begin altering their daily routines in ways that would gradually develop into the practice of agriculture. Growing wild in the semi-arid grasslands of this region was an unusual variety of edible wild grains that were ideal for domestication . . . . These included several kinds of cereals: two kinds of wheat (emmer and einkorn), barley, and rye, all of which are easily gathered sources of carbohydrate. Also available were peas and lentils (beans), both good sources of protein. This range of natural bounty was unique to this one region.

Part of this bounty was a result of the semi-arid climate of the Near East. In regions with long dry seasons, where most of the annual vegetation dies each year, plants use their energy to produce seeds for reproduction, and those edible seeds are convenient food sources for humans. By comparison, the much larger amount of vegetation stored in well-watered forests provides little that is edible, and deserts are nearly barren. As a result, people in the Fertile Crescent could derive a portion of their food needs from crops but still augment their nutrition intake by gathering wild seeds, hunting, and in places fishing.[112]

Ruddiman goes on to explain more about the development of agriculture in the Near East region about 11,000 years ago:

Another major advantage of the Fertile Crescent region was that it was the natural home to many types of wild animals that proved easy to domesticate. As Jared Diamond noted in *Guns, Germs, and Steel*, many more types of mammals exist than can easily be domesticated. Some are by nature too wild or skittish or solitary, and some are simply too small to be of much use. The Fertile Crescent was again fortunate in having the wild predecessors of goats, sheep, pigs, and cattle . . . [Therefore,] [n]utritionally, the Fertile Crescent provided everything necessary for humans to adapt a new kind of life based entirely on agriculture.[113]

A central point that Ruddiman is making relates to "bioregionalism" – the obvious but often disregarded reality that different ecoregions around

---

112 Ruddiman, *supra* note 68, at 65–66.
113 *Id.* at 68–69.

the Earth have different characteristics. It is a point that dates back at least to Alexander von Humboldt (1769–1859), whose research into the distribution of species and ecosystems in geographic space and through geological time were expanded on by the Russian German climatologist Wladimir Köppen; the Köppen climate classification system that he first published in 1884 and modified in later decades rests on the concept that native vegetation is the best expression of climate.[114]

Ruddiman explores bioregionalism further in another passage. Having explained that agriculture "originated independently in several regions" – the Fertile Crescent, the Yellow River Valley of northern China, the Central American lowlands, the high terrain around the Peruvian Andes, and the tropics of Africa and New Guinea – he offers this observation: "Studies of remote and isolated tribes who have maintained Stone Age cultures into the modern era have shown them to be world-class botanists within their own habitat. Most can distinguish hundreds to thousands of plants . . . [and their] motivation for acquiring this knowledge is obvious: they needed to gather food for survival, and they could not afford to make mistakes."[115]

In discussing other aspects of pre-"industrial" agriculture, Frederick Kirschenmann focuses on two particular elements. The first is "return" – a concept he describes by explaining how it is now largely absent from modern agriculture:

> Modern American agriculture largely ignores the *law of return*. With the introduction of fossil fuels in the 1950s, we shifted entirely from a nutrient cycling to an input/output system of production. We now rely exclusively on exogenous inputs to supply basic nutrients; farms are specialized, standardized, and simplified like factories. Diverse cropping systems and crop/livestock integration, required for efficient nutrient cycling, are no longer necessary. These production systems meet the needs of an industrialized food system, but they depend on fossil-fuel inputs for fertility, pest control, processing, packaging, and shipping.[116]

Nutrient cycling was common until the mid-nineteenth century. Most human and animal wastes were recycled as food for the plant kingdom. As society became urbanized in the mid-nineteenth to the mid-twentieth centuries, human waste recycling became difficult, so we began depositing it in landfills and sewers. As agriculture became

---

114 The more recent Trewartha climate classification system, developed by American geographer Glenn Thomas Trewartha in 1966 and updated in 1980, is a modified version of the 1899 Köppen system, created to answer some of the deficiencies of that earlier system.

115 Ruddiman, *supra* note 68, at 65. For other references to "bioregionalism", see note 27 in Chapter 4, *infra*.

116 Kirschenmann's emphasis on the significance of "fossil-fuel inputs" corresponds to the emphasis that Tim Crews of The Land Institute places on "fossil fuel slaves". See text accompanying note 53 in Chapter 2, *supra*.

industrialized in the late twentieth century and crop and livestock production systems became isolated from each other, recycling animal wastes became difficult and costly, creating unintended consequences such as soil loss, nutrient pollution, and imperiled water systems.[117]

The other feature of pre-"industrial" agriculture that Kirschenmann draws attention to is "restraint". He discusses that feature, and explains how it figured into a sophisticated system of food production in North America before the European invasion starting in the late fifteenth century:

> The *agriculture of restraint* that natives practiced before the arrival of Columbus was likely developed in response to fifteen thousand years of experience. Until quite recently, we had dismissed their agriculture as "primitive" because it seemed so contrary to the "clean cultivation" mosaic familiar to the European mind, and because it didn't appear to fully exploit the rich resources available here. Today we [realize otherwise] . . . . In the northern plains, for example, the Ankara, Mandan, and Hidatsa tribes had evolved principles of agriculture that [relied on] . . . [d]iversity, recycling, restrictive cultivation, moisture conservation, and selectivity, based on bioregionalism [as their] key strategies . . . . All of these principles are now very familiar to advocates of sustainable agriculture.[118]

Kirschenmann's brief description of some key features of agriculture as practiced by Native Americans prompts us to consider that subject in somewhat more detail. After all, North America is – for better or for worse – the birthplace of "industrial" agriculture; so it is particularly appropriate to consider how this same continent fared in terms of food production under pre-"industrial" agriculture.

IIIB. The Native American agricultural experience

Put in rather simple terms, some of the key characteristics of today's "industrial" agriculture that were *not* part of food production as practiced by Native Americans in North America were these: large scale operations, heavy reliance on fossil carbon for high energy input, an emphasis on monocropping, extensive nutrient/chemical pollution of waterways, and intense productivity.[119] The characteristics that *did* feature prominently in

---

117 Kirchenmann, *supra* note 45, at 180 (emphasis added).

118 *Id.* at 264 (emphasis added).

119 These and other components of industrial agriculture – or what I have more comprehensively labeled "industrial fossil-carbon-based enhanced extractive agriculture" – are discussed in subsection IB of Chapter 1, *supra*, and are reflected in summary form in Figure 1.1 in that same chapter.

Native American food production in that region – and in many other parts of the world as well in the pre-"industrial" agriculture era – are fallowing, crop diversity, flooding, burning, adaptations to local ecology, and remarkable energy-efficiency based on current (solar) power. Other features included the requirement for large tracts of land to support smaller population groups.

Today's agriculture, therefore, looks quite different from the agriculture practiced by Native American groups before the arrival of European immigrants. Gregory McIsaac and Edward Williams offer an account that gives special emphasis to the "agricultural knowledge" possessed by the Native Americans in temperate zones of North America:

> [Their] agricultural knowledge [was] reflected in a diverse mix of crops planted in intercropped fields, extensive approaches to maintaining soil fertility through fallowing, reliance on human labor and simple tools, a relationship between the spiritual well-being of the group and the farming enterprise, division of labor by gender, and some combination of farming, hunting, fishing, and gathering. Tillage was usually based on intercropped maize, beans, sunflower, pumpkins, sweet potatoes, and squash. These were planted in mounds, which retarded erosion more effectively than contemporary approaches to farming that feature monocropped parallel rows. . . [In particular,] [c]rop diversity was very important. Efforts were made to maintain wild varieties of domesticated plants so they could cross-pollinate those in the gardens . . . . Frequently, multiple varieties of maize would be intercropped in the same field. Maize varieties were well adapted to local environments. Hidatsa flint corn, for example, was grown in the semi-arid Northern Plains climate, matured in sixty days, and was resistant to hail and frost. Its short stalk was also resistant to the wind.[120]

The significance of these observations on the diversity of crops used by Native Americans becomes clearer when contrasted to the "industrial" agriculture of today. Now 90 percent of the world's food comes from 30 crop species, even though about 7,000 crop species exist.[121]

---

120 Gregory McIsaac and William R. Edwards, eds., SUSTAINABLE AGRICULTURE IN THE AMERICAN MIDWEST: LESSONS FROM THE PAST, PROSPECTS FOR THE FUTURE 37–38 (1994). For another account of Native American agriculture, highlighting many of the same points emphasized by McIsaac and Edwards, see generally R. Douglas Hurt, INDIAN AGRICULTURE IN AMERICA (1987). Several of the techniques that McIsaac and Edwards describe – such as the early development of squash, the use of mound planting, and the importance of crop diversity – had origins in Mesoamerica dating back at least to about 3000 BCE, and by 900 BCE "Mesoamerican farmers were superb plant breeders". Hurt, *supra*, at 7–8.

121 Clay, *supra* note 27, at 49. For further details on this set of figures (90%, 30 crop species used, of 7,000 crop species that exist), see note 27 in Chapter 2, *supra*.

McIsaac and Edwards summarize the key characteristics that Native American agriculture had in North America before contact with Europeans:

> Land was to be used to sustain the economic and spiritual life of a community. Large tracts of land were required by a community to maintain its practice of farming, hunting, fishing, and gathering of wild foods. Land tenure and use was controlled by groups, usually a community, that could sanction individuals whose practices broke social norms. Individuals or lineages had the right to use land for farming or other subsistence activities. Individuals did not, however, have the right to claim permanent title or to transfer title or ownership of land. [122]
>
> [In addition, this system of food production was energy efficient.] In Native American farming systems, five to fifty food calories were obtained for each calorie of energy invested. [123]

These accounts of Native American agricultural methods are not intended to suggest that they always succeeded, or that there was no degradation of the land, no collapse of production, no famines. What we know of Native American history indicates many such problems and disasters.

For instance, Charles Mann has explained that "between about 1100 and 1300 A.D. [CE], cataclysms afflicted Indian settlement from the Hudson Valley to Florida", often because they had, upon adopting maize, "set aside millennia of tradition, started burning and clearing thousands of acres of land, and experienced floods and mudslides". [124]

These instances of failure, though, in the Native American agricultural experience do not undercut the main point I wish to emphasize here: in many respects, in many areas, for many centuries, food production in North America was highly successful based on criteria that we would surely consider important today. It featured the "law of return" that Kirschenmann has referred to, it involved a high degree of sophisticated agricultural knowledge; it drew from a broad range of crops; it minimized erosion; it was energy efficient in terms of the number of calories of energy produced from a calorie of energy expended.

Indeed, Mann explains in his account of Native American agriculture that after the period of "cataclysms" he cited, lessons were learned and corrective actions were taken. That is, after about 1300 CE, evidence of widespread erosion peters out; Native Americans were still growing a lot of maize, but

---

122 McIsaac and Edwards, *supra* note 120, at 38, citing a 1988 article on Native American agriculture by J. T. Milanich.

123 McIsaac and Edwards, *supra* note 120, at 38, citing a 1975 article by J. S. Steinhart and C. E. Steinhart on energy use in the US food system, as published by the American Association for the Advancement of Science.

124 Charles C. Mann, 1491: New Revelations of the Americas Before Columbus 301–302 (2011).

they changed their methods. They began systematically replanting large belts of woodland, thereby decreasing erosion. Moreover, the trees planted in these woodland areas were carefully selected, creating veritable orchards of fruit and nut-producing trees such as hickory, beech, acorn, butternut, hazelnut, pecans, walnuts, and chestnuts. The result was a vast patchwork of farmland and orchards, with enough forest to also allow for hunting – or, as Mann expresses it, "[t]he result was a new balance of nature."[125]

IIIC. The Green Revolution and "industrial" agriculture

If North America saw the sort of "new balance of nature" that Mann refers to – and that was based on the sort of "bioregionalism" noted above – it did not last long. In the early 1800s the industrial revolution took root in North America. The industrial revolution popularized ideas that were different from those of specialization and bioregionalism: they were the ideas of generalization and globalization. Goods could be produced in great quantities in factories operating with machine power (not just human muscle) and using assembly lines, interchangeable parts, and the like, and then shipped elsewhere.

Moreover, all this could be done with unprecedented speed. In his book *About Time*, Adam Frank explains what a change the industrial revolution brought to the way people saw time, and clocks that kept time – especially for purposes of measuring working hours for laborers hired to operate the machines and to run the assembly lines that started dominating the manufacture of goods.[126] Frank asserts that perspectives changed also on productivity – so that with the industrial revolution came greater emphasis on producing more items within limited time.[127]

---

125 *Id.* at 302.

126 See generally Adam Frank, ABOUT TIME: COSMOLOGY AND CULTURE AT THE TWILIGHT OF THE BIG BANG (2011). Others have also written, of course, about the significance of clocks. In the view of Lewis Mumford, writing in the early twentieth century, the mechanical clock – as developed by monks in the Middle Ages and subsequently adopted by the rest of society – was more important than the steam engine: "The clock, not the steam-engine, is the key-machine of the modern industrial age. . . .The clock . . . is a piece of power-machinery whose 'product' is seconds and minutes . . ." Lewis Mumford, TECHNICS AND CIVILIZATION 14–15 (1934).

127 Frank credits (or blames) Ambrose Crowley, founder of an ironworks facility near Newcastle, England, in the late seventeenth century, for having brought a heavy reliance on clocks (and on wardens to ensure that the laborers came to work on time) in the early days of the industrial revolution. Frank, *supra* note 126, at 90–91. Frank claims that Crowley's set of rules to manage the facility's operations – *The Law Book of the Crowley Ironworks* – set forth "a first vision of the new industrial time". *Id.* at 90. Frank also emphasizes the importance of the *minute hand* on clocks, saying that the minute "was a small but workable unit of time" that "announced a new kind of time that would govern the home [and] the workplace." *Id.* at 91. The new emphasis on time, and clocks with minute hands to mark its passing, "changed . . . each worker's 'imperatives of time' as they now needed to fit their lives into the factory's time and the contours of the factory's day." *Id.*

These changes in perspective relating to productivity and time had only limited application to agriculture at first. The development of the mechanical cotton gin in the 1790s represented an early adoption of industrialization in the processing of agricultural products, although not in their planting, cultivation, and harvesting.[128] Other reflections of the industrial revolution appeared before long, though, that dramatically affected many aspects of American agriculture.

In, 1819, for instance, Jethro Wood patented an iron plow with interchangeable parts, following the development by several persons, including Thomas Jefferson, of moldboard plows with reduced resistance and therefore greater efficiency. In 1834, the McCormick reaper was patented, and in 1837 an effective threshing machine was patented. (That was the year also, as mentioned earlier, that John Deere and Leonard Andrus began manufacturing steel plows.) A grain drilling machine for planting seeds was patented in 1841, and a mowing machine to facilitate cutting grasses for hay was patented in 1844. The next quarter-century saw the development and patenting of a range of other agricultural implements and equipment, including the self-governing windmill, the two-horse straddle-row cultivator, modified versions of plows, the spring-tooth harrow (for seedbed preparation), and the early forms of steam tractors.[129]

These developments continued in the 1880s, with the widespread marketing of twine binders used in harvesting, the introduction of horse-drawn combines (also for harvesting), and the production of hybridized corn. By 1890, it took on average "40–50 labor-hours . . . to produce 100 bushels (5 acres) of wheat with gang plow, seeder, harrow, binder, thresher, wagons, and horses", whereas just 60 years earlier it had required "[a]bout 250–300 labor-hours . . . to produce 100 bushels (5 acres) of wheat."[130] Similar productivity increases were made for corn: by 1890, it took "35–40 labor-hours . . . to produce 100 bushels (2½ acres) of corn", whereas just 40 years earlier it had taken "[a]bout 75–90 labor-hours . . . to produce 100 bushels (2½ acres) of corn".

The year my mother's father was born in farm country in northeast Missouri – 1892 – the first gasoline-powered tractor was produced by

---

128 The mechanical cotton gin dramatically increased the speed with which cotton fibers could be separated from the seeds. In addition to causing a massive growth in the production of cotton in the USA – expanding from 750,000 bales in 1830 to nearly three million bales in 1850 – and therefore creating an increasing economic dependence of the American South on slavery, the development of the mechanical cotton gin is thought by some historians to mark the beginning of the industrial revolution, at least in America. See, e.g., Paco Underhill, *The Cotton Gin, Oil, Robots, and the Store of 2020*, 20 Display & Design Ideas 48 (2008).

129 Details in this paragraph and the following two paragraphs are drawn from several sources, but primarily from *Growing a Nation: The Story of American Agriculture* (2014), www.agclassroom. org/gan/timeline/farm_tech.htm, which itself "is based upon work supported by the National Institute of Food and Agriculture (NIFA)" of the USDA.

130 *Id.* (entries for 1830 and 1890).

John Froelich, and by the year my mother was 10 years old – 1930 – gasoline-powered tractors had gone through numerous modifications and were being marketed widely, along with a range of farm implements that those tractors would pull and operate. The year after I was born in 1953, my father purchased a new Allis-Chalmers tractor that had a power take-off drive, extensive hydraulics – these will raise and lower a plow, a rear-mount blade, or a post-hole digger – a hand clutch as well as a foot clutch, an automatic starter, and adequate power to pull and operate planters, rakes, hay elevators, wagons, corn-pickers, grain combines, and a range of other implements. My brother and I still use that tractor occasionally, but not for any cropping operations.

Also by 1950, because of those technological developments, the average effort required to produce 100 bushels of corn had dropped to 10–14 labor hours (from 35–40 hours in 1890) and now required only two acres, not two and a half acres. The average effort required to produce 100 bushels of wheat had dropped to six and a half hours (from 40–50 hours in 1890) and required only four acres, not five as in 1890.[131]

Perhaps the largest influence that the industrial revolution had on agriculture, though, was felt in the middle part of the twentieth century with the advent of the Green Revolution. It is to that phenomenon that I now turn.

It might seem odd to have come nearly to the end of a third chapter defining and evaluating modern extractive agriculture without having devoted much sustained attention to the so-called Green Revolution that gained such notoriety and achieved such success on its own terms beginning in the 1940s. Up to this point, I have referred to the Green Revolution only indirectly in discussing (i) certain economic aspects of the increased crop yields that it brought and (ii) how it prompted losses in biodiversity and increases in agricultural chemical use).[132]

Now, however, I wish to turn directly to the Green Revolution, in order to set the stage for the general proposition that I aim to explain and support as I close this chapter. The proposition is this: *"Industrial" agriculture – that is, the remarkably advanced form of modern extractive agriculture that the Green Revolution has ushered in with such flourish and acclaim over the past seventy years – has done deep and lasting harm to society by transforming farming, rural life, and food production in ways (i) that unwisely discard some values and efficiencies that for thousands of years have been central not only to our production of food but also to our role and identity within the ecosphere and (ii) that have, by linking agriculture*

---

131 *Id.* (entries for 1945 and 1955).

132 In some of those preceding subsections, I have noted that the term "Green Revolution" can be construed narrowly to comprise only one element of a larger set of developments that created the form of agriculture prevalent in many countries today – but that in some contexts the term is construed very widely. Most of the observations from Mark Tauger, quoted below, reflect that wider construction of "Green Revolution".

with industry (and the fossil carbon on which industry depends), made agriculture today economically unsustainable, environmentally unsustainable, and dangerous to human health.

Mark Tauger offers a somewhat "sanitized" description of how the Green Revolution began in the mid-1940s and unfolded in its early years:

> [The Green Revolution was triggered in part by the fact that] during the Second World War, Mexico had three successive failures of its wheat crop caused by rust, a fungal plant disease, and in 1943 the Mexican government appealed to the U.S. government for aid. The U.S. government referred the Mexicans to the Rockefeller Foundation, which turned to the University of Minnesota, where one of the top specialists on rust, Elvin Stakman, was based. Stakman organized a team of specialists to go to Mexico to deal with this problem. One member of his team was Norman E. Borlaug, who completed a PhD under Stakman in 1942.
>
> This team set up a plant breeding program that also trained Mexican scientists. In a few years they developed wheat varieties resistant to rust, suited to Mexican conditions and higher yielding. Borlaug, however, saw the potential to achieve something better. The U.S. in 1945-46 had obtained a [wheat] plant with a large head. It was a descendant of the Rono varieties developed in the nineteenth century but with admixtures from American and Russian varieties. Borlaug and his team crossed Norin with the new rust-resistant varieties for seven years until they finally developed varieties with the genetic characteristics they sought.
>
> By the 1950s his program was breeding dwarf wheats that responded to heavy doses of fertilizers by growing much more grain on a sturdy stem that would not lodge – collapse – from the weight of the head full of grain . . . . The program organized Mexican landlords to produce seed, distributed this to farmers widely, and by 1956 Mexico became self-sufficient in wheat. The key to this success was a "package": the seed, fertilizers, and adequate irrigation, and the researchers insisted that it would work on any scale of farm with those inputs.[133]

I referred above to this account of the Green Revolution as somewhat "sanitized" because it leaves out important political influences that US authorities – including in particular Vice-President Henry A. Wallace – brought to bear on the research program in Mexico. I explore those interesting political influences later.[134] For present purposes, the overall chronology and consequences of the Green Revolution are most important. Tauger offers this assessment of its results:

---

133 Mark B. Tauger, Agriculture in World History 153 (2010).
134 See text accompanying notes 90–93 in Chapter 6, *infra*.

The Green Revolution brought a long-term increase in agricultural productivity and food supplies. The area planted in high-yielding rice and wheat in developing countries increased from 41,000 hectares in 1965–66 to 50.5 million hectares by 1970-71. Perhaps two billion people in the world by 2000 could not have been kept alive [otherwise]. This success was also in part the result of increased fertilizer production.[135]

The point Tauger makes in that last sentence is worth further attention: the success of the Green Revolution in creating a dramatic increase in crop yields was "in part the result of increased fertilizer production." As explained earlier, great quantities of ammonia (NH_3) – a combination of hydrogen and nitrogen – have been used ever since the Second World War to create synthetic nitrogen fertilizers to boost crop yields.[136] Indeed, roughly four-fifths of all ammonia produced around the world is devoted to use as agricultural fertilizer. Production of the ammonia, in turn, requires great quantities of energy, which in the USA and many other countries is drawn from natural gas – which is of course a form of fossil carbon that is extracted from beneath the surface of the Earth. This is one way in which the sort of agriculture that has emerged in the past 70 years is tied into the same dependence on fossil carbon that characterizes the industrial revolution more generally.

After all, it was fossil fuel that provided power for the industrial revolution to take hold. The first of the fossil fuels used for this purpose was coal. Although coal had been used since prehistoric times, it was – as one source has expressed it – "the overwhelming need for energy to run the new technologies invented during the Industrial Revolution that provided the real opportunity for coal to fill its first role as a dominant worldwide supplier of energy."[137] Then came the discovery and development of other forms of fossil fuel – principally oil[138] and natural gas – to continue powering the industrial revolution and transforming society.[139]

---

135 Tauger, *supra* note 133, at 154.

136 See notes 19 and 20 in Chapter 1, *supra*, and accompanying text.

137 US Department of Energy, *A Brief History of Coal Use*, www.fossil.energy.gov/education/energylessons/coal/coal_history.html.

138 For a description of the discovery of oil by Edwin Drake in Pennsylvania in 1859, and of the effect it had on industrial mechanization (particularly the internal combustion engine), see text immediately following note 15 in Chapter 1, *supra*.

139 For a fascinating synopsis of some key trends of environmental history over the past 200 years, with special attention to the development and use of fossil fuel, see John R. McNeil, *Global Environmental History in the Age of Fossil Fuels (1800–2007)* (2011), www.cartografareilpresente.org/article254.html. John McNeil, a professor of history at Georgetown University, writes that although "[c]oal was king for the span of two human generations", oil carries twice the energy per ton as does coal, so oil quickly gained favor. "Between 1800 and 2000, total worldwide energy use grew by 80–90 fold, the most revolutionary process in human history since domestication [of plants and animals roughly ten thousand years ago]. Fossil fuels accounted for almost all the growth, and today make up about 77% of all energy use." *Id.*

The Green Revolution, then, may be seen as dramatically strengthening the linkage between the development of fossil carbon fuels and the character of agriculture. The linkage includes not only nitrogen fertilizers as described above but also the use of fossil fuels for the mining of phosphorus and potassium (also used as fertilizers) and for the powering of farm equipment and machinery used in processing and transporting agricultural products.[140]

In sum, industry and agriculture now have converged to a very great degree, in large part because of their shared use of fossil fuels. This is an important way in which the term "industrial agriculture" makes sense.

Tauger offers these observations on the convergence or interdependence of industry and agriculture – and on the ways in which this development is risky and unwelcome:

> Most studies, economic theories, and state policies have long viewed agriculture and industry as different sectors and often attributed to agriculture the role of supporting industrial development. [Although since the Second World War] agriculture and industry have steadily grown more interdependent [and is today considered] . . . for the most part one sector of a global industrial economy . . . [nevertheless] agriculture has certain features that differ from all other industries, and [this] makes the recent dependence of food on industry extremely risky for the world . . . . [After all,] [l]ife forms are much more complex and less fully understood than most raw materials used in industrial production. The magnitudes of world population and agricultural production mean that changes in it can have serious and unanticipated effects on the environment and subsistence.[141]

The concern Tauger voices over industrial agriculture is echoed, of course, by many others. Kirschenmann, for instance, elaborates on two specific ways in which industrial agriculture poses problems: by fostering a "consumption ethic" and by making "factory food" a commonplace reality. Kirschenmann explains first his "consumption ethic" point:

> Consumerism grew out of the industrial era. Erich Fromm suggests that it was the industrial era's illusion that we could fully control nature . . . that led us to believe in "unlimited production and, hence, unlimited consumption." This [illusion] . . . transformed us into consumers, turned

---

140 For a useful survey of these, see Grace Communications Foundation, *Energy and Agriculture*, www.gracelinks.org/118/energy-and-agriculture.

141 Tauger, *supra* note 133, at 162. As I emphasize in section II of Chapter 8, it is in part this special character of agriculture, and the dangers inherent in treating it like industry for purposes of governance, that agriculture – or, more generally, agroecological integrity – should fall not within the jurisdiction of traditional single-territory states but rather under the responsibility of a different global framework of governance and protection based on "eco-zones".

inward, intent on usurping nature to satisfy our own individual desires, even at the expense of our neighbors. As consumers we became individual siphons on society, instead of cooperating members of a local neighborhood . . .

We can never achieve a people/food/land equilibrium without both a new consumption ethic and a new production ethic. The new consumption ethic must be based on satisfying the needs of local communities within the constraints of local ecosystems, rather than on the desires of individuals driven by illusions of unlimited resources. In other words, commercial transactions must be driven by community *custom* designed to enrich the fabric of the entire community, rather than individual satisfaction driven by whatever subjective need we may feel. Such an ethic would transform us from *consumers* into *customers*.[142]

Against these concerns over the long-term implications of industrial agriculture as it has emerged from the Green Revolution is this simple fact: the Green Revolution created, in an astonishingly short period of time, a dramatic increase in the food-producing capacity of a growing world. A quick reference to the Wikipedia entry for "Green Revolution" reveals this terse description: "[This set of] initiatives, led by Norman Borlaug, the "Father of the Green Revolution" [is] credited with saving over a billion people from starvation."[143]

This is true, and it is important. However, for the reasons I have tried to present in this chapter, the Green Revolution – along with the larger industrialization-of-agriculture context that it fits into – has had a great many very unfortunate consequences. As one obituary emphasized on Norman Borlaug's death in 2009, many "commentators [on Borlaug's work] pointed to the problems that [came] in the wake of his 'Green Revolution' . . . [especially as] concerns steadily mounted over the long-term sustainability of the chemical-based farming practices involved . . . [and over] the profound social and ecological changes that the revolution heralded among peasant farmers".[144] Some of those consequences can only be overcome or reversed

---

142 Kirschenmann, *supra* note 45, at 58. A ground for concern in these observations by Kirschenmann is this: Although it seems reasonable, and even essential, that *local* circumstances drive decisions about the proper "people/food/land equilibrium" (to use Kirschenmann's term), the ecological and social shortcomings of modern extractive agriculture seem to require *global* changes in priorities. It is fundamentally problematic to expect that local determinations and implementation of national and global policies will work well. It is with this in mind that I will turn in Chapters 6 and 7 of this book to national and global legal reforms.

143 See http://en.wikipedia.org/wiki/Green_Revolution, citing Peter B. R. Hazel, THE ASIAN GREEN REVOLUTION 1 (2009) (published by the International Food Policy Research Institute). The Wikipedia entry goes on to explain that the term "Green Revolution" was first used in 1968 by former United States Agency for International Development (USAID) director William Gaud.

144 Christopher Reed, *Norman Borlaug: Agricultural Scientist Who Averted Famine with a Controversial 'Green Revolution'"*, GUARDIAN, Sep. 13, 2009.

now with tremendous effort. This book can be seen in part as a contribution to that effort – looking specifically at fundamental changes to be made in perspectives, priorities, policies, laws, and regulations necessary to transform agriculture.

It should be clear from the earlier sections of this chapter that my views on the economic, environmental, and health implications of modern extractive agriculture are shared by a wide variety of thoughtful observers. Some of those observers have not focused directly on the Green Revolution itself as a target of criticism, but others have done so. To see an example of the latter, consider the website of the University of Arkansas National Agricultural Law Center, which offers this bullet-point summary of how the Green Revolution and its technological changes have contributed to a form of agriculture that is not viable over the long term:

- Monocropping and pesticides: Green Revolution agriculture relies on the extensive use of pesticides, which are necessary to limit the high levels of pest damage that inevitably occur in monocropping – the practice of producing or growing one single crop over a wide area.
- Biodiversity: Green Revolution agriculture reduces agricultural bio-diversity by relying on a few high-yield varieties of each crop. This decrease in biodiversity makes the food supply susceptible to pathogens that cannot be controlled by agrochemicals and is responsible for the permanent loss of valuable genetic traits bred into traditional varieties. Additionally, deforestation in order to develop new farmland causes the loss of wild biodiversity.
- Soil erosion: Increased erosion caused by monocropping causes siltation of waterways and carries fertilizer, pesticides, and herbicides from the soil into foreign watersheds and ecosystems.
- Water scarcity: Green Revolution agriculture's heavy reliance on irrigation has added to the global problem of water scarcity.
- Carbon emissions: [The Green Revolution's] [h]eavy reliance on fossil fuels has transformed the agriculture industry into a heavy emitter of untenable amounts of carbon into the atmosphere.[145]

My aim in drawing critical attention to the Green Revolution, in terms of its economic, environmental, and human-health aspects, is not to detract from the scientific breakthroughs that Norman Borlaug and others achieved.

---

145 See *Sustainable Agriculture – An Overview*, on the website of the University of Arkansas National Agricultural Law Center, at www.nationalaglawcenter.org/assets/overviews/sustainableag. html. For another explanation of several political, socioeconomic, and nutritional criticisms that have been leveled directly at the Green Revolution, see Daniel Pepper, *The Toxic Consequences of the Green Revolution*, available on the website for U.S. News & World Report at www.usnews. com/news/world/articles/2008/07/07/the-toxic-consequences-of-the-green-revolution. (July 7, 2008).

Those breakthroughs unquestionably had short-term benefits for some farmers and many non-farmers. Instead, my aim is to suggest that the Green Revolution, in representing a climax of the extractive system of agriculture – the most modern, the most resource-intensive, the most production-oriented ever – represents *also* the epitome of a system that has profound shortcomings.

In short, I have tried to develop in this chapter the foundations for the proposition that I announced in my introduction of the Green Revolution, a few pages prior to this one. I asserted there that "industrial" agriculture, to which the Green Revolution has been a particularly prominent contributor, has done deep and lasting harm to society by transforming farming, rural life, and food production in ways (i) that unwisely discard some values and efficiencies that for thousands of years have been central not only to our production of food but also to our role and identity within the ecosphere and (ii) that also have, by linking agriculture with industry (and the fossil carbon on which industry depends), made agriculture today economically unsustainable, environmentally unsustainable, and dangerous to human health.

As for the first of those points, I have in this final section of Chapter 3 emphasized how odd, historically and socially speaking, today's agriculture has become. It runs, I have asserted, "against the grain" of human development up through about the end of the eighteenth century. It has disregarded the law of return that Kirschenmann described; it has abandoned the close knowledge of nature that bioregionality implies; it has abetted the development of a consumption ethic; it has changed our relationship to land by commodifying it; it has had a similar effect on our relationship to community and shared resources and destinies; and it has become almost breathtakingly energy-*inefficient* through its addiction to fossil carbon. Agriculture has been changed in all these ways in the past two centuries.

Why does this matter, if indeed it does? *Not* because traditional rural life, farming systems, and food production have been unequivocally happy, healthy, peaceful, abundant, and satisfying in earlier eras – far from it. Many aspects of rural and farm life in past generations and millennia have been hard, harsh, and even horrible, especially by standards that most people would wish to apply today. (Of course, the very same thing could be said of urban life as well in many places and at many times before the modern era, and still today for millions of people worldwide.) The concern I have emphasized in this section (III) is that the profound changes that have occurred in agriculture – changes that the Green Revolution accelerated and intensified so dramatically – threaten to blind us to the fact that the benefits of those changes come at the *expense* of some values and ethics and efficiencies that were developed and proven effective over several millennia of human agricultural experience. We must not, I think, abandon those values lightly for benefits of "industrial" agriculture, especially given the *other* risks and shortcomings that have accompanied this epitome of modern extractive agriculture. Foremost among those other risks and shortcomings are the ones I have emphasized above: economic unsustainability (addressed in Chapter 2), environmental destruction (section I), and dangers to human health (section II). Taken in aggregate, these features of modern extractive agriculture make it sensible and indeed urgent to seek viable alternatives.

Part III

The promise of agroecological husbandry

4 What is agroecological husbandry?

I have presented in the preceding Part II of this book a realistic and grim assessment of modern extractive agriculture. Its economic unsustainability, its degrading effects on the environment, the dangers it poses to human health – these problems have all deepened in recent decades, partly as a result of the Green Revolution and the industrialization of food production more generally. Indeed, in concluding my account of modern extractive agriculture at the end of Chapter 3, I asserted that the advances emerging from the Green Revolution have come at the expense of some fundamental values regarding the human role and identity within the ecosphere.

What I wish to do in Part III is to explain how a different form of food production, one based on agroecological husbandry, is both possible and preferable. I begin this chapter with a general description of agroecological husbandry as an alternative to the ages-old extractive approach to agriculture, and then I turn in Chapter 5 to a more detailed account of the advantages that agroecological husbandry has over the form of agriculture that prevails in the world today, and especially in the USA and other high–agricultural–output regions.

In describing (in this chapter) the main contours of agroecological husbandry, I emphasize the extent to which this alternative approach differs from the approach it would replace. Even the term "agroecological husbandry" – initially awkward though it might seem – helps to underscore just how novel it would be to embrace this alternative approach to food production (focusing on grains, legumes, and pulses in particular), especially because of certain terms and connotations that the new label of "agroecological husbandry" avoids. However, I also try to emphasize below that although the label is new, some features of agroecological husbandry itself are not really new. In important ways, it reflects perspectives and values – particularly those that relate to how humans fit in, and depend upon, the natural world – that have merely been temporarily suppressed by the industrialization of agriculture and now need reinvigoration. Fortunately, some key scientific developments of just the past few decades make it possible now to forge a way *out* of the grip of modern extractive agriculture and to claim, or reclaim, the benefits that can come from a natural-systems approach to food production.

I. Etymology, science, perspective, and history

IA. *An etymological approach*

As should be clear from earlier portions of this book, especially Chapter 1, I construe the term "agriculture" rather broadly for purposes of examining its effects and contemplating its future. Indeed, the term "farming" – which traditionally has encompassed both the raising of crops and the raising of livestock – can usually be substituted for "agriculture" throughout the preceding discussions, because my examination of the economic, environmental, and other effects of modern extractive agriculture has included some aspects of livestock production, not just cropping.

In order to explain my term "agroecological husbandry", however, I wish to draw close attention to the word "agriculture". It originates in the Latin word *agricultūra*, from *ager*, "a field", and *cultūra*, "tillage of the soil".[1] Thus,

---

1 This etymological point can be found on innumerable websites. For an interesting debate about the Latin roots of the term "agriculture", and specifically the difference between "*cultūra*" and "*colere*", see the comments following Jason Godesky, *Agriculture or Permaculture: Why Words Matter*, at: www.rewild.info/anthropik/2007/06/agriculture-or-permaculture-why-words-matter/. One of the comments offers this explanation:

> *Cultura* is, admittedly, a fairly complex Latin word, and very tightly bound into Roman conceptions of agriculture, civilization, *pietas* and progress, so those meanings all appear in Latin at some point, though the earliest meaning specifically referred to tilling soil. Later, this was used to suggest tilling soil for civilization and culture in a person, the same way the soil is tilled for wheat; later still, worship itself became a kind of ideological tilling. They're all correct meanings, and all reveal the mythic framework of the Romans, who saw *imperium*, *pietas* and their glorious, totalitarian regime as the logical culmination of *cultura*.

a literal definition of the word "agriculture" could be "tillage of the soil in a field" or simply "tillage of fields".

Over the course of the past few decades, the most drastic form of physical tilling of a field – that is, turning over the top layer of soil by using a steel plow – has waned in popularity in the USA and in many other countries, in favor of "low-till" or "no-till" farming. However, the fundamentally disruptive approach that plowing represents remains central to agriculture around the world; it simply appears in a different form in some countries. Instead of physically turning and churning the soil as a literal matter through plowing, disking, cultivating, and the like – thus leaving it "exposed as a flayed skin laid open"[2] – modern production agriculture uses chemical pesticides to suppress all plant life other than the single desired crop grown in monoculture. Then each year at harvest time, a form of "clear-cutting" occurs that removes that monoculture from the ground, leaving bare soil stripped of plant life, and thereby stifling or starving the normal ecological activity that would otherwise occur there to enrich the soil.

What is that "normal ecological activity"? In Wes Jackson's words, quoted in Chapter 1, the soil "teem[s] with small organisms".[3] In Don Worster's words, it involves "the humus, the organic residue of roots, carrion, feces, bone, and leaves mixed through the mineral components", and it also involves "an incomprehensibly large number of bacteria working away, decomposing the dead, fixing nitrogen, forming nitrates to feed the living".[4]

In short, conventional agriculture, even as practiced in its most advanced forms, relies either on tilling the ground as a literal matter or on tilling the ground in a figurative manner, by a chemical means of disturbing or even sterilizing the soil. One way of explaining the meaning of agroecological

---

Id. Frederick Kirschenmann likewise refers to the Latin word *colere* as an important element to the term "agriculture":

> The term "agriculture" is made up of two words, *ager* from the Greek (meaning "field" or "land") and *colere* from the Latin (meaning "to cultivate"). *Colere*, however, is multifaceted; both "cult" and "culture" are derived from *colere*. Embedded in its meaning is the notion of a community caring for its own refinement. *Colere* presumes a transcendent ethic guiding the community in its efforts to enhance its quality of life. We may therefore assume that to ancient people the word agri*culture* meant cultivation of plants and domestication of animals in the context of a caring community committed to the sacred obligation of caring for the land.

Frederick L. Kirschenmann, CULTIVATING AN ECOLOGICAL CONSCIENCE: ESSAYS FROM A FARMER PHILOSOPHER 50 (2010).

2 John Opie, *Ecology and Environment*, appearing as the third chapter in THE GREAT PLAINS REGION 82–84 (Amanda Rees, ed., 2004), a contribution to THE GREENWOOD ENCYCLOPEDIA OF AMERICAN REGIONAL CULTURES (describing treatment of the soil leading up to the Dust Bowl days on the American Great Plains).

3 See text accompanying note 30 in Chapter 1, *supra*.

4 Donald Worster, THE WEALTH OF NATURE: ENVIRONMENTAL HISTORY AND THE ECOLOGICAL IMAGINATION 81–82 (1993).

husbandry from a simple etymological perspective is that it retains the reference to *ager* (fields) but removes the reference to *cultūra* (tilling). It introduces the notion of ecology and highlights the importance of the ecosystem to a "natural-systems" agriculture. (As I will explain shortly, in subsection IC, agroecological husbandry also involves the notion of "*husbandry*" to distinguish it from modern extractive agriculture's emphasis on maximizing *production*.)

IB. The ecosystem as the standard

Tilling the soil, whether in a literal way or a figurative way through chemicals, is a natural and necessary component of conventional agriculture because such agriculture focuses almost exclusively on annual crops, not perennial crops. Annuals thrive on disturbance – that is, on a temporary disruption, and removal if possible, of competing plant life. By contrast, perennials thrive on continuity and non-disturbance.

I highlight this pair of associations – disturbance in the case of annuals and non-disturbance in the case of perennials – because it illustrates a key distinction between agroecological husbandry and conventional agriculture. Agroecological husbandry takes natural ecosystems as the model or standard on which to base a new approach to the process of growing food in soil. The specific type of natural ecosystem that I have in mind is that of the native grasslands that formerly covered vast areas of the world and that have, over thousands of years, been converted to agricultural use.

As I have explained in my recent book on grasslands, the flora in those areas consist primarily of perennials, not annuals.[5] For instance, what gives the Great Plains of North America their distinctive character – at least in those rare segments of them that have not come under the plow or been damaged by livestock grazing – is the abundant presence of big bluestem, little bluestem, Indian grass, grama grass, buffalo grass, switchgrass, and other species. All of these are perennials, as are the many legumes that typically proliferate in some grasslands and help provide the grasses with much-needed nitrogen.[6] By contrast, the form of maize (corn), wheat, rye, oats, barley,

---

5 See John W. Head, Global Legal Regimes to Protect the World's Grasslands 31 (2012) [hereinafter Grasslands] (drawing from a Canadian survey of grasslands to explain that "[g]rasslands are usually dominated by perennial grasses over the annual and biennial types").

6 In the case of the native grasslands of the American Great Plains, for instance, a wide variety of leguminous plants provide, through nitrogen-fixing bacteria in their rhizomes, a significant portion of the nitrogen that serves as natural fertilizer for the grasses. (The lead plant, *Amorpha canescens*, is an abundant leguminous plant in northeast Kansas where my wife and I live.) Just how and where other nitrogen is produced in the complex biotic community under the surface of grasslands – in what is sometimes called the "rhizosphere" (from the Greek word *rhiza*, meaning "root") – is still the subject of important research. For a summary of the significance of the rhizosphere, see David H. McNear, *The Rhizosphere: Roots, Soil, and Everything Between* (Nature Education Knowledge 4(3):1, 2013), www.nature.com/scitable/knowledge/library/the-

rice, and other grains that dominate today's agricultural production are annuals. They must be planted anew each year, and the soil must be made to accommodate them, and to eliminate their competitors, by *disturbance* of the soil – either by literal tilling or by the application of various forms of biocides to poison or prevent other life forms in the soil.

If we were, then, to envision a technique of growing grains and other food crops in a way that would mimic the natural order and resilience and ecological economy of grasslands, that technique would focus attention on perennials, not annuals. It would also focus on a broad mixture of plants living in a diverse community. Grasslands display such diversity. Note, for example, this description offered by an expert explaining the nature of the tall grass prairies of North America:

> Three herb layers are apparent in the tall grass prairie, each characterized by relatively high species diversity. Both sod-forming and bunch grasses are present. Perennial forbs [such as coneflower, for instance,] are abundant and varied; different species bloom at different times during the growing season[,] contributing to an ever-changing palette of colors. Bluestems comprise the uppermost herb layer. Other, shorter upright grasses and forbs form an intermediate layer. Recumbent species such as the grama grasses (*Bouteloua* spp.) make up the lowest, ground-hugging layer.[7]

The broad diversity described in that passage is absolutely at odds with the *lack* of diversity found in traditional agriculture. Expressed differently: grassland ecology is the epitome of diversity, of polyculture; by contrast, modern extractive agriculture of the sort described and criticized in Chapters 1 through 3 of this book is the epitome of monoculture. Even the customary practice of crop rotation – following corn with soybeans, for example, in order to take advantage of the nitrogen-fixing function that soybeans carry out, thus helping the fertility of the soil – has largely been abandoned in the American Midwest in favor of a single monocrop, usually corn, on the same field year after year.

Therefore, a technique for growing food that departs drastically from modern extractive agriculture and that instead takes the ecology of grasslands as its model would involve perennials grown in polycultures, as opposed to annuals grown in monocultures. These elements lie at the center of

---

rhizosphere-roots-soil-and-67500617. As defined there, the rhizosphere comprises three parts: the endorhizosphere (inside the root mass), the rhizoplane (the zone adjacent to the root including the root epidermis and mucilage), and the ectorhizosphere (extending from the rhizoplane out into the bulk soil).

7 Susan L. Woodward, *The North American Prairies* (1996), appearing on the website of Radford University, at www.radford.edu/~swoodwar/CLASSES/GEOG235/biomes/tempgrass/prairie.html.

agroecological husbandry, and I explore them more fully in subsection IE. Before doing that, however, we should keep our focus on the general character of agroecological husbandry – and particularly why it focuses on "husbandry".

IC. Husbandry over production

In the description I have offered thus far of agroecological husbandry as a new technique for growing food, I have carefully avoided using three particular words. The first word is "production". Technically, of course, the growing of food is the "production" of food. However, the emphasis of conventional agriculture, built as it is on the tradition of using annual grains grown in monocultures, has revolved almost entirely around increasing yields in order to meet rising demand, emerging mainly from the explosion of global human populations. "Production agriculture" is a term often used to capture this emphasis on ever-increasing yields.[8] In the preceding paragraphs, I have avoided using the term "production" in my attempts to describe agroecological husbandry, so as to avoid any undue association with the dominant contemporary approach of raising food (and feed and fiber).

The term I offer instead of "production" is "husbandry". The latter term derives, of course, from "husband", which according to the *American Heritage*

---

8 This term appears widely in the language used by groups involved in and supporting agriculture. For instance, it is possible at the Delaware Technical and Community College to earn an Associate's Degree in "production agriculture", which that institution's website says "involves the growing and marketing of plants and livestock" and requires a "thorough knowledge of marketing, management, and finance combined with production skills". See *Production Agriculture*, at www. dtcc.edu/owens/programs/pdf/production_agriculture.pdf. Although this description might seem innocuous enough, some authors have focused on the ethical aspects of emphasizing *production* at the expense of other values. A book review of the 2006 work *Agriculture's Ethical Horizon* offers this synopsis:

> *Agriculture's Ethical Horizon* . . . argues that agricultural *productivity* has been the quintessential value of agriculture that has trumped other concerns such as sustainability, environmental preservation, and social justice (e.g., fair commodity prices, welfare of migrant farm laborers). Increasing world population demands greater productivity, but *Agriculture's Ethical Horizon* argues that the prevailing production ethic is insufficient to address the myriad issues that 21st-century agriculture faces. Modern practices bent on ever-greater production per acre have created externalities such as soil erosion, pesticide resistance, and groundwater depletion, with woefully inadequate attention about their long-term consequences. Social policies and economies of scale favor ever-larger farms, resulting in loss of family farms and dwindling of rural communities. A new, more encompassing ethic is needed to guide agriculture that places other values on par with production.

See Louis S. Hesler, *Book Review*, 55 AMERICAN ENTOMOLOGIST (Summer 2009) (emphasis added) (reviewing Robert L. Zimdahl, AGRICULTURE'S ETHICAL HORIZON (2006)), available on the website of the Entomological Society of America, at: www.entsoc.org/pubs/bookreviews/book-review-agriculture-ethical-horizon.

Dictionary traces its roots back to the Old Norse term *hūsbōndi*, itself composed of the roots *hūs* ("house") plus *bōndi* or *būandi*, the present participle of *būa*, ("to dwell"). Hence, a husband in this narrow etymological sense (and irrespective of a person's gender) is "a householder". The term "husbandry", in turn, therefore carries the narrow denotation of "management of a household". And yet "husbandry" carries a broader connotation as well, to encompass such notions as conservation, frugality, economy, and the prudent or judicious use and nurturing of resources.[9]

I mean "husbandry"[10] in that latter sense, to refer to the understanding, conserving, and nurturing of the long-term viability of an ecosystem[11] for its own sake because of its own value. That value can include benefits that inure to humans, of course, but the benefit accruing to humans would not naturally – and *should* not – be the main reason for husbandry. In this respect, the concept of husbandry shares some important characteristics with the concept of an equitable trust, especially as that concept has developed in English law. As I have explained in another context, the concept of the trust, tracing its roots back to Roman law, involves an equitable obligation that legally binds a person (the trustee) who has legal title and control over certain property (the trust property) to manage that property *not* for his or her own direct benefit but rather for the benefit of a specified group of persons named as the beneficiaries of the trust.[12] In Chapter 7 I will give some further attention

---

9 One online etymology source notes that the term "husbandry" was in use as early as about 1300 CE, to signify "management of a household" and by the late fourteenth century to signify "farm management", and that it derives from the word "husband" in "a now-obsolete sense of 'peasant farmer'". The same source offers this definition for the word "husband" when it is used as a verb: "manage thriftily," from the early fifteenth century, deriving from "husband" in "an obsolete sense of 'steward'". See www.etymonline.com/index.php?term=husbandry and http://www.etymonline.com/index.php?term=husband&allowed_in_frame=0, respectively. Another source offers this explanation of "husbandry": "farming, especially when regarded as a science, skill, or art; [the] management of affairs and resources". See www.memidex.com/husbandry. The same source offers these as synonyms or related terms: conservation; frugality; economy; parsimony; providence; "the control or judicious use of resources"; and the "science and practice of producing crops and livestock from the natural resources of the earth". The term "animal husbandry" is still widely used to convey this latter meaning in respect of livestock.

10 In using "husbandry", of course, I am not suggesting that farmers are or should be *male* (which the term "husband" might initially be thought to mean). Indeed, throughout most of history, across many cultures, substantial responsibility for food production has rested with women. A 2007 agriculture census in the USA showed that nearly one-third of all farm operators in this country are women and that women serve as "principal operators" in control of over 300,000 farms accounting for over 64 million acres of farmland. USDA, *2007 Census of Agriculture*, "Women Farmers", at 1 and Table 50, www.agcensus.usda.gov/Publications/2007/Online_Highlights/Fact_Sheets/Demographics/women.pdf.

11 For some observations about the meaning of "long-term viability" in this context, see note 34 in Chapter 1, *supra*.

12 See generally John W. Head, *Sketching a Global Agroecology Eutopia: The Land Institute in Directional Context* [hereinafter *Sketching*], appearing as Chapter 9 in in Paul V. Stock, Michael Carolan, and Christopher Rosin, FOOD UTOPIAS – REIMAGINING CITIZENSHIP, ETHICS AND COMMUNITY

to the role I see for the notion of trusts and trustees in this context; I will also explain briefly there how the concepts of "public trust doctrine" and "constructive trusts" as developed in national and international law differ in important ways from those that underlie US trust law.[13]

A second word I have avoided thus far, in addition to "production", is "green". Again, my purpose in doing so is to shy away from any risk of guilt by association. Specifically, I wish to distinguish agroecological husbandry as a proposed new approach to growing food – concentrating on perennials grown in polycultures – from the so-called Green Revolution that gained such notoriety and achieved such success, at least on its own terms, in the latter part of the twentieth century, as described in subsection IIIC of Chapter 3.[14]

A third word I have avoided in describing agroecological husbandry is "sustainable". As with the word "production", the word "sustainable" is, as a technical matter, entirely appropriate as a description of a technique of growing food that uses a grassland ecosystem as its standard or model. Such a technique would, after all, be eminently sustainable in character. Yet the word "sustainable", despite its noble rise to prominence in the environmental-protection movement[15] and its use in the names of numerous admirable initiatives and organizations,[16] does carry some heavy and unattractive baggage in this context: the term "sustainable agriculture" is widely used to describe merely a mitigated version of extractive agriculture – continuing, that is, to rely on annuals, on monocropping, and on fossil carbon, but merely with some increased safeguards intended to slow the pace of damage and deterioration that modern agriculture inflicts.[17]

(2015). References to the concept of the trust, and its relation to the concept of "husbandry", appear at pages 162–163 of that volume.

13 See subsection IIB3, and especially text accompanying notes 43–62 in Chapter 7, *infra*.

14 I should note that although I have avoided in the preceding paragraphs the word "green" – as in "Green Revolution" – I do not shy away from the word "revolution". A shift from conventional agriculture to a new approach to growing food by using perennials in polycultures would unmistakably qualify as a revolution, as noted by the editors of the periodical *Scientific American*: "If the plant scientists succeed, the achievement would rival humanity's original domestication of food crops . . . and be just as revolutionary." Michael A. Parks, *Grass Roots*, SIERRA, Nov.–Dec. 2010, at 50, 52–53 (quoting a 2007 *Scientific American* story).

15 An early use of the term occurred in 1987 with the Brundtland Report, titled *Our Common Future*, which celebrated the cause of "sustainable development", for which it offered this definition: "Sustainable development is development that meets the needs of the present without compromising the ability of future generations to meet their own needs." The 1992 Rio Conference, referred to in subsection IA1 of Chapter 7, prompted the UN General Assembly to create the Commission on Sustainable Development, details of which can be seen on its website at www.un.org/esa/dsd/csd/csd_aboucsd.shtml.

16 An example is the National Sustainable Agriculture Coalition, where my friend Ferd Hoefner has worked for years on legislative and policy reform that would be beneficial to farmers and to the soil on which we all depend.

17 For example, the elements of "sustainable agriculture" as enumerated on Wikipedia (probably an adequate source for this general purpose) are agroforesty, crop rotation, mixed farming, and

In the preceding few paragraphs, I have exaggerated certain word choices to make a point. The point is this: agroecological husbandry differs fundamentally from conventional agriculture – especially in conventional agriculture's modern fossil-carbon-based industrial enhanced extractive manifestation. Moreover, agroecological husbandry aims to take as its model the very sort of grassland ecology that modern extractive agriculture has largely displaced and destroyed around the world. In sum, to underscore these crucial distinctions, I have gone to rather extreme lengths – including avoiding the words "production" (because of its association with "production agriculture"),[18] "sustainable" (because that word can be hijacked to celebrate half-measures),[19] and "green" (for its possible association with the Green Revolution) . . . and indeed even avoiding the word "agriculture" itself in this context.

I have not tried to be so strict with my terminology elsewhere in this book. After all, it is a little difficult to steer clear of the word "agriculture" or the term "sustainable agriculture" in this context. Indeed, in some respects I embrace the main elements of "sustainable agriculture" as it has been

---

multiple cropping – none of which involves either perennial crops or polycultures. Similarly, the legislation that governs the USDA's National Institute of Food and Agriculture, in defining the term "sustainable agriculture", envisions it as merely a modestly modified form of traditional agriculture by prescribing that "sustainable agriculture" will "[m]ake the most efficient use of nonrenewable resources and on-farm resources and integrate, *where appropriate*, natural biological cycles and controls" – hardly a rejection of fossil carbon in favor of renewable resource use. See the USDA's website at: www.nifa.usda.gov/nea/ag_systems/in_focus/sustain_ag_if_legal. html (emphasis added), citing 7 USC §3103(19).

18 The distinction between "production" and "husbandry" can be seen, I believe, in Carl Sauer's use of the term "husbandry" in one of his most famous essays: "The more farming becomes *industry* and *business*, the less remains of the older *husbandry* in which man [or woman] lived in balance with his [or her] land." Carl Sauer, *The Agency of Man on the Earth*, in SELECTED ESSAYS 1963–1975 (Bob Callahan, ed., 1981), at 359 (emphasis added).

19 In contemplating the meaning(s) of the term "sustainable agriculture", Wes Jackson has noted the dangers of this hijacking for political purposes:

[I]t is becoming increasingly clear that "sustainable" is a complex political word. Political terms are especially vulnerable to cooptation, to the point that the term could be used as a weapon by *proponents* of large-scale industrialized agribusiness, people who want agriculture to go on as it is now . . . . We cannot . . . allow it to be defined by such people or, for that matter, only by farmers. More accurately, the word comes from the few people in the common culture who are frustrated with the extractive economy and the desecration of the land and water that sustain agriculture.

Wes Jackson, *Making Sustainable Agriculture Work*, appearing in NATURE AS MEASURE – THE SELECTED ESSAYS OF WES JACKSON, at 88 (2011) (emphasis added). For a commentary on the proper definition of "sustainability" in agriculture – and insisting that a sustainable agriculture does *not* deplete water resources, jeopardize human health, cause extinctions of species, or include economic profitability as a necessary condition – see Timothy E. Crews, Charles L. Mohler, and Alison G. Power, *Energetics and Ecosystem Integrity: The Defining Principles of Sustainable Agriculture*, 6 AMERICAN JOURNAL OF ALTERNATIVE AGRICULTURE 146–149 (1991).

described by some sources.[20] However, at least for purposes of the preceding paragraphs, I have avoided those four words – "agriculture", "green", "sustainable", and "production" – in order to emphasize the fundamental difference between modern extractive agriculture and what I have given the rather odd label of "agroecological husbandry".

I hasten to add that the first part of that term "agroecological husbandry" is by no means my own creation. That is, even though I think the term "agroecological husbandry" originates with me, the term "agroecological" – or its noun form, "agroecology" – dates back several decades, and it reflects a rich heritage of alternative views regarding agriculture, ecology, and humanity.[21]

ID. The historical roots of agroecological husbandry

Frederick Kirschenmann explains some of the background to the term "agroecology". He offers these explanations in an essay that quotes several other writers as well:

> The word agroecology came into our vocabulary around 1970 and refers to a way of practicing agriculture that attempts to balance environmental and economic risks of farming while maintaining productivity over the long term. Agroecology refers to farming practices that "have the general effect of making the agro ecosystem more like the natural ecosystem and less like the urban-industrial system, and hence a less disorderly and a more harmonious component of our total landscape." Agroecology also refers to an alternative food system that supports ecologically sounds farming practices. In fact, it is reasonable to believe

---

20 For an overview of "sustainable agriculture" and a look at some domestic US initiatives that have been taken toward it, see the website of the University of Arkansas National Agricultural Law Center, at: www.nationalaglawcenter.org/assets/overviews/sustainableag.html and at: www.nationalaglawcenter.org/readingrooms/sustainableag/. As indicated on that website, "[w]hile the term 'sustainable agriculture' is not clearly defined, most definitions include four basic goals: adequate food and fiber production; sound environmental stewardship; economic viability; and development of strong rural communities. Sustainable agriculture may be understood as a process of striking a balance among these goals, viewing them as complementary rather than competing." See also Committee on Twenty-First Century Systems Agriculture, National Research Council, Executive Summary, in TOWARD SUSTAINABLE AGRICULTURAL SYSTEMS IN THE 21ST CENTURY (2010) (carrying the report of the Committee on Twenty-First Century Systems Agriculture, convened by the National Research Council to assess scientific evidence for the strengths and weaknesses of different production, marketing, and policy approaches to improving the sustainability of American agriculture).

21 Routledge carries over a dozen titles, for instance, with "agroecology" as part of the title. See https://www.routledge.com/products/search?keywords=agroecology. Some authors use a hyphen, as in "agro-ecology", and I have followed that usage in some of my earlier writings. See, e.g., John W. Head, International Law, Agro-Ecological Integrity, and Sovereignty – Proposals for Reform 63 THE FEDERAL LAWYER 56 (June 2016).

that without alternative food system infrastructures, it may not be possible to farm in an ecologically sound manner: "When agricultural products become valued more as market commodities to be sold to the highest bidder rather than as food to nourish us, and when short-term yields are maximized at the expense of long-term sustainable production, then the agro ecosystem becomes more of a drain than a contribution to the life-support environment."[22]

Kirschenmann goes on to explain how "during the past several decades we have succeeded in reducing agri*culture* to agri*business*" and what role he sees for agroecology in reversing that trend:

> Our word "business" comes from the Old English *bisig*, meaning to be constantly occupied. In the sixteenth and seventeenth centuries, it also carried the connotation of impertinence and mischievousness. The word *agribusiness* was presumably introduced as a way of urging farmers to think about farming as a business "like any other" and *not* as a way of life. Since the 1970s, agribusiness propaganda has been warning farmers that if they want to survive, they had better get on the agribusiness bandwagon.
>
> While the etymological roots of these terms are instructive, perhaps even more illuminating is the connotation they carry in the agricultural community. "Agri*culture*" is now generally shunned because it refers to old-fashioned and ineffective farming practices of the past. Agri*culture* is considered synonymous with poor management and nostalgic adherence to farming methods that do not meet the challenges of modern market realities. Land-grant universities, consequently, have distanced themselves from agri*culture*. Farm magazines almost universally prefer the word agri*business*. In mainstream agricultural circles, agri*business* is synonymous with high-tech, modern, efficient farming that uses best management practices, while agri*culture* is a vestige of the past. The barely recognized agro*ecology* is hardly ever used in farm magazines.[23]

According to Kirschenmann, the shift from agriculture to agribusiness "was designed to legitimize the industrial agricultural paradigm", and this has in turn introduced "at least three sets of problems: agronomic, social, and biospheric". Agronomic problems, he says, include soil loss, groundwater depletion, pesticide resistance, and deterioration of soil and water quality – problems that I examined briefly in section I of Chapter 3. Social problems, he says, "include deteriorating rural communities, the loss of family farms,

---

22 Kirschenmann, *supra* note 1, at 51 (citing Eugene Odum, Paul Thompson, and other sources on "agroecology" and related concepts).
23 *Id*. at 51–52.

increased health risks, the concentration of wealth and power in agribusiness corporations . . . , and community distress over factory farms". Biospheric problems include "depletion of fossil fuel energy resources, global warming, wildlife destruction, reduced biodiversity, and ecological imbalances".[24]

Kirschenmann then urges the adoption of a new "ethic", one that "will recognize the seamless connections between healthy soil, healthy ecological neighborhoods, and vibrant human communities." This is the same approach I am urging with agroecological husbandry. Kirschenmann asserts that agriculture should be "designed to fit into local ecologies" and should be carried out by "local people [who] live in local ecosystems long enough and intimately enough to know how to manage them in an ecologically sound manner".[25]

In sum, some key elements of agroecological husbandry – and even the term "agroecology" and the natural-systems approach it reflects – are not new. Indeed, what Kirschenmann refers to as "ecological agriculture" can be traced back to the 1930s, when Herbert C. Hanson (an expert on grasslands) "stressed the importance of ecological approaches" to agriculture, and yet further back to the early 1900s:

> Liberty Hyde Bailey, dean of agriculture at Cornell University [in the late 1800s and early 1900s], wrote that "a good part of agriculture is to learn how to adapt one's work to nature . . . . To live in right relation with his natural conditions is one of the first lessons that a wise farmer or any other wise man learns." At about the same time, agricultural visionaries in other parts of the world were stressing a similar path. Sir Albert Howard of Great Britain, Rudolf Steiner in Austria, and Mokichi Okada in Japan all emphasized the importance of farming in concert with nature.[26]

One particular element of agroecological husbandry – reflected again in observations made by Frederick Kirschenmann – is its regional character.[27]

Designing an agriculture that is "in adjustment with the environment" is uniquely regional. Agricultural practices that achieve ecological goals in

---

24 *Id.* at 52–53.

25 *Id.*

26 *Id.* at 121–122. For a fascinating account of the life and work of Liberty Hyde Bailey, see Harlan P. Banks, *Liberty Hyde Bailey*, in 64 Biographical Memoirs, Office of the Home Secretary, National Academy of Sciences (1994).

27 For a recent discussion of "bioregionalism" in the context of food systems, see Joshua Lockyer, *"We Should Have a Culture Around Food": Toward a Sustainable Food Utopia in the Ozark-Ouachita Bioregion*, appearing as Chapter 4 in Paul V. Stock, Michael Carolan, and Christopher Rosin, Food Utopias – Reimagining Citizenship, Ethics and Community (2015). See also note 115 and accompanying text in Chapter 3, *supra*.

one region will likely not work in another region. Recognizing ecological and social differences will be crucial to success. Efforts to "harmonize" agriculture on a global scale will probably fail. While [a national] government can set national goals, it cannot manage natural resources. Because nature is dynamic and every regional ecosystem is unique, proper management can only be executed by local managers who understand local systems and who can modify management strategies to accommodate local changes in nature.[28]

In some respects, the move toward the new form of agriculture I am describing here – agroecological husbandry – can be regarded as a natural progression from some earlier reforms made in the crisis periods of the twentieth century. Lester Brown, the president of Earth Policy Initiative, offers this narrative of how some of those reforms in the USA from about 80 years ago gave careful attention to the need for soil conservation:

> The 1930s Dust Bowl that threatened to turn the U.S. Great Plains into a vast desert was a traumatic experience that led to revolutionary changes in American agricultural practices, including the planting of tree shelterbelts (rows of trees planted beside fields to slow wind and thus reduce wind erosion) and strip cropping (the planting of wheat on alternate strips with fallowed land each year) . . . . [Later programs built on those early efforts.] In 1985, the U.S. Department of Agriculture, with strong support from the environmental community, created the Conservation Reserve Program (CRP) to reduce soil erosion and control overproduction of basic commodities . . . . Under this program, farmers were paid to plant fragile cropland in grass or trees. The retirement of [many millions of] acres under the CRP, together with the use of conservation practices on 37 percent of all cropland, reduced annual U.S. soil erosion from 3.1 billion tons to 1.9 billion tons between 1982 and 1997. . .[29]

---

28 Kirschenmann, *supra* note 1, at 128 (emphasis in original). In connection with another passage quoted from Kirschenmann's work – see note 142 in Chapter 3, *supra* – I highlighted a point of concern regarding his emphasis on local decision-making and management regarding agriculture and land use. I would reiterate that point here. For national or global issues, local government might well be too fragmented, too subject to the impacts of competition with other governmental entities, too subject to local interpretation, and too subject to being provincial to be counted on to make decisions that constitute an effective national or global policy. This is a reality that explains, among other things, the need for intensive training of farmers responsible for engaging in agroecological husbandry, a point I discuss in subsection IB4 of Chapter 7, *infra*.

29 Lester R. Brown, WORLD ON THE EDGE: HOW TO PREVENT ENVIRONMENTAL AND ECONOMIC COLLAPSE 142–144 (2011). Brown then discusses low-till and no-till agriculture, in which "farmers simply drill seeds directly through crop residues into undisturbed soil." He notes that in the USA, "the no-till area went from 17 million acres in 1990 to 65 million acres in 2007", and no-till practices have spread to other agricultural areas as well, especially in Brazil, Argentina,

From the basis of that account, however, the direction Lester Brown takes in his reasoning departs from the direction of agroecological husbandry. Instead of envisioning a revolutionary transformation in the approach that agriculture should take, Brown urges what might be regarded as "more of the same, but smarter". In Brown's view, there is still untapped potential in some of the same forms of increasing land productivity that were seen during the Green Revolution. He urges, for instance, "expand[ing] the land area that produces more than one crop per year".[30] By contrast, the primary approach that agroecological husbandry would take is *not* further intensification or extensification of conventional agricultural practices but rather a shift to natural-systems practices involving perennials grown in polycultures – mimicking the ecology of the grasslands that were destroyed to conduct modern extractive agriculture.

II. Exploring perennial polycultures more closely

Having introduced the general contours of agroecological husbandry, I now turn to a closer examination of one of its key elements – that is, the way in which it mimics natural grassland ecosystems by using perennials grown in polycultures.

IIA. Searching for the missing combination

For over 35 years, Wes Jackson has directed The Land Institute headquartered in Salina, Kansas. In that capacity, he has been the principal proponent in a campaign to develop perennials that would be grown in polycultures – a campaign that has earned Jackson much recognition and numerous awards, including a MacArthur "Genius" grant. I quoted Jackson in Chapter 1 for his broad historical narrative, in which he described the history of agriculture by compressing the last 750 million years into a single calendar year.[31] Beyond offering this historical contextualization, however, Jackson has also explained the broad biological context in which it is sensible to develop perennial polycultures as a revolutionary new form of agriculture. In his book *New Roots for Agriculture*, Jackson identified 16 categories of plants from which food might theoretically be produced for human consumption. Those 16 categories emerged from four pairs of characteristics. He identifies them as follows:

Australia, and Canada. *Id.* He does not, however, emphasize the fact that no-till practices depend heavily on the use of synthetic pesticides.

30 *Id.* at 169.

31 As noted earlier, the significance of 750 million years is that this represents the most recent one-sixth of the total age of the Earth; it was in this period, Jackson explains, that the "explosive emergence of higher life" occurred. See note 28 in Chapter 1, *supra*.

[W]e may . . . contrast . . . annual versus perennial, . . . monoculture versus polycuture, . . . woody versus the herbaceous condition and whether the human interest is in the fruit/seed product or [in] the vegetative part of the plant. When we consider these four contrasting considerations, in all possible combinations, we have sixteen categories for assessment.

We can eliminate four of these sixteen categories for they involve woody annuals, a phenomenon unknown in nature. This leaves us with twelve categories for consideration, as listed in the Table.[32]

Jackson then presents the table appearing in Figure 4.1. In doing so, he gives special emphasis to one specific missing category or "missing combination": herbaceous perennial plants grown in polycultures for their fruit or seed.

Polyculture vs. Monoculture	Woody vs. Herbaceous	Annual vs. Perennial	Fruit/Seed vs. Vegetative	Current status
1 Polyculture	Woody	Perennial	Fruit/Seed	Mixed orchard (both nut and fleshy fruits)
2 Polyculture	Woody	Perennial	Vegetative	Mixed woodlot
3 Polyculture	Herbaceous	Annual	Fruit/Seed	Mixed cropping (corn-beans in the tropics)
4 Polyculture	Herbaceous	Annual	Vegetative	Dump heap garden, companion planting
5 Polyculture	Herbaceous	Perennial	Fruit/Seed	—————————————
6 Polyculture	Herbaceous	Perennial	Vegetative	Pasture and hay (native or domestic)
7 Monoculture	Woody	Perennial	Fruit/Seed	Orchard (both nut and fleshy fruits)
8 Monoculture	Woody	Perennial	Vegetative	Managed forest or woodlot
9 Monoculture	Herbaceous	Annual	Fruit/Seed	High-producing agriculture (wheat, corn, rice)
10 Monoculture	Herbaceous	Annual	Vegetative	Ensilage for livestock
11 Monoculture	Herbaceous	Perennial	Fruit/Seed	Seed crops for category 12
12 Monoculture	Herbaceous	Perennial	Vegetative	Hay crops (legumes and grasses) and grazing

Figure 4.1 The array of possible agricultural productivity pairings[33]

In order to present a more visual image of these 12 possibilities, as well as the four non-possible combinations (among the 16 total conceivable ones that emerge theoretically from Jackson's four pairs of considerations), I have prepared the diagram appearing in Figure 4.2.[34] Like Figure 4.1, Figure 4.2

---

32 Wes Jackson, NEW ROOTS FOR AGRICULTURE 96–97 (1980, 1985 reprint) [hereinafter NEW ROOTS].

33 Although Jackson does not identify specifically which categories in his list include the wide array of foods that are commonly referred to as "vegetables" – that is, potatoes, tomatoes, peppers, squash, peas, carrots, and other crops (some of which are produced for commercial sale in "truck farms") – most of these would fall within categories 9 and 10.

34 The diagram in Figure 4.2 is based on the array of possible pairings enumerated in Figure 4.1, *supra*. It identifies all 16 *conceivable* pairings of the four variables identified by Jackson: polyculture

reflects the key insight that distinguishes Jackson's 1980 work: *one key combination is missing*. Eleven of the combinations are harvested in the real world to serve human purposes. The other combination is not. Again, it is item number 5 (both in Figure 4.1 and Figure 4.2): a polyculture of herbaceous perennials primarily for production of fruit or seed.

The main focus of the research that Wes Jackson has managed for more than 35 years at The Land Institute is to fill in that blank for item number 5: herbaceous perennial seed-producing polycultures. Jackson emphasizes that agriculture as practiced for thousands of years has focused enormous attention on the combination carrying the number 9 (herbaceous *annual* seed-producing *mono*cultures) but has paid virtually no attention to the combination carrying the number 5.

Jackson's mission, therefore, stands completely apart from the efforts that most agricultural researchers have engaged in for years, and indeed for centuries. Jackson regards it as a mistake to disregard or pooh-pooh the possibility of developing perennial polycultures for food production. The main rationale for Jackson's work is that this mistake needs to be corrected – and that it can change the world.

There is a further rationale, however, that Jackson and his colleagues at The Land Institute have offered regarding their efforts to develop a new form of cereal grains, legumes, and oilseeds: it is now *possible,* as never before, to do so. In a paper he has co-authored with two other scientists, Jackson explains why it would have been difficult for humankind to have tried developing perennials (in a way that would be useful for feeding human populations, that is) 10,000 years ago:

> [U]nder the original set of conditions available to early cultivators, perennial plants were not conducive to domestication . . . . Humans had long gathered and eaten seed of many herbaceous perennial species, especially grasses, but the domestication step that could have generated perennial grain crops never happened . . . . [The reason] is that in the domestication process, the obvious advantages enjoyed by annual species were denied to perennials.

versus monoculture; woody versus herbaceous; annual versus perennial; and whether the plant is grown primarily (in terms of human food use) (i) for its fruit or seed or (ii) for its vegetative part. I have marked with an arrow the specific combination of these four factors that is missing from this 16-pairing array, in the sense that it does not represent a combination that humans have developed as a source of food. That "missing combination", as noted above in text, is the item numbered 5 – herbaceous perennial plants grown in polycultures for their fruit or seed. It is this that researchers at The Land Institute and associated institutions are intent on developing. In addition, I have used black background to identify those four combinations that do not exist in nature and given them letters (A, B, C, and D) instead of numbers, in order to be consistent with the numbering provided by Jackson as shown in Figure 4.1.

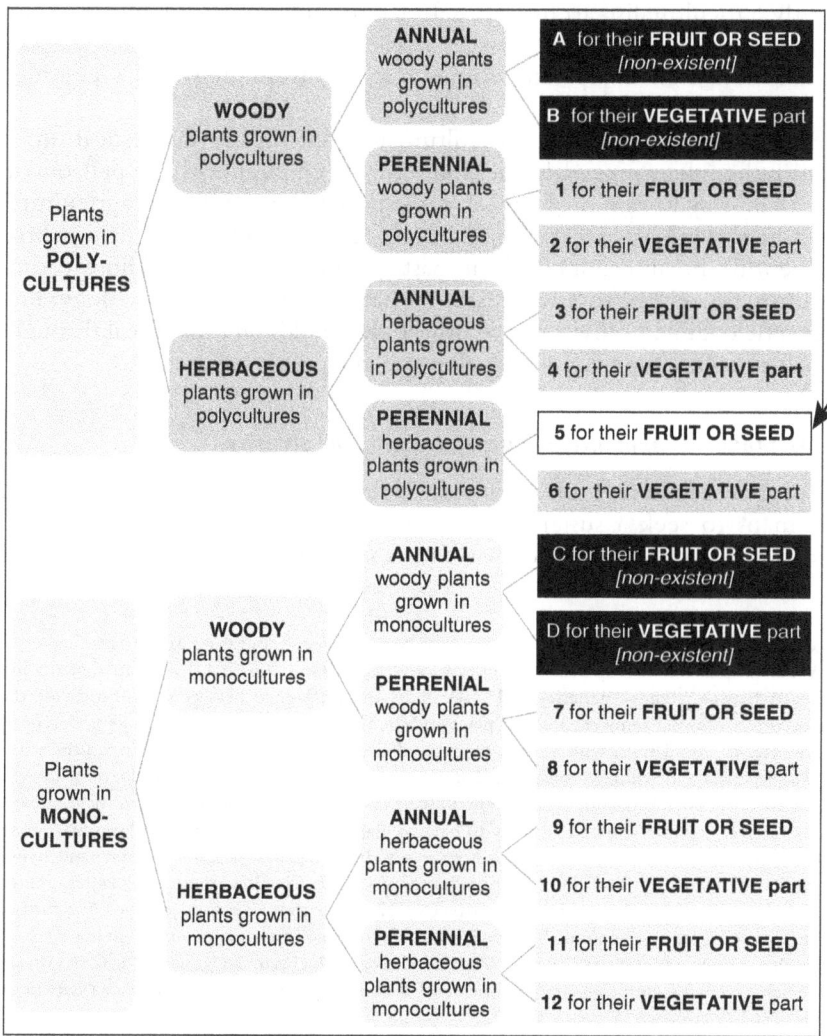

Figure 4.2 Diagramming the conceivable agricultural productivity pairings

For one thing, the perennial plants from which people gathered seed always re-grew much more strongly from vegetative buds at or below the soil surface than they would have from dropped or sown seed. Also, whereas the first annual species to be domesticated – including wheat, barley, and rice – form their seed primarily through self-pollination, the most severe form of inbreeding, most perennial species are largely incapable of inbreeding, and when they do, lethal mutations wipe them out. Their ability to inbreed gave annuals a distinct advantage over perennials in the domestication sweepstakes, by making it possible for early agriculturalists to quickly select rare mutant plants that hold onto their seed after ripening so that it can be harvested easily – the key characteristics that separates domestic species from their wild ancestors.[35]

Having explained why early agriculture focused on the domestication of *annuals* rather than *perennials*, Jackson and his colleagues then spell out in detail a "new synthesis" in science, with special applicability to agriculture. Such a new synthesis, Jackson and his co-authors explain, builds on three key scientific breakthroughs of the past century and a half. Those earlier breakthroughs related to advances in (i) natural selection, (ii) ecology, and (iii) genetic coding.[36] Now a new synthesis that builds on those breakthroughs makes it possible, they say, to develop perennial polycultures.[37]

IIB. Metaphor A: humanity's journey toward food security

In short, new scientific tools now make it feasible, as it never was until now, for humans to seek a sustainable global food supply through a "natural-systems" form of agriculture that focuses on perennial grains grown in

---

35 Wes Jackson, Stan Cox, and Tim Crews, *The Next Synthesis*, appearing as the last chapter in Nature as Measure – The Selected Essays of Wes Jackson 206 (2011). Perennial plants also tend to have a higher "genetic load", which is the difference between the fitness of the theoretically optimal genotype for that plant and the fitness of the observed average genotype in a population. Populations with a low genetic load (as is typical of annuals) will have fitness that is less dispersed and will have offspring that are more likely to survive to reproduction. On the other hand, populations with a high genetic load (as is typical of perennials) will have greater variance in fitness and will be less likely to have as many organisms at or near the optimal fitness level. Early agriculturalists would have been easily discouraged with the variation (and hence unpredictability of outcome) in offspring of perennial plants that they might have experimented with in search of genotypes that had attractive characteristics as a food source. For further discussion of factors serving as disincentives to the domestication of perennial grains in early agricultural development, see David L. Van Tassel, Lee R. DeHaan, and Thomas S. Cox, *Missing Domesticated Plant Forms: Can Artificial Selection Fill the Gap?*, Evolutionary Applications ISSN 1752–7461 (2010).

36 Jackson and his co-authors refer to these as (i) the synthesis of 1859 ("uniting the life sciences"), centering on Charles Darwin's *The Origin of Species*, published in that year, (ii) the synthesis of 1937 ("genetics and the origin of species"), centering on the book *Genetics and the Origin of Species* published in that year by Theodosius Dobzhansky, and (iii) the synthesis of 1953 ("the molecular coding of genetics and evolution"), centering on the work of Watson and Clark in elucidating the structure of the genetic code. *The Next Synthesis*, *supra* note 35, at 201–203.

37 Significantly, the authors emphasize that unlike those earlier breakthroughs, this "new synthesis" would not take a reductionist approach – focusing on ever-narrower sub-disciplines of study – but would instead focus on the ecosystem level of the natural order in order to develop new strains of grains and legumes that would replace annuals grown in monocultures. According to Jackson and his co-authors, "[s]o powerful has been molecular biology that it was natural for scientists to look increasingly downward, toward smaller scales, in their attacks on problems in every area of biology. This reductionist approach was not new" and can in fact, they say, be traced to the Enlightenment. *Id.* at 203.

polycultures. I find it helpful to use a metaphor – Metaphor A, represented as a drawing in Figure 4.3 – to depict the feasibility of seeking such a new form of food production. The next few paragraphs explain this metaphor, and I will return to it in Chapter 8. In fact, it is at that point that I will use a second metaphor (Metaphor B), to suggest that laws and institutions should be developed simultaneously with the emergence of new forms of food production, especially grains and legumes.

In the context of Metaphor A, humankind might be imagined as a tribe or band of travelers with various needs and goals. As a matter of survival, one of their goals is food security – that is, a method of establishing a safe

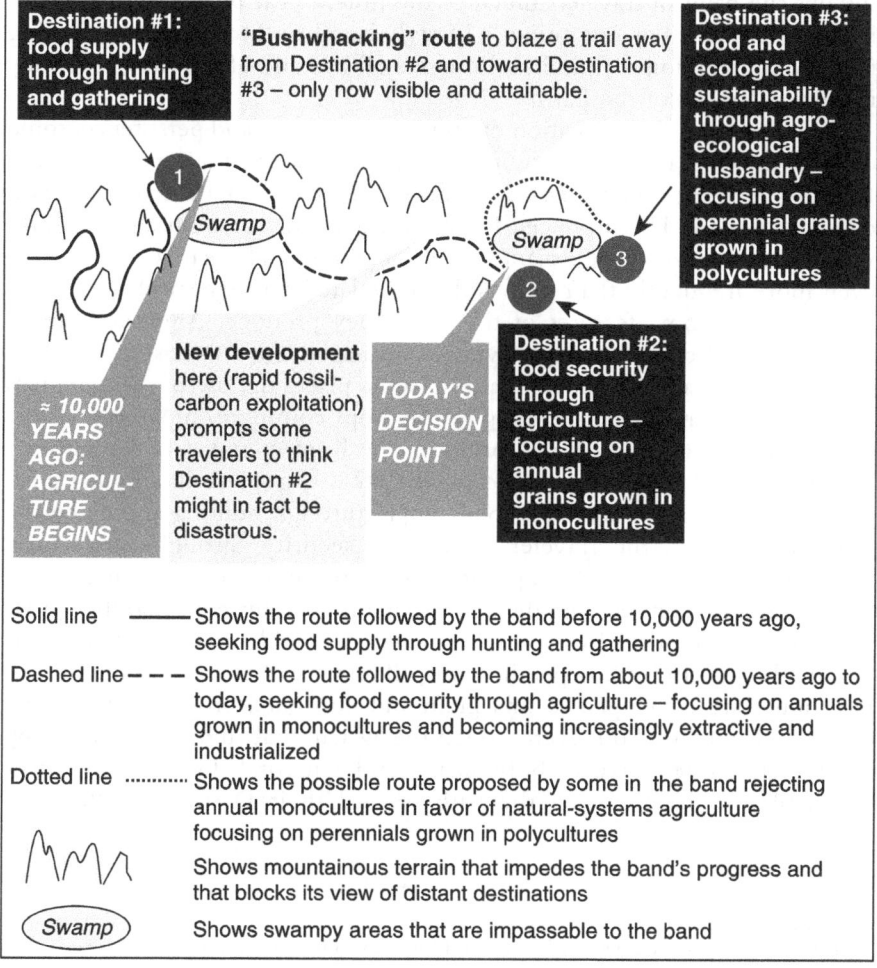

Figure 4.3 (Metaphorical) journey of a band of travelers seeking food supply, security, and sustainability

and reliable food supply adequate to meet the travelers' needs for physical health and energy on an ongoing basis. For untold numbers of centuries, the band of travelers relies on hunting and gathering food that they find available in the landscapes (and seascapes) that they share with other species on the planet. During those many centuries, the (figurative) *destination* that this band of travelers seeks to reach through their efforts is *food supply through hunting and gathering.* Their progress toward that goal or destination – which we might call *Destination #1* – involves the development of ever more efficient weapons and strategies for catching, killing, and cooking other animals as well as an ever more sophisticated understanding of what plants are edible, when and where they can best be gathered, and how their growth can be encouraged. This technique, drawing from the bounty of the ecosphere, sustains the band of travelers much of the time, but it proves unsatisfactory in some respects. For one thing, a lifestyle based on hunting and gathering for food supplies imposes limits on the size of the band of travelers, especially in certain regions of the Earth.

Then, through a combination of accident, curiosity, and persistence, some of them develop a new technique for securing food. It holds great promise. Instead of relying on *available* food, they start relying on a narrow selection of *produced* food. The produced food consists mainly of annual seed grains, which they realize can be planted and harvested each year, or sometimes even more frequently than that, and manipulated through careful selection to accentuate some features of the seeds they produce. Another form of produced food comes through livestock management – that is, the domestication of animals for food, and also in some places the use of these animals for power in planting and tending to the crops coming from seeds.

Once this system has shown promise, the band of travelers establishes a new (figurative) destination – *Destination #2* – to replace its original goal: Instead of seeking an adequate food supply through hunting and gathering (Destination #1), the travelers seek food security through agriculture, concentrating on annual crops of grains grown in monocultures. The travelers do not confine their diets *only* to grains, of course, nor do they grow their food crops *only* in a "single-species-per-field" manner (that is, as monocultures). Indeed, depending on soil and weather conditions, various techniques are developed to mix grain and non-grain crops, such as corn, beans, and squash.[38] Still, even though the agricultural landscapes created by the travelers often involve both spatial and temporal diversity, the most dominant characteristics of the path toward their Destination #2 is its

---

38 For reference to the efficiency of "traditional Mexican corn-bean-squash farms", see note 53 in Chapter 2, *supra.* For a reference to other forms of intercropping involving additional species, see text accompanying note 120 in Chapter 3, *supra.* Crop rotation represents another technique designed to maintain at least some diversity (consecutive, not simultaneous) in cropping patterns.

concentration on annual grain crops grown in monocultures and stored as seeds for food or future planting.

Over the course of many centuries, the band of travelers makes gradual progress toward Destination #2, navigating through mountainous terrain and circumventing impassable swamps. Suddenly, however, the group's journey toward that new destination – food security through agriculture – changes dramatically in two related ways. First, the journey experiences a breathtaking acceleration with the discovery and development of new resources, particularly fossil-carbon resources that can be extracted from those regions under the surface of the Earth where they were laid down eons ago, before humans existed on the planet. (These resources are put to use not only for grain agriculture, of course.)

These new fossil-carbon resources also create a second type of change: as agriculture is de-coupled from current solar power as its primary energy source – replacing solar energy in large part with energy derived from fossil fuel – the enormous increase in the power that can be applied to agriculture allows it to become even more mechanized and even more concentrated in its attention to monocultures and high production. In short, because of this new "extractive" or "industrial" form of agriculture, the band of travelers is now progressing much faster toward Destination #2 – with its eyes fixed more than ever before on annual grain crops grown in monocultures as a means of achieving (hoped-for) food security through agriculture.

Then the tables begin to turn. Some members of the group realize that the new fossil-carbon resources not only (i) are limited in availability and are non-renewable on a human time scale but also (ii) are responsible for creating extremely serious long-term changes in the physical attributes that make the Earth livable for humans. In addition, other shortcomings of this new industrial agriculture and its effects become more evident to the band of travelers. For instance, most members of the band are not needed for participating in the food-production process, even if they want to, because its reliance on fossil-carbon resources requires food production to be centralized in large-scale operations controlled by only a few participants. Moreover, even the large-scale operators cannot succeed at food production without drawing heavy economic support from the rest of the group. Hence, even as their Destination #2 (food security through agriculture) seems tantalizingly close, some members of the band of travelers now consider that destination to be thoroughly repulsive and dangerous as its true character comes more clearly into view.

The band of travelers now splits into several factions. Some deny the evidence that Destination #2 is flawed; they urge others in the band of travelers to press forward toward food security through ever more industrialized and concentrated forms of agriculture. Some other members of the band of travelers sink into depression, seeing in the turn of events the hand of fate or an angry and vengeful god bringing despair to sinners in the form of a painful end to a doomed journey. Surprisingly, some members of the band welcome the end of the supposedly doomed journey; they express confidence that their own faith in some supernatural power will physically transport them out of the ecosphere whose limits they have cheerfully disregarded. One or two members of the group even flirt with the idea of retracing their steps – that is, returning to the Destination #1 of hunting and gathering.

Yet another faction in the band of travelers takes an entirely different perspective on the situation. Members of this faction do not focus their attention on either Destination #1 or on the now-repulsive Destination #2 but rather on a new *Destination #3* that now has come into view and seems attainable. However, it will be attainable *only* if the resources of the group can be mustered effectively and devoted to blazing a trail toward that new destination, since further mountainous terrain must be navigated to get there. Destination #3 uses the natural world – and particularly the functioning of the native grasslands that once covered vast parts of the Earth's surface – as its model or standard for production of food, and it has as its centerpiece the development of perennial grains grown in polycultures.

As such, Destination #3 amounts to a partial endorsement and a partial rejection of *both* destinations that the band of travelers had earlier considered to hold such promise. On the one hand, Destination #3 endorses Destination #1's recognition that humanity is and must be integrated with and mutually dependent on all of the components of the ecosphere. In that respect, Destination #3 rejects the supposition inherent in Destination #2, that humankind stands above or outside the natural word and its natural limits on the use (or abuse) of the Earth's resources. On the other hand, Destination #3 endorses Destination #2's confidence that humans can, through very smart and cautious action, succeed in bringing a prudent hand of management to the production of food – and in that respect, Destination #3 rejects the supposition inherent in Destination #1 that humankind must rely entirely on the natural world as it happens to exist at any particular time, and that humans cannot modify or manage it at all.

Those members of the band of travelers who urge their colleagues toward Destination #3 – food and ecological sustainability through a natural-systems approach that uses the natural world as its model and therefore focuses on perennials grown in polycultures – have one especially potent argument. They assert that until recently, Destination #3 was unattainable as a practical matter, and that in fact it was hardly even visible.[39] This may be regarded as

---

39 This detail in Metaphor A reflects the point I made earlier: Wes Jackson and some of his colleagues at The Land Institute have explained why it would have been so difficult, for technical reasons, for humankind to have tried developing perennials (in a way that would be useful for feeding human populations, that is) 10,000 years ago. These reasons include such matters as mechanisms of plant regrowth, self-pollination, and "genetic load". See note 35, *supra*, and accompanying text. Perhaps another reason is economic and political in character: until recently, the political systems operating in the world have not been strong enough to manage economic

a saving grace for having ventured so close to Destination #2. That is, according to this faction of the group, the only reason that they can now blaze a trail toward Destination #3, through a process of "bushwhacking"[40], is that the travelers now have at their disposal certain recent advances in science, including some advances just made possible by drawing from the non-renewable fossil-carbon resources whose headlong use in recent years now threatens to bring the travelers' journey (and indeed their planet) to a tragic end. In short, *it is only now that an annual-monocultures approach to agriculture* (Destination #2) *could be discarded in favor of a perennial-polycultures approach to food and ecological sustainability* (Destination #3).

Let me start bringing this Metaphor A narrative to a close for now. The "band of travelers" in the narrative is humanity as a whole, and its ability now to "blaze a trail . . . through a process of bushwhacking" toward a different destination of food and ecological sustainability refers to the adoption of a natural-systems form of agriculture that I have described here as agroecological husbandry, which focuses on perennials grown in polycultures. What I have done in Chapters 2 and 3 of this book is to explore certain fundamental flaws in the path that humanity is now on, toward an increasingly concentrated, extractive, industrialized form of agriculture. Now, in this chapter and the next chapter, I am examining how a profoundly different form of natural-systems food production, focusing especially on grains and legumes, is now becoming both possible and preferable. In Chapters 6, 7, and 8, I will describe how a national and international legal framework can be designed and built to facilitate a global transition to this novel form of food production.

The diagram I offer in Figure 4.3 illustrates the Metaphor A that I have introduced here. This diagram will also appear in Chapter 8, where I trace the route of a similar "journey" of a band of travelers (human society) seeking political security through legal concepts, including the concept of sovereignty. For that, I will introduce a Metaphor B, with a corresponding diagram. Both metaphors envision a form of "bushwhacking" toward a new destination only made possible by recent developments.[41]

---

affairs well enough both (i) to prevent profit motives from driving some members of society to disregard externalities of the sort discussed in Chapter 2 – especially in the text accompanying notes 53–66 in that chapter – and (ii) to mobilize broad-based sources of public funds and other resources well enough to facilitate a transformation from an ages-old form of agriculture to a new one.

40 One online dictionary defines "bushwhacking" as "to clear a path through thick woods by cutting down bushes and low tree branches". Another offers this definition: "to force one's way through a forested or overgrown area where no path exists."

41 As explained in Chapter 8, the two parallel metaphors have explanatory power. Just as humanity has travelled for roughly 10,000 years toward a particular destination of *food* security (centered on annuals grown in monocultures), only to find now that this destination is in fact deeply unattractive, likewise humanity has traveled for the past four centuries toward a particular destination of *political* security that also now seems deeply unattractive, for reasons that are widely

An obvious question to be addressed before undertaking the strenuous "bushwhacking" exercise called for in Metaphor A – that is, before pouring energy into an effort to make a wholesale abandonment of modern extractive agriculture in favor of agroecological husbandry based on perennial polycultures – is this: would such an exercise be worthwhile? I aim to address that question in two ways. First, in the remaining paragraphs of this Chapter 4, I offer a projection of an optimal global setting in which a new form of food production would operate. Then, in Chapter 5, I get down to basics by examining in some detail both (i) what would be *preferable* about shifting to agroecological husbandry and (ii) whether it is *realistically practical* as a scientific matter to do so on a large scale.

III. Projections for an agroecological future[42]

The difference between a "utopia" and a "eutopia", as I understand it, is that between a "no place" and a "good place". By most accounts, the word *utopia* was coined by Sir Thomas More for his 1516 book *Utopia*, describing a fictional island society in the Atlantic Ocean. Drawn from the Greek words οὐ ("not") and τόπος ("place"), it can be construed as "no place". By contrast, the English homophone *eutopia* draws from the Greek words εὐ ("good" or "well") and τόπος ("place"), and can therefore be construed as "good place". The fact that *utopia* and *eutopia* have identical pronunciations makes for a rather clever double meaning: any place that we can imagine as being a "good place" is commonly regarded as also being impossible to create in reality.

In this last portion of Chapter 4, I offer a set of projections for an agroecological future. I start with these observations about the difference between *utopia* and *eutopia* in order to emphasize my view that *both* terms are appropriate in this context. On the one hand, it is true that there is "no place" *currently* in the world that has the features I discuss in the following pages. In that sense my projections might seem *utopian* in character. On the other hand, recent advances and tangible successes in "natural-systems"

---

known in general terms but are seldom discussed in the context of food production and ecological protection. I believe that this agroecological context offers a fresh perspective for recognizing the defects in the international legal system and considering how those defects can be overcome. Some of this effort involves more sophisticated and proven forms of cooperation among states and non-state actors, and some of it involves examining what I refer to as the "monolithic sovereignty" that sits at the center of that system. I will describe the "monolithic sovereignty" concept in section I of Chapter 8 and then will explain why a "bushwhacking" exercise is needed in order to arrive at a different destination of political and ecological stability through new forms of eco-regional multilateral action and through a blended and multilayered "pluralistic sovereignty" – one that is in fact already practiced in some settings around the world and is ripe for more nuanced development and application in this context as well.

42 The following paragraphs draw in part from a chapter that I contributed recently to a book on "food utopias". See Head, *Sketching, supra* note 12.

agriculture do offer grounds for optimism that the trajectory of environmental degradation that we see humanity following today might not be an inevitable path. For those individuals and groups who wish to keep resisting that degradation – I count myself among them – it strikes me as both essential and realistic to have some vision of such a "good place", a *eutopia*, firmly in our sights.

I will explore in Chapter 5 why it is *realistic* from a scientific perspective to think that agroecological husbandry, featuring perennial grains grown in polycultures, can be developed in the foreseeable future based on the advances already made. But why do I posit (immediately above) that it is also *essential* to have some overall vision of such a future? Because bridging the gap between current reality and a future agroecological eutopia ("good place") can be better accomplished if we have some vision of where the far end of the bridge rests.

It makes sense, therefore, to ask: "What are the features of a eutopia that would be worth working toward in the days ahead?" That is the general question I wish to consider here briefly. In doing so, I will defer for now any substantial discussion of the political obstacles that will almost surely make it difficult to make these eutopian projections a reality. I am just as aware as anyone else of how significant those difficulties might be, and I address several of those difficulties in later chapters, especially Chapters 6 and 7 and in the remarks with which I conclude the main text of this book in Chapter 8.

IIIA. "New roots" projections – globalized and updated

I am hardly alone in thinking about the "big picture" of what sort of utopian (or eutopian) world agroecology would fit into if it could be achieved. For instance, there is undeniably a strong theme of eutopianism running through the work of The Land Institute, and particularly in the writings of Wes Jackson. Indeed, in Chapter 9 of his 1980 *New Roots* book, Jackson sets forth a brief description of "one utopian farm" and the community and landscape into which it fits. That description is set in the future – the year 2030 – in central Kansas, with some references also to other parts of North America. Jackson provides details not only of the *physical* layout of farms and farm communities, including references to farm families relying on solar panels and wood stoves as they lived and worked on land held through a "Land Trust" system,[43] but also of the *ethical and social* landscape as it would appear in 2030.

Jackson notes, for instance, that "much of the social pathology in the last century [looking back, that is, from 2030] could be attributed to the fact that so many people were engaged in meaningless work or in work that seemed

---

43 For Jackson's vision of farm communities in 2030, see Jackson, *supra* note 32, at 118–132.

meaningless in an urban setting. Gradually a great untruth that formerly had the power of a myth began to be exposed."[44] The untrue myth Jackson refers to is that industrial agriculture, based on unsustainable use of fossil carbon, was, over the long term, both (i) efficient and effective for food production, and (ii) acceptable as a social matter. Neither of these had ever been true, of course, when seen accurately in a long-term view, so that by 2030 society had, in Jackson's projection, changed in a variety of fundamental ways as the realities of food production and rural life had changed.

In the remainder of this chapter, I hope to lay foundations for an updated projection into the future of the sort that Wes Jackson offered in 1980. However, because I believe any such updated projection must be global in its reach, my own observations will extend into areas that were not encompassed in Jackson's view. Some of them lie beyond my direct expertise, which is concentrated most heavily in the areas of international law and institutions. Still, I offer this list of what I believe many observers would consider the most important elements in such an agroecological eutopia (good place) that we can project for the future based on recent developments and new knowledge:

- an emphasis on *land husbandry* that would place highest value on restoration and preservation of Earth's natural resources – especially the health of its soil – and on climate stability, climate resilience and water security (see subsection IIIB);
- a *distribution of agricultural plant and animal species* that would balance indigenousness and preservation with productivity for nourishment of humans and other species (see subsection IIIC);
- a *global human population* that is both (i) large and diverse enough to provide adequate creative productivity, so as to sustain cultural energy and innovation and yet (ii) not so large as to overburden the reasonable sustainable carrying capacity of the Earth and therefore subject large numbers of people to malnourishment or other forms of deprivation (see subsection IIID);
- a system of *energy production and distribution* based on renewable sources – not fossil-carbon sources – and designed to guard against pollution and high transmission costs (see subsection IIIE);
- an *economic system* that (i) aims for "production equilibrium" – that is, an equilibrium between production and resource availability, as distinct from a system that gives primacy to growth and consumption – and that (ii) requires all externalities to be accounted for and reflected in the pricing of goods, particularly agricultural goods (see subsection IIIF); and
- a *legal and political framework* that facilitates and sustains these various systems – relating to land husbandry, species biodiversity, human population,

---

44 *Id.* at 125.

energy, and economics – in ways that reflect certain fundamental values of fairness.

I will turn to the last of those elements – the pertinent legal and political framework – in Chapters 6, 7, and 8 of this book. In the following subsections I will provide, for each of the first five points enumerated on the preceding page, some elaborations that I have constructed with help from several colleagues[45] and organized in the same order as the above bullet points.

IIIB. Land husbandry

A key element of this projection for an agroecological future would be sustainable land husbandry. Perhaps it is redundant to use the modifier "sustainable" with the noun "husbandry", since there can be no such thing as unsustainable husbandry. As I explained earlier in this chapter, I use the term "husbandry" to encompass the understanding, conserving, preserving,[46] and nurturing of the long-term viability of an ecosystem for its own sake because of its own value. (By using the phrase "because of its own value", I intend to answer in the affirmative Christopher Stone's famous question from 1972 – "Should trees have standing?"[47]; in other words, "husbandry" involves living *with* the world, not just *on* it or through its depreciation.)

The value of an ecosystem can include benefits that inure to humans, of course, but the benefit to humans would not naturally – and *should* not – be the main reason for husbandry, just as any benefit that might be enjoyed by

---

45 I wish to acknowledge with gratitude the especially helpful contributions made to my consideration of these issues, and to the preparation of these elaborations, by Caleb Hall, Miriam Friesen, and Tim Crews. Especially valuable research and drafting for subsections IIIB, IIIE, and IIIF came from Mr. Hall and Ms. Friesen.

46 I have intentionally used both "conserving" and "preserving" here, despite the fact that they carry different connotations in the context of environmental protection, as explained on the National Park Service website:

> Two opposing factions had emerged within the environmental movement by the early 20th century: the conservationists and the preservationists. The conservationists . . . focused on the *proper use of nature,* whereas the preservationists sought the *protection of nature from use.* Put another way, conservation sought to regulate human use while preservation sought to eliminate human impact altogether. [Such preservationist views often revolve around what is sometimes called "deep ecology",] a philosophy that believes in an inherent worth of all living beings, regardless of their instrumental utility to human needs.

See www.nps.gov/klgo/forteachers/classrooms/conservation-vs-preservation.htm.

47 See Christopher D. Stone, *Should Trees Have Standing? – Toward Legal Rights for Natural Objects,* 45 SOUTHERN CALIFORNIA LAW REVIEW 450 (1972). For information about the Global Alliance for the Rights of Nature, which calls for a "recognition that our ecosystems – including trees, oceans, animals, mountains – have rights just as human beings have rights", see http://therights ofnature.org/what-is-rights-of-nature/.

a trustee would not naturally, and should not, be the main reason for a trustee to discharge his or her fiduciary duty. I will return briefly to the idea of trustees and trusteeships in Chapter 7 of this book, especially subsection IIB3 of that chapter.

Hence some key aims of land husbandry – situated at the center of a natural-systems form of food production – would be (i) to restore to their *status quo ante* condition a great many ecosystems around the Earth that human activities have damaged, (ii) to set aside extensive areas of the Earth's surface where wilderness should prevail with little or no human interference, and (iii) to engage in food production in the non-wilderness areas that places higher priority on soil health than on maximizing short-term productivity. This last point would require the development of agro-ecosystems that are synchronized with, and not exceeding, the operating rates of the critical ecosystem processes on which productivity depends.

I realize that these are generalities. But we can fill in some specific land-husbandry features of a world based on agroecology if we draw on the work of several observers who have given direct attention to matters of restoring and preserving natural resources, especially soil health.

One of these is Wendell Berry. In his 1977 book *The Unsettling of America*, Berry explains how commodification of farm products and the "fence row to fence row" mantra of agribusiness in the USA pushed most farmers off of their land, and how now-consolidated business interests enrich themselves at the expense of both the soil and the remaining rural communities. These realities have continued, of course, to this day. What can be taken from Berry's work in projecting a future based on agroecological husbandry is that farming should be anchored in biology rather than economics.[48] In those areas where land is to be used for food production, land use decisions should be made according to what the land can produce sustainably, not what it can be forced to produce in the short term to the detriment of its long-term health and productivity.

Another source of inspiration in envisioning the land-husbandry details of a global agroecological future would be Peter Singer's 1975 book *Animal Liberation*,[49] in which he describes the central tenet of the history of human progress as the belief that humans are superior to other animals and hence entitled to do whatever we want to them. Singer rejects this belief as irrational. To the extent that an agroecological eutopia would include live-stock, it might borrow that conclusion from Singer and then argue (i) that although animal suffering may be impossible to completely prevent, suffering should be kept to a minimum, and (ii) that no individual or group has absolute rights to dispose of animals or a landscape selfishly.

---

48 Wendell Berry, THE UNSETTLING OF AMERICA 137 (1977).
49 Peter Singer, ANIMAL LIBERATION (1975).

Another source of inspiration could be found in the "re-wilding" proposals urged by J. B. MacKinnon,[50] and in a specific 1987 proposal by Frank and Debra Popper for the construction of a "Buffalo Commons" – a vast nature preserve created by returning 139,000 square miles of the drier portion of the Great Plains to native prairie, and by reintroducing the American Bison that once grazed there. The proposal would affect ten midwestern and western US states: Montana, Wyoming, Colorado, Oklahoma, New Mexico, Texas, North Dakota, South Dakota, Nebraska, and Kansas.[51]

The Buffalo Commons proposal is one that many Americans would view as preposterous. However, if this view reflects their assumption that the proposal is *unattainable* as a practical matter because it would interfere with private property rights and would involve the exercise of sweeping powers of eminent domain and confiscation, then a possible reply is this: the legal system that exists in the USA concerning private "property" in land has been shaped by peculiar historical and cultural influences that are likely to disappear in coming years as the economic, ecological, and social factors I have explained in Chapters 2 and 3 become more clearly evident. If that does occur, then the Buffalo Commons proposal – or others like it that are designed to restore and preserve damaged ecosystems – will probably look attractive and in fact essential. Moreover, as a legal matter, the implementation of such proposals could be facilitated by the use of concepts that already exist and have been widely used, such as the concept of the constructive trust as developed in English law. As noted previously, I will return to these matters in Chapter 7.

Yet another source of inspiration for the land-husbandry aspects of an agroecological future could be found in David Montgomery's 2007 book *Dirt: The Erosion of Civilizations*. That book catalogues the various forms of soil degradation – erosion, poisoning, and the like – that have occurred in various civilizations over thousands of years and that have accelerated in our own over recent decades. Despite this, Montgomery holds out some prospect of reversing these trends in his reference to labor-intensive (as opposed to capital-intensive) agriculture, which he says could not only provide a way out of hunger and poverty for "the third of humanity that lives on less than two dollars a day" but also "help rebuild the planet's soil".[52]

In his 1949 masterpiece *A Sand County Almanac*, Aldo Leopold urges the adoption of a "land ethic", which he says "enlarges the boundaries of the community to include soils, waters, plants, and animals" and thereby

---

50 See generally J. B. MacKinnon, The Once and Future World: Finding Wilderness in the Nature We've Made (2013) (asserting that having an effectively functioning planet requires making more space for nature and for wild things in the human world rather than simply separating the two).

51 Deborah Epstein Popper and Frank J. Popper, *Great Plains: From Dust to Dust,* Planning, December 1987.

52 David Montgomery, Dirt: The Erosion of Civilizations 245 (2007).

"changes the role of *Homo sapiens* from conqueror of the land community to plain member and citizen of it." [53] For purposes of projecting forward to an agroecological future, a guiding principle we can take from Leopold is that "a thing is right when it tends to preserve the integrity, stability, and beauty of the biotic community. It is wrong when it tends otherwise." [54] On this reasoning, the world we can project for the future would preserve the ecological integrity of the landscapes from which food is derived.

Yet another source of inspiration in fashioning the land-husbandry aspect of a world based on agroecological husbandry is Willa Cather. I think it is not too much of a stretch to place her among the ranks of the environmentalists – and particularly those for whom natural areas should be respected and preserved, not consumed and conquered. In her novel *Death Comes to the Archbishop*, Cather describes one of the many trips made by the protagonist Father Latour across the landscape (or was it a skyscape?) of the southwestern United States and of the observations Latour made about "the white man's way to assert himself" anywhere he travels:

> The ride to Santa Fé was something under four hundred miles. The weather alternated between blinding sand-storms and brilliant sunlight. The sky was as full of motion and change as the desert beneath it was monotonous and still – and there was so much sky, more than at sea, more than anywhere else in the world. The plain was there, under one's feet, but what one saw when one looked about was the brilliant blue world of stinging air and moving cloud. Even the mountains were mere ant-hills under it. Elsewhere the sky is the roof of the world; but here the earth was the floor of the sky. The landscape one longed for when one was far away, the thing all about one, the world one actually lived in, was the sky, the sky!
>
> Travelling with [his Navajo guide] Eusabio was like travelling with the landscape made human. He accepted chance and weather as the country did, with a sort of grave enjoyment. He talked little, ate little, slept anywhere, preserved a countenance open and warm, and like Jacinto he had unfailing good manners. The Bishop [i.e., Father Latour] was rather surprised that [his guide Eusabio] stopped so often by the way to gather flowers . . .
>
> When they left the rock or tree or sand dune that had sheltered them for the night, the Navajo was careful to obliterate every trace of their temporary occupation. He buried the embers of the fire and the remnants of food, unpiled any stones he had piled together, filled up the holes he had scooped in the sand . . . . Father Latour judged that, just as it was the white man's way to assert himself in any landscape, to change it,

---

53 Aldo Leopold, A SAND COUNTY ALMANAC 239–240 (1966) (1949).
54 *Id.* at 262.

make it over a little (at least to leave some mark or memorial of his sojourn), it was the Indian's way to pass through a country without disturbing anything; to pass and leave no trace, like fish through the water, or birds through the air . . .

In the working of silver or drilling of turquoise the Indians had exhaustless patience; upon their blankets and belts and ceremonial robes they lavished their skill and pains. But their conception of decoration did not extend to the landscape. They seemed to have none of the European's desire to "master" nature, to arrange and re-create. *They spent their ingenuity* in the other direction; *in accommodating themselves to the scene in which they found themselves.* This was not so much from indolence, the Bishop thought, as from an inherited caution and respect. It was as if the great country were asleep, and they wished to carry on their lives without awakening it; or as if the spirits of earth and air and water were things not to antagonize and arouse. When they hunted, it was with the same discretion; an Indian hunt was never a slaughter. They ravaged neither the rivers nor the forest, and if they irrigated, they took as little water as would serve their needs. *The land and all that it bore they treated with consideration; not attempting to improve it, they never desecrated it.*[55]

Still more inspiration for envisioning the land-husbandry aspect of an agroecological future can be drawn from Frederick Kirschenmann. In the following paragraphs he emphasizes the need for closeness to the land, and to the workings of the natural world, in order for a productive agrarian life to be feasible.

Recognizing problems in a timely manner through close, personal observations is crucial to making a sustainable system work. If one allows a weed problem to grow so that everyone in the neighborhood can recognize it, it is probably too late to find a remedy. Weed and insect problems can only be solved without pesticides by utilizing preventive management strategies. Consequently, if a farmer wants to reduce or eliminate reliance on fertilizers and pesticides, it will be necessary to walk the fields, scrutinize weed problems, carefully analyze growing plants and give close attention to the entire soil profile. That means the farmer has to live on the farm. Absentee sustainable farming will not work. Sustainable practices probably also will not work on farms too large to be personally managed by the owner-operator.[56]

---

55 Willa Cather, Death Comes for the Archbishop 234–237 (1927) (emphasis added).

56 Kirschenmann, *supra* note 1, at 37. In a preceding passage, Kirschenmann cites the sixteenth-century Italian political theorist Niccolo Machiavelli for the proposition that "a wise king always lives in the province he rules, because that is the only way to recognize problems in a timely manner . . . [and] if a ruler lacks close, personal knowledge of his realm, evils [like weeds for a farmer] 'are allowed to grow so that everyone can recognize them, there is no longer any remedy

[Niles] Eldredge argues that one of the reasons it is difficult for modern humans to recognize their dependence on local ecosystems is that we are so disconnected from the natural world, "we don't even know the most basic details of our own food production." Such alienation separates us from fundamental cycles of life and death, birth and decay, production, and waste recycling. We are deprived of the opportunity to experience the drama of species connectedness. Hence it is "small wonder people have a hard time getting concerned that a particular species of pine tree in the Pacific Northwest or a particular seabird species, let alone some, as yet, undiscovered species of beetle in the depths of the Amazonian rain forest, is under imminent threat of extinction. How can the demise of any of those species possibly have an impact on us?" If we redesign agriculture to make us more aware of the "most basic details of our own food production," then agriculture might help us become more aware of our dependence on local ecosystems and thereby motivate us to restore and maintain them.[57]

In sum, projecting a future based on agroecological husbandry could draw on these and other thoughtful observers to identify details of the role that *land husbandry* would play in such a future. What types of details do I have in mind? They would answer such questions as the following:

- What *degree* of restoration – of ecosystems, of soil quality, and of values that Aldo Leopold and Wendell Berry and others say have been lost – should we strive for, and how could this effectively be quantified? For instance, does it make sense to strive for an 80–80–80 goal (my own invention), so that 80 percent of the world's ecosystems would be restored to 80 percent of their pre-industrial-age health within 80 years?

---

to be found.'" *Id.*, quoting from Hedley Bull's translation of Machiavelli's *The Prince*. For some observations as to whether Machiavelli's views or Kirschenmann's views fully reflect today's reality, see note 57, *infra*.

57 Kirschenmann, *supra* note 1, at 50 (citing Niles Eldredge). As I have emphasized earlier, Kirschenmann's emphasis on relying on local government for decision-making and management might be subject to question. See note 142 in Chapter 3, *supra*, and note 28, *supra*. Similarly, his emphasis (inspired in part by Machiavelli) on the need for local presence – including his assertion, for instance, that "[a]bsentee sustainable farming will not work" – might be subject to question in light of technological advances that permit remote sensing. See, e.g., *Dragnet* THE ECONOMIST, Jan 24, 2015 (explaining how new satellite-based surveillance systems can closely monitor global fishing operations in order to identify law-breakers). Moreover, with advanced farm-management and farm-operation services, the "personal management by the owner-operator" urged by Kirschenmann might not always seem necessary, or even sensible. Still, those advances seem unlikely to address (and indeed might exacerbate) the concerns Kirschenmann and others highlight regarding the *alienation* that results from human separation from the natural world. Local presence and involvement in agriculture can, as Kirschenmann states, "help us become more aware of our dependence on local ecosystems and thereby motivate us to restore and maintain them".

If too many ecosystems have already been irreversibly degraded (as through global climate change), is a 70–80–90 goal more feasible?

- Within the time allocated to restoring land and land husbandry (establishing what Aldo Leopold has called a "land ethic") – whether that time allocation is 80 years or some longer or shorter period – how should we *prioritize* the efforts to restore to a sustainable degree of health the many ecosystems that have been damaged by human activity? Should we start with those ecosystems that have been most severely damaged and therefore are most at risk of total collapse, such as certain large river deltas, coral reefs, and old-growth forests? Should we instead concentrate first on those ecosystems that have been degraded in a way that can be reversed with fairly modest changes in policy and practice, such as putting an end to the availability of public lands in the western US states for private exploitation through livestock grazing and minerals extraction?

- What about the sources of *funding* for these (land restoration) purposes: should the necessary funds come from national and local governments, from existing international institutions such as the World Bank, or from a newly-established institution devoted primarily to ecological restoration and agricultural renovation? It needs to be borne in mind that in legal cultures that currently have strong private-property-ownership protections, substantial funding will probably be needed to compensate existing holders of property rights in land to be affected by restoration efforts. I return to some of these issues in Chapter 7 of this book.

- How much of the Earth's surface should be set aside as *wilderness* areas off limits to human interference – 100 times as much acreage as is presently set aside? 10,000 times as much acreage? What would be the minimum sizes of such set-aside wilderness areas, in order to facilitate the operation of ecosystems on an appropriately grand scale? For instance, might the Poppers' plan for setting aside 139,000 square miles in the American Great Plains actually be too small to support a Buffalo Commons? Would it be more realistic to set aside half the earth (both terrestrial and maritime areas) as a permanent preserve, as the famous biologist E. O. Wilson proposes in his most recent book?[58]

- Combining some of the perspectives summarized previously from Wendell Berry and David Montgomery, just what is the *level of sustainable productivity* that we can reasonably expect from those areas devoted to farming once the sort of soil restoration referred to beforehand has been completed? The answers to this question would of course differ by ecoregions, by soils, by the degree and type of livestock production,

---

58 See Edward O. Wilson, HALF-EARTH: OUR PLANET'S FIGHT FOR LIFE (2016). For a discussion of this book (Wilson's 32nd) and the sense of urgency it conveys, see Claudia Dreifus, *A Plea, While There's Still Time*, THE NEW YORK TIMES, Mar. 1, 2016, at D5 (explaining that Wilson's proposal "means creating something equivalent to the U.N.'s World Heritage sites that could be regarded as the priceless assets of humanity").

by the degree to which climate change alters potential productivity, by the particular balance of mechanized versus non-mechanized labor involved in the farming operations, and by a range of other factors. Significant among those factors is the extent to which "sustainable productivity" is defined to encompass not just production of grain or meat or other commodities but also the social and cultural "goods" that farming and rural communities create.

IIIC. *Plant species and animal species*

I referred in the preceding subsection to the need for plants and animals to be "indigenous or appropriately adapted" to the ecosystem(s) in which they are found. In respect of those areas of the Earth that would be devoted to farming in a future that insists on an agroecological approach to food production, the "appropriate adaptation" of *plant* species would encompass the management of selected species of plants – probably perennials grown in polycultures – developed expressly to produce food for human and non-human use. What this would mean in practice is that the Earth's land cover would include both (i) vast ranges of territory that feature indigenous plants – not alien and invasive species – and (ii) smaller but still substantial ranges of territory devoted to "appropriately adapted" plants used for production of food, feed, and fiber.

The "appropriate adaptation" of *animal* species would reflect several needs. One would be the need to facilitate land husbandry described in subsection IIIB – for instance, by using cattle, bison, and other large herbivores in field, prairie, and grassland management and by nurturing small invertebrates (worms, insects, etc.) to facilitate soil tilth and fertility. A second would be the need to assure reasonable safety of humans against attack or disease caused by animals. Others would be the need to facilitate climate stability and resilience and water security and the need for a modest level of livestock operations to produce meat for human consumption. In keeping with the latter two of those points, the grazing of cattle and other large herbivores for human consumption of their meat would occur at drastically lower levels than those that prevail today. Why? Both (i) to control emission of methane, one of the most potent greenhouse gases, and (ii) to reflect the need for more healthful, less meat-intensive diets than those that are currently followed in some countries (and that are in fact rising quickly in some countries, including China). Furthermore, the raising of any domestic livestock – not just cattle, but also poultry, sheep, and hogs – would, in a global agroecological future, follow ethical standards of treatment that are currently disregarded to a large degree in modern industrial animal-based food production.

In respect of both plants and animals, some difficult questions might arise in determining which species are "indigenous" to an area. In fact, these questions might be especially important in respect of wilderness areas. Human transformation of the natural world has been so great, especially in the past two to four centuries, that indigenousness is a fuzzy concept; different species may arguably be "indigenous" to an ecosystem as of different dates. Therefore, different "as of" dates might be selected for different ecosystems; but in most cases an "as of" date for these purposes might be no more recent than about 1700, since the last 300 years have seen such a vast increase in the introduction of non-indigenous species into many ecosystems. This suggests that much of the land cover on the Earth will, in an agroecological future, look dramatically different from how it looks today; it will reflect a massive habitat-repair project reversing the damage inflicted by humans over recent centuries.

A restored vitality of indigenous plant and animal populations would reflect the research and conclusions of such experts as Alan Savory on the interaction of herbivores, predators, and grasslands. The emphasis that Savory and others place on grasslands – whose destruction he claims is "the immediate cause of poverty, social breakdown, violence, cultural genocide . . . and a significant contribution to climate change"[59] – is especially significant in projecting a global agroecological world because the principal cause of grasslands destruction and degradation is agriculture, especially if defined to include livestock grazing.[60] Likewise, a restored vitality of indigenous plant and animal populations might also reflect work being done by such experts as Stewart Brand and Andrew Torrance on "de-extinction" – that is, the restoration of certain species that humanity wiped out.[61]

Again, these are generalities. In order to elaborate further on the plant-and-animal-species features of an agroecological future, we could usefully refer to various experts on restoring and preserving biodiversity. For instance, considerable work on global diversity has been carried out by the World Wildlife Fund (WWF), especially in its development of the notion of "ecoregions". The WWF website offers this explanation of "ecoregions":

> Biodiversity is not spread evenly across the Earth but follows complex patterns determined by climate, geology and the evolutionary history of the planet. These patterns are called "ecoregions". . . . WWF defines an ecoregion as a "large unit of land or water containing a geographically distinct assemblage of species, natural communities, and environmental conditions". The boundaries of an ecoregion are not fixed and sharp,

---

59 *Allan Savory Works to Promote Holistic Management in the Grasslands of the World*, www.ted.com/speakers/allan_savory. A "TED Talk" by Savory can be accessed from that website.

60 See GRASSLANDS, *supra* note 5, at 41–45.

61 For the work of Stewart Brand, see *The Dawn of De-Extinction: Are You Ready?*, www.ted.com/talks/stewart_brand_the_dawn_of_de_extinction_are_you_ready. A "TED Talk" by Brand can be accessed from that website. For a summary of the work of Andrew Torrance on this topic, see *Professor Examining Potential Laws to Regulate De-extinction* (University of Kansas press release, Aug. 21, 2013), https://law.ku.edu/news/professor-taking-lead-potential-laws-de-extinction.

but rather encompass an area within which important ecological and evolutionary processes most strongly interact.[62]

By identifying 867 distinct terrestrial ecoregions on the Earth, the WWF classification system provides a baseline for determining what specific efforts are needed to restore biodiversity to those ecoregions – and what each of those ecoregions would look like upon the successful completion of those efforts.

Biodiversity has a special place in the professional literature for those areas of the Earth that would – in an agroecological future – be devoted to farming. For instance, in his 2005 article *Renewing Husbandry*, Wendell Berry offers these observations about the importance of local adaptation of farm crops:

> Our recent focus on productivity, genetic and technological uniformity, and global trade – all supported by supposedly limitless supplies of fuel, water, and soil – has obscured the necessity for local adaptation. But our circumstances are changing rapidly now, and this requirement will be forced upon us again by terrorism and other kinds of political violence, by chemical pollution, by increasing energy costs, by depleted soils, aquifers, and streams, and by the spread of exotic weeds, pests, and diseases. We are going to have to return to the old questions about local nature, local carrying capacities, and local needs. And we are going to have to resume the breeding of plants and animals to fit the region and the farm.[63]

In sum, the envisioning of an agroecological future could draw on a range of thoughtful observers to identify details of the role that *species diversity*, both of plants and of animals, would play in such a world. What types of details do I have in mind? Those that would answer such questions as the following:

* In order to maintain adequate species diversity and adaptation on Earth, how much of the planet's land would be used for food production? Might it be 20 percent, compared to the current figure of 40 percent

---

62 *Ecoregions*, at http://worldwildlife.org/biomes. Another WWF definition gives somewhat more emphasis to climate: "Ecoregions are large areas of relatively uniform climate that harbour a characteristic set of species and ecological communities." *About Global Ecoregions*, wwf.panda. org/about_our_earth/ecoregions/about/. For a more detailed reference to WWF ecoregions, see subsection IIA of Chapter 8, *infra*. As noted earlier, geographical climate classification systems date back a couple of hundred years. See text accompanying note 114 in Chapter 3, *supra*.

63 Wendell Berry, *Renewing Husbandry*, in Orion (Sep.–Oct. 2005), www.orionmagazine.org/ index.php/articles/article/160/ (hereinafter Berry, *Husbandry*).

and the 1700 figure of 7 percent ?[64] This would depend partly, of course, on population projections discussed in subsection IIID and on the question posed in the next bullet-point.

- How much yield can be expected from a new inventory of grain and legume crops – such as those perennial strains being developed at The Land Institute and elsewhere for growing in polycultures – and what does the answer to that question tell us about how much of the Earth's surface would need to be devoted to growing of food crops, taking into account population projections and also root crops, fish, and other non-grain foods?
- Likewise, how would the allocation of land for wilderness, for farming, and for other designations be driven by the need to adjust to climate change that is already occurring and to facilitate climate adaptation and resilience for the future, especially through carbon sequestration, which I have described elsewhere[65] – as well as certain other environmental concerns such as water security?
- What degree of protection from human interference would be needed in wilderness areas in order to maintain the plant and animal diversity in those areas? Would those areas need to be off limits to all human entry, or just to human alteration and development?

As with other questions raised above, the questions I have posed here share a pair of common underlying issues that I have deferred addressing for purpose of this discussion: (i) by what procedure, and by whom, are such questions to be answered, and (ii) how will those answers be actually *implemented*? I address those underlying issues in Chapters 6 and 7.[66]

---

64 See James Owen, *Farming Claims Almost Half Earth's Land, New Maps Show*, NATIONAL GEOGRAPHIC NEWS, Oct. 28, 2010, http://news.nationalgeographic.com/news/2005/12/1209_051209_crops_map.html (reporting that "an area roughly the size of South America is used for crop production, while even more land – 7.9 to 8.9 billion acres . . . – is being used to raise livestock").

65 See GRASSLANDS, *supra* note 1, at 63 (summarizing some of the climate-change implications of retaining and restoring large tracts of grassland ecosystems, and noting that the overall potential of carbon sequestration by grasslands compares favorably, because of their extensive root systems, with the potential for carbon sequestration by rain forests). See also note 7 in Chapter 5, *infra*, and accompanying text (explaining the carbon-sequestration potential of perennial-polyculture crops).

66 In particular, see sections II and III of Chapter 6 (regarding the elements of a 50-year farm bill that would mandate changes in land use practices and biodiversity protections) and sections I and II of Chapter 7 (enumerating duties of countries to adopt legislation setting aside nature reserves and prohibiting certain agricultural practices – all governed by a set of principles regarding natural-systems agriculture and subject to oversight and management by an international institution designed to overcome shortcomings that afflict current global organizations. For additional attention to the last of these points, regarding international institutional design, see John W. Head, A GLOBAL CORPORATE TRUST FOR AGROECOLOGICAL INTEGRITY: MANAGING A NEW AGRICULTURE IN A WORLD OF ECO-STATES (forthcoming).

IIID. Global human population

A third key element involved in projecting a future that features agroecology would focus on human population. As with the elements I have addressed in the preceding paragraphs – land husbandry and species diversity – my observations here about human population will not dwell on the breathtaking recent increases in global population figures (which recently moved past the 7 billion mark), or on projections that have been offered for the growth in those figures if unabated by some global calamity, or even on the debate over whether those projections are cause for concern. I have given some attention to those topics in another context.[67] Instead, for present purposes I wish to consider what the human population on Earth *would* be in a future that replaces modern extractive agriculture with agroecological husbandry.

This issue turns in part on the concept of carrying capacity, which might be regarded as addressing this question: What would be the *maximum* size of the human population on Earth, based on the resources available to support such a population? Beyond that carrying-capacity question, though, is another one that seems more important in projecting an agroecological world: What would be the *preferred* or *optimal* size, not the maximum size, of the human population on Earth? After all – looking at the issue from a food-production perspective – just because it might be technically possible to grow, and even to distribute, enough food to keep alive a global human population of, say, 9 billion or 10 billion does not mean that it would be ethically acceptable to have that large a human population on the planet.

As for maximum *carrying capacity*, a 2001 UN report noted that two-thirds of the estimates for the Earth's carrying capacity (for humans) fall in the range of 4 billion to 16 billion people, with a median of about 10 billion.[68] Some more recent estimates are much lower than this, particularly if resource depletion and increased consumption are taken into account.[69]

---

67 GRASSLANDS, *supra* note 5, at 182–193 (examining human population growth in the context of the degradation of the Earth's grasslands regions). For some further observations about global human population and the claim that modern extractive agriculture is necessary in order to "feed the world", see the discussion in section III of Chapter 2, *supra*.

68 See UN Department of Economic and Social Affairs, Population Division, *World Population Monitoring 2001: Population, Environment, and Development* (2001), ST/ESA/SER.A/203 www.un.org/esa/population/publications/wpm/wpm2001.pdf, at 31.

69 See, e.g., W. F. Ryerson, *Population, The Multiplier of Everything Else*, in Richard Heinberg and Daniel Lerch, THE POST CARBON READER: MANAGING THE 21ST CENTURY SUSTAINABILITY CRISIS 3 (2010), www.postcarbon.org/report/131587-population-the-multiplier-of-everything-else. See also Lester R. Brown, WORLD ON THE EDGE 16 (2011). For an explanation of the connection between global human population and these other factors (resource depletion and increased consumption), with particular emphasis on carbon emissions and global climate change, see Eduardo Porter, *Reducing Carbon by Curbing Population*, THE NEW YORK TIMES, Aug. 6, 2014, pp. B1, B5 (citing a study asserting that "reducing the burning of fossil fuels might be easier if there were fewer of us consuming them").

What about the *preferred* or *optimal* size of the human population on Earth from an ethical standpoint? It is worth noting that some observers would reject the concept of an optimal population, partly due to their confidence that humans would be able to adapt to virtually any population level,[70] and others would reject the notion of optimal population for more practical reasons – that it would never be possible to reach consensus on the factors to be considered in calculating an optimal world population.

Still, some efforts have been made to identify an optimal human population by taking into account such factors as these: (i) the availability of adequate wealth and resources to everyone; (ii) the offering of basic human rights to everyone; (iii) the preservation of cultural diversity; (iv) the allowance of intellectual, artistic, and technological creativity; and (v) an adequate preservation of biodiversity. It is noteworthy that several of these factors have an important ethical dimension: Most observers would regard it as unethical to permit or promote a global human population that is so great as to subject large portions of it to malnourishment or other forms of deprivation such as unhealthy air, contaminated water, or inadequate sanitation.[71]

Using the five factors enumerated in the preceding paragraph, the author who in 1968 wrote *The Population Bomb* offered in the 1990s an optimal global human population figure of about 1½ billion to 2 billion people.[72] Other sources suggest figures of 2 to 3 billion depending on varying assumptions.[73] (For comparison purposes, note that a global human population of 2 billion was reached around 1928, and a global human population of 3 billion was reached in about 1961.)[74]

---

70 In this regard, the writings of Julian Simon come to mind. Simon argued that increasing populations add creative capacity to society. Julian Simon, THE ULTIMATE RESOURCE 221 (1981).

71 It is worth noting that the implementation of some religions – such as the form of Catholicism applied by church authorities in the Philippines – seems to give relatively little heed to the obvious consequences of their policies on matters of population control.

72 See Gretchen C. Daily, Anne H. Ehrlich, and Paul R. Ehrlich, *Optimum Human Population Size*, 15 POPULATION AND ENVIRONMENT (no. 6) (July 1994). These authors acknowledge that "[i]nnumerable complexities" are involved in calculating an "optimal" global human population, but they insist that "[i]t is nonetheless instructive to make a tentative, back-of-the-envelope calculation of an optimum on the basis of present and foreseeable consumption patterns and technologies." See also Paul R. Ehrlich, THE POPULATION BOMB xi (rev. ed. 1975); and Paul R. Ehrlich and Anne H. Ehrlich, *The Population Bomb Revisited*, 1 ELECTRONIC JOURNAL OF SUSTAINABLE DEVELOPMENT 63, 67 (2009).

73 See, e.g., Ken Small, *Global Population Reduction: Confronting the Inevitable*, in 17 WORLD WATCH MAGAZINE (no. 5) (Sep.–Oct. 2004), www.worldwatch.org/node/563 (calling for "a global population shrinkage of at least two-thirds to three-fourths, ... [to a] 'population optimum' of not more than 2 to 3 billion"). See also Martin Desvaux, *Towards Sustainable and Optimum Populations* (2008), available on the website of the organization Population Matters at http://populationmatters.org/documents/towards_populations.pdf (offering various figures depending on differing per-person "footprints" of consumption but specifying 2.7 billion for "a 'modest' world footprint" with some margin for maintaining biodiversity).

74 GRASSLANDS, *supra* note 5, at 184. The subsequent "billion-person milestones" were reached in about 1974 (4 billion), 1987 (5 billion), 1999 (6 billion), and 2011 (7 billion). *Id.* Frederick

In addition to identifying an optimal overall total human population, an agroecological future would also feature a reimagined *distribution* of that population. Such distribution of world population would almost surely be different from that of the present day in that it would comport with the availability of various resources, including of course those needed for food production (rainfall, groundwater, arable soil, etc.) but also those needed for production of essential manufactured goods. Moreover, the distribution of human population would turn on such factors as indigenousness (of people) in one region or another, ethnicity, and other social and cultural factors.

I would emphasize again that many of the projections and imaginings I am exploring in these paragraphs – on a reduced and redistributed global human population, on the determination of what portions of the Earth's land would be used for food production and what portions for biodiversity-enhancing natural reserves, and so forth – seem far-fetched in today's world. How could any such issues be decided? If decisions are made, how could they be implemented?

If, as I have suggested in Chapters 2 and 3, modern extractive agriculture is unsustainable on economic, ecological, and other grounds, then it stands to reason that changes *will occur* whether humans take action or not – just as changes in sea level *will occur* in the next few decades along many coasts, creating millions of climate refugees. If no attempt is made to envision a future that might be sustainable, then the changes that come will occur through natural forces (extinctions, for instance, or substantial destabilization of the Earth's climate) that are likely to be unfavorable to human survival. Accordingly, it seems reasonable to make some effort to envision an agroecological future that *does* address even such sensitive and difficult issues as global land use and human population. Doing so might provide a sense of where to locate what I referred to earlier as "the far end of the bridge" between today's reality and a future eutopia ("good place") based on agroecological husbandry.[75]

IIIE. Energy

In sharp contrast to the *current* global system of energy generation, distribution, and use – which has become heavily dependent on fossil-carbon fuels controlled by a handful of commercial entities – a future focusing on agroecological husbandry would feature systems of energy production and distribution based on current, renewable (non-fossil-carbon) sources. These

---

Kirschenmann offers this further perspective on the growth of global human population: "After agriculture was invented, the human population jumped from 10 million to about 50 million in just 3,000 years. Thereafter, the human population doubled at an ever-increasing rate." In fact, "[t]he most recent doubling took just thirty-six years, from 1969 to 1996." Kirschenmann, *supra* note 1, at 143–144.

75 See the third paragraph at the beginning of this section III.

would include solar power, wind power, geothermal power, biofuel, and hydropower sources, all designed and operated to guard against pollution and high transmission costs. Importantly, the overall use of electrical power would be dramatically less than in today's world, as would the transmission of electrical power over long distances. That is, power generation would be locally-sourced to the greatest possible extent.

For some inspiration in constructing the energy component of an agroecological future, consider the efforts made in Denmark, where the government has set a goal of achieving complete independence from fossil fuels by 2050. Its plans include (i) highly efficient energy consumption, (ii) electrification of heating, industry, and transportation, (iii) use of more wind-power energy, (iv) efficient use of biomass resources, (v) use of biogas (originating in large part from livestock manure), (vi) photovoltaic solar modules and wave power as supplements for wind power, and (vii) an intelligent system regarding the timing and coordination of energy use.[76] This project suggests that it can be possible for an entire country[77] to shift focus toward an energy system that integrates and efficiently uses all potential renewable energy sources, including "waste" products from industries – with special emphasis on the agricultural sector.

Energy sustainability at the *local* level (as distinct from the country-wide level) has also attracted attention elsewhere in Scandinavia. A recent study based in Sweden examines combinations of various renewable energy sources, in comparison with fossil fuel sources, to determine how energy use can be incorporated into small-scale agricultural operations to optimize farms' productive capacity per unit of energy use.[78]

Likewise, small-scale and localized energy generation serves as the center-piece of work by Hermann Scheer, whose 2006 book *Energy Autonomy*[79]

---

76 See *Energy Strategy 2050: From Coal, Oil and Gas to Green Energy* (March 2011), www.ens.dk/en/info/publications/energy-strategy-2050-coal-oil-gas-green-energy.

77 In case it is thought that Denmark is a small and irrelevant country, it is worth noting that its total GDP is in the top 60 countries even though its human population is less than 6 million (giving it a per capita GDP in the top 40 countries), it has a very low Gini index (signifying relative equality in economic circumstances of its population), its low current account deficit puts it in the top 20 countries in terms of balance of payments, the value of its exports puts it in the top 40 countries, its foreign reserves put it in the top 25 countries, its outbound foreign investment also puts it in the top 25 countries, and its infant mortality rate is among the very lowest in the world. See CIA World Factbook Denmark Page, www.cia.gov/library/publications/the-world-factbook/geos/da.html.

78 See Sheshti Johansson, Kristina Belfrage, and Mats Olsson, *Impact on Food Productivity by Fossil Fuel Independence – A Case Study of a Swedish Small-Scale Integrated Organic Farm*, 63 Acta Agriculturae Scandinavica Section B Soil & Plant Science 123 (2013).

79 Hermann Scheer, Energy Autonomy: The Economic, Social and Technological Case for Renewable Energy (2006). Hermann Scheer, who died in 2010, was president of the European Association for Renewable Energy, chairman of the World Council for Renewable Energy, the winner of the Right Livelihood Award (regarded as the alternative Nobel Prize) in 1999, and a member of the German parliament.

touts the feasibility of renewable sources of energy and of distributed, decentralized energy generation. Scheer, whose other books include *A Solar Manifesto*,[80] insists that relying on renewable energy is technologically, commercially, and politically possible – and manifestly more sensible than relying on either fossil or nuclear energy sources. Scheer's conclusions are echoed by numerous others, including the authors of the 2011 book *Climate Capitalism*,[81] which highlights some recent innovations in the energy, construction, transportation, and agriculture sectors – innovations that, according to the authors, illustrate how changes towards the use of renewable resources contribute both to business profitability and to social and economic stability.

In sum, the envisioning of an agroecological future could draw inspiration from a range of thoughtful observers to identify and specify the central role to be played by renewable energy sources, decentralized in their generation and distribution to the extent possible. Relying on such energy sources – and, in particular, substituting solar energy for fossil-fuel energy on which farming has become so thoroughly dependent in the past several decades – will contribute to a form of farming and agricultural production that promises economic and ecological stability.

IIIF. Economics

It is essential to see how the two elements of stability that I referred to at the end of the preceding subsection, where I referred to "economic and ecological stability", fit together. Ecology and economics might well be seen as the twin pillars supporting the agroecological future that I am projecting in this chapter.

As an illustration of this, consider Kenneth Boulding's 1966 essay on "spaceship Earth".[82] In that essay Boulding urges us to think of our planet as

---

80 Hermann Scheer, A Solar Manifesto (2005).
81 Hunter Lovins and Boyd Cohen, Climate Capitalism: Capitalism in the Age of Climate Change (2011). Hunter Lovins is president of Natural Capitalism Solutions, a Colorado-based NGO to educate senior decision makers in the business case for a regenerative economy. Trained as a sociologist and a lawyer, she is a professor of sustainable business management at Bard College and Denver University. Boyd Cohen is the director of innovation and a professor of Entrepreneurship, Sustainability, and Smart Cities at the Universidad del Desarrollo in Santiago, Chile.
82 Kenneth Boulding, *The Economics of the Coming Spaceship Earth*, in Valuing the Earth 297, 297–309 (Herman Daly and Kenneth Townsend eds., 1993). The term "Spaceship Earth" gained currency when in 1965 Adlai Stevenson gave a speech at the United Nations with this passage in it:

> We travel together, passengers on a little space ship, dependent on its vulnerable reserves of air and soil . . . [and] preserved from annihilation only by the care, the work, and, I will say, the love we give our fragile craft. We cannot maintain it half fortunate, half miserable, half confident, half despairing, half slave [and] half free in a liberation of resources undreamed of

a closed system, as a spaceship traveling through the universe with fixed resources. Since the Earth, like such a spaceship, is a closed system, entropy can only increase as we continue to extract resources from the natural world, and therefore endless growth is simply impossible. Boulding argues that economics should take this into account and stop preaching the ideology of endless growth.

Growth – or a questioning of its role in economics – is at the center also of Herman Daly's 1993 essay on the "steady-state economy".[83] According to Daly, such an economy has a fairly constant supply of population and a fairly constant supply of exosomatic capital (such as sunlight). Both of these (population and exosomatic capital) are plentiful enough in such an economy to support a sustainable life for people, but both are used conservatively – that is, at a relatively low "throughput" rate.

Daly's work in developing the notion of a "steady-state economy" draws from numerous other authors, including John Stuart Mill, and it has built considerable momentum quite recently, as reported by Richard Heinberg in this account from 2011:

> The past three decades, and especially the past three years, have seen an explosion of discussion about alternative ways of thinking about economics. There are now at least a score of think tanks, institutes, and publications advocating fundamentally revising economic theory in view of ecological limits. Many alt-economics theorists question either the possibility or advisability of endless growth.
>
> The fraternity of conventional economists appears to be highly resistant to these sorts of challenging new ideas . . . . [Those economists start] with certain basic premises that are clearly, unequivocally incorrect: that the environment is a subset of the economy; that resources are infinitely substitutable; and that growth in population and consumption can continue forever. In conventional economics, natural resources like fossil fuels are treated as expendable income, when in fact they should be treated as capital, since they are subject to depletion. As many alternative economists have pointed out, if economics is to stop steering society into the ditch it has to start by reexamining these assumptions.[84]

---

until this day. No craft, no crew can travel safely with such vast contradictions. On their resolution depends the survival of us all.

See the account at www.bartleby.com/73/477.html.

83 Herman Daly, *The Steady-State Economy: Toward a Political Economy of Biophysical Equilibrium and Moral Growth*, in VALUING THE EARTH, *supra* note 82, at 325–363. Herman Daly is an ecological economist and professor at the School of Public Policy of the University of Maryland. He has received the Right Livelihood Award (regarded as the alternative Nobel Prize) and the National Council for Science and the Environment Lifetime Achievement Award.

84 Richard Heinberg, THE END OF GROWTH: ADAPTING TO OUR NEW ECONOMIC REALITY 246–247 (2011). Richard Heinberg is the senior fellow at the Post Carbon Institute and the author of a dozen

Two of the points included in the passage quoted from Heinberg deserve special attention. First, he notes that conventional economics assumes "that the environment is a subset of the economy". In another portion of his book on "the end of growth", Heinberg offers this further explanation:

> The subsuming of *land* within the category of *capital* by nearly all post-classical economists [has] amounted to a declaration that Nature is merely a subset of the human economy – an endless pile of resources to be transformed into wealth. It also [means] that natural resources could always be substituted with some other form of capital – money or technology. The reality, of course, is that the human economy exists within and entirely depends upon Nature, and many natural resources have no realistic substitutes. This fundamental logical and philosophical mistake . . . [of conventional economics has] set society directly on a course toward the current era of climate change and resource depletion, and its persistence makes conventional economic theories – of both Keynesian and neoliberal varieties – utterly incapable of dealing with the economic and environmental survival threats to civilization in the 21st century.[85]

A second point I would emphasize from the first passage quoted above from Heinberg, concerns *consumption*. According to Heinberg, conventional economics is deeply flawed in assuming – indeed, requiring – "that growth in . . . consumption can continue forever". Reality and prudence, Heinberg insists, require us to place limits on consumption, both of renewable and of nonrenewable resources.[86]

A quickly-growing literature echoes this concern about over-consumption. Tim Jackson's 2006 "reader" on sustainable consumption,[87] Daniel Miller's 2012 text on "consumption and its consequences",[88] and a 2002 book on

---

books. Another prominent advocate for developing the notion of a "steady-state economy" is Lisi Krall, a professor of Economics at the State University of New York, Cortland. One biographical sketch notes that Krall's work "is driven by one overarching concern: How to reconcile the global market economy with the biophysical limits of the planet in a way that encourages humans to once again become a species embedded in the Earth". See the website of the International Forum on Globalization, at http://ifg.org/lisi-krall. Krall is the author of many books and articles, including *Proving Up: Domesticating Land in U.S. History* (exploring the interconnections of economy, culture, and land in US history), and is a member of the Center for the Advancement of the Steady State Economy, the website for which can be found at http://steadystate.org/.

85 Heinberg, *supra* note 84, at 39–40.
86 See quoted text accompanying note 84, *supra*, drawing from *id.* at 246–247.
87 Tim Jackson, ed., The Earthscan Reader on Sustainable Consumption (Routledge 2006). Tim Jackson is a professor of Sustainable Development at the University of Surrey. He currently holds a Professorial Fellowship on Prosperity and Sustainability in the Green Economy. He has served as Economics Commissioner on the UK Sustainable Development Commission.
88 Daniel Miller, Consumption and Its Consequences (2012). Daniel Miller holds a PhD in Anthropology from Cambridge and has written or edited over three dozen books.

"confronting consumption"[89] – all of these works challenge the view (common in conventional economic theory) that consumption is the main driver, and indeed the purpose, of an economic system. According to these and other authors, efforts to make *production* more sustainable or "green" will never succeed in putting global society on an ecologically and socially sustainable path; instead, these authors place primary emphasis on sustainable *consumption* – and on the impossibility of achieving this form of sustainability unless we reduce consumption substantially from current levels.[90]

I believe a carefully constructed vision of an agroecological future must grapple with these issues – and must almost surely opt ultimately for an economic model that does *not* give primacy to the need for growth in consumption.[91] Likewise, such a vision will almost surely need to incorporate some elements of the work being carried out by Bill McKibben and others regarding what he calls the "deep economy".[92] McKibben's central thesis is that a "deep economy" would place greater reliance on local producers and consumers, rather than constituting a predominantly centralized system where a relatively few large producers do business with almost all consumers. His *Deep Economy* book, drawing on evidence from China, India, and New England, argues that this economic framework can realistically be adapted elsewhere in the world, creating thereby a system that is resilient and flexible, and thus less susceptible to market failures.

Moreover, an agroecological future would embrace an economic model in which all externalities are properly accounted for in the pricing of goods – particularly agricultural goods. In Chapter 2, I emphasized how our modern extractive agriculture fails to do so: the current system inflicts

---

89 Thomas Princen, Michael Maniates, and Ken Conca, eds., CONFRONTING CONSUMPTION (2002).

90 For another recent collection of essays related to sustainable consumption, including agriculture's role in shifting production/consumption patterns, see WORLDWATCH INSTITUTE, STATE OF THE WORLD 2010: TRANSFORMING CULTURES: FROM CONSUMERISM TO SUSTAINABILITY (2010). Another perspective on the need to achieve sustainable *consumption* focuses on the need to place limits on *investment*. John Fullerton, the founder and president of the Capital Institute and a director of New Day Farms, Inc., asserts that "[a] transition to a sustainable economy requires not only population stabilization, breakthroughs in resource productivity, and checks on material consumption, but also constraints on aggregate investment." He explains that "the goal of optimizing relatively short-term reliance on investment" is central to modern finance, and that when such investment is successful, it "induces exponential growth in the aggregate stock of financial capital" – thus spurring "ever-increasing demands for natural resources." Hence, Fullerton argues, there must be a fundamental transformation of finance. John Fullerton, *Limits to Investment: Finance in the Anthropocene* (April 2014), appearing on the website of the Great Transition Institute, www.greattransition.org.

91 For other explanations of such an economic model, see Kent A. Klitgaard and Lisi Krall, *Ecological Economics, DeGrowth, and Institutional Change*, 84 ECOLOGICAL ECONOMICS 247–253 (2012).

92 Bill McKibben, DEEP ECONOMY: THE WEALTH OF COMMUNITIES AND THE DURABLE FUTURE (2008). Bill McKibben is an author and environmentalist who was awarded the Right Livelihood Prize (regarded as the alternative Nobel Prize) in 2014. In 1989 he wrote *The End of Nature*, widely regarded as the first book for a general audience about climate change.

environmental damage – "negative externalities" – that currently are not fully accounted for, giving the illusion of economic sustainability where in fact it is absent.[93]

Prominent among the authorities focusing on this point are authors of a 2008 book *Natural Capitalism*.[94] They assert that once all externalities of ecosystem services are properly priced and accounted for, businesses involving renewable energy will flourish. A similar point is made by Lester Brown, who offers this explanation of full-cost accounting and pricing in his 2011 book *World on the Edge*:

> The key to restructuring the economy is to get the market to tell the truth through full-cost pricing . . . . If we can get the market to tell the truth, [and hence] to have market prices that reflect the full cost of burning gasoline or coal, of deforestation, of overpumping aquifers, and of over-fishing, then we can begin to create a rational economy [and particularly a rational] world energy economy. Phasing in full-cost pricing will quickly reduce oil and coal use. Suddenly wind, solar, and geothermal will become much cheaper than climate-disrupting fossil fuels.[95]

In sum, the envisioning of a world featuring agroecological husbandry could draw inspiration from a range of thoughtful observers to establish an alternative structure of economic theory – one in which (i) endless growth is not an essential element, (ii) consumption is recognized as being subject to limits, not celebrated as a central aim of the economy, and (iii) all externalities are properly accounted for in the pricing of goods, particularly agricultural goods.

---

93 See text accompanying notes 53–66 in Chapter 2, *supra*.

94 Paul Hawken, Amory Lovins, and L. Hunter Lovins, NATURAL CAPITALISM: CREATING THE NEXT INDUSTRIAL REVOLUTION (1999).

95 Lester R. Brown, WORLD ON THE EDGE: HOW TO PREVENT ENVIRONMENTAL AND ECONOMIC COLLAPSE 183–185 (2011). Brown uses the cost of gasoline in the USA as an illustration: "When added together, the many indirect costs [of gasoline] to society – including climate change, oil industry subsidies, oil spills, and treatment of auto exhaust-related respiratory illnesses – total roughly $12 per gallon." That amount of $12 is an externality which full-cost accounting would incorporate into the price of gasoline. *Id.* at 183–184. For an earlier work on the need for full-cost pricing through proper accounting for externalities see Robert Ayers and Allen V. Kneese, *Production, Consumption, and Externalities*, 59 AMERICAN ECONOMIC REVIEW 282 (1969) (asserting that "[t]he current production and consumption economy does not take into account the value of common goods such as air and water resources; nor does it account for the external costs – potentially in the tens of billions of dollars per year – of discharging residual waste from production and consumption"). See also Armon Rezai, Duncan K. Foley, and Lance Taylor, *Global Warming and Economic Externalities*, 49 ECONOMIC THEORY 329 (2012) (explaining that "despite worldwide policy efforts such as the Kyoto Protocol, the emission of greenhouse gases . . . remains a negative externality" undercutting efforts to mitigate global warming).

5 Realizing the advantages of agroecological husbandry over modern extractive agriculture

Having explained in Chapter 4 what agroecological husbandry is in general terms, and how it could fit into a larger fabric of ecological, social, and economic reform, I now to turn to specifics. In particular, I examine here (i) *how agroecological husbandry addresses key problems* inherent in modern extractive agriculture (section I) and (ii) *what progress has been made* in designing perennial polycultures as a central element of agroecological husbandry (section II).

In titling this Chapter 5 "Realizing the advantages of agroecological husbandry over modern extractive agriculture", I am therefore using the word "realizing" in two senses. First, I aim to offer explanations that will help "realize" agroecological husbandry's advantages in the sense of *understanding* them. Secondly, I aim to explore how those advantages can be "realized" in the sense of *bringing them to reality*.

I. What might be preferable about agroecological husbandry?

What specific benefits does agroecological husbandry promise, thereby making it preferable to modern extractive agriculture? In other words, why should any effort be made to develop agroecological husbandry, and specifically to create herbaceous perennial seed-producing polycultures of the sort that The Land Institute and other research centers are developing?

I offer two kinds of answers. The first kind of answer reflects the content found in Chapter 2 and most of Chapter 3, in which I catalogued the many problems – especially economic, ecological, and human-health problems – posed by modern extractive agriculture. As I explain in section IA, agroecological husbandry overcomes many of those problems. The second kind of answer reflects the content of section III of Chapter 3, which posited that modern extractive agriculture has created an ever-widening gap between humans and the rest of nature, and that for this reason it runs "against the grain" of human development.

IA. Addressing economic, ecological, and health-related problems

IA1. Ecological issues

Perennial polycultures stand at the center of agroecological husbandry. There is a host of important benefits that perennial polycultures could bring – if they were effectively developed and used in food production – to overcome various economic, environmental, and health problems cited in Chapters 2 and 3. These benefits accrue because perennial polycultures mimic ecological processes of natural grasslands in many ways. Consider the following:[1]

---

1 For further details on some of these points, see John W. Head, GLOBAL LEGAL REGIMES TO PROTECT THE WORLD'S GRASSLANDS (2012) [hereinafter GRASSLANDS], at 216–217, 221–222. See also Jerry D. Glover, Cindy M. Cox, and John P. Reganold, *Future Farming: A Return to Roots?*, SCIENTIFIC AMERICAN, Aug. 2007, pp. 66–73. Several items in the following bullet-point list are also highlighted on the website of Mission 2015: Biodiversity (associated with the Massachusetts Institute of Technology), at http://web.mit.edu/12.000/www/m2015/2015/perennial_agriculture.html.

- Perennial polycultures can dramatically reduce the required amount of agricultural fertilizer and chemical pesticides.[2] These draw heavily from fossil carbon. The quantity of fossil carbon is limited – at least on a human-based time-scale[3] – and its extraction and use bring detrimental change to the air and the water through emissions and run-off.
- Perennial polycultures can also dramatically reduce the fossil-carbon fuels[4] needed to power farm equipment – partly because fewer passes over a field are necessary, and partly because less equipment is needed. This further reduces the draw on non-renewable fossil-carbon deposits and reduces greenhouse-gas emissions that contribute to climate change.[5]
- Perennial polycultures can arrest the degradation that traditional agriculture causes to soil through erosion, damage to soil structure, and reduction in soil organic matter.
- Perennial polycultures can reduce the loss of water compared with annual grain crops, since the deeper roots of perennials "intercept, retain, and utilize more precipitation" when it falls.[6]

---

2 On the reduced need for pesticides in perennial polycultures, Wes Jackson and his colleagues offer this explanation: "Perennial species have evolved to outlast pests over the long term"; accordingly, a diverse polyculture featuring many types of perennials will have the ability to outlast a great many pests, since "[w]here there is species diversity there is chemical diversity" – which would require "a tremendous enzyme system on the part of an insect or pathogen to produce an epidemic". Wes Jackson, Stan Cox, and Tim Crews, *The Next Synthesis*, appearing as the last chapter in Nature as Measure – The Selected Essays of Wes Jackson 218 (2011).

3 There is a mismatch, in other words, between the pace at which fossil-carbon deposits – coal, oil, and natural gas in particular – were formed and the pace at which they are being used. Although the pace at which these fossil-carbon deposits are being *used* is on an ever-quickening human time-scale, fossil-carbon-deposits were *created* on an entirely different time-scale – that is, over such vast stretches of time that they are, from a human perspective, entirely non-renewable. David Pimentel, whose work on dangers of pesticide use, soil erosion, and overuse of water I referred to in Chapters 2 and 3, is frequently cited as using the figure of 100,000 for this purpose, suggesting that fossil carbon deposits are being used 100,000 times faster than they were formed – a figure echoed on the website of the Geological Association of Canada. See www.gac.ca/PopularGeoscience/factsheets/Set_FossilFuels_e.pdf.

4 The opposite of fossil-carbon fuels would be contemporary fuels, such as sunlight, on which ecosystems operate in their natural state.

5 A recent study estimates that a perennial crop could reduce the fossil-fuel energy required for production by as much as 90 percent compared to conventional no-till annual wheat. Jerry D. Glover *et al.*, *Harvested Perennial Grasslands Provide Ecological Benchmarks for Agricultural Sustainability*, 137 Agriculture, Ecosystems and Environment 3–12 (2010). Applied to all US grain, this savings would reduce commercial energy use in the USA by 3 percent and annual carbon dioxide emissions by 162 million tons. To put this in perspective, all Toyota Prius hybrid vehicles produced in that model's first ten years prevented 9 million tons of carbon dioxide from entering the atmosphere. See *Toyota Celebrates 10 Years of Prius* (2010), available on the Toyota website at: http://pressroom.toyota.com/article_display.cfm?article_id=2033. I am grateful to Tim Crews of The Land Institute for bringing these points to my attention.

6 Jerry D. Glover *et al.*, *Increased Food and Ecosystem Security via Perennial Grains*, 328 Science 1638 (2010). As noted in that scientific paper, "[a]nnual grain crops can lose five times as much water . . . as perennial crops".

- Perennial polycultures can, because of their diversity, better resist attacks by pests and pathogens.
- Perennial polycultures can sequester carbon,[7] thus (i) recapturing a significant amount of the carbon that was released from the soil in the past several decades[8] and (ii) contributing to the resilience and stability of the climate.[9]
- In addition to sequestering carbon, perennial polycultures would probably reduce emissions of nitrous oxide, which is more potent as a

---

7 For details on the carbon-sequestration potential of perennial polycultures of the sort that until the nineteenth century covered much of the American Great Plains, see Thomas H. DeLuca and Catherine A. Zabinski, *Prairie Ecosystems and the Carbon Problem*, 9 FRONTIERS IN ECOLOGY AND THE ENVIRONMENT 407, 413 (2011) (discounting the carbon-sequestration effectiveness of no-till agriculture and calling instead for more research into "[a]lternative agricultural production systems that more closely emulate the biodiversity, phenology, and biogeochemical processes associated with native prairie ecosystems"). See also John H. Davidson, *North America's Great Carbon Ocean: Protecting Prairie Grasslands Keeps Carbon in the Soil and Slows the Pace of Climate Change*, SAVING LAND, Winter 2010, pp. 18–23 (insisting that "in today's world, prairie needs to be recognized for its capacity to help reduce climate change by sequestering heat-trapping carbon from the atmosphere"). Although much attention has been paid recently to tropical rainforests as "carbon sinks" – that is, as a form of ground cover that can capture carbon released into the atmosphere from a range of sources – the same is true of grassland areas, and indeed the overall potential of carbon sequestration by grasslands compares favorably with the potential for carbon sequestration by rainforests. See GRASSLANDS, *supra* note 1, at 62–63 (citing works published by the FAO).

8 One especially well-respected observer has estimated that between 100 and 200 gigatons of carbon have been lost from the land (that is, released into the atmosphere) due to land use change in terrestrial ecosystems since 1850. Most of this, he asserts, came from vegetation conversion of forests to crops but that about a quarter of it (so 25 to 50 gigatons) came from loss of soil organic matter due to cultivation for agriculture. R. A. Houghton, *Historic Changes in Terrestrial Carbon Storage*, appearing in R. Lal, K. Korenz, *et al.* (eds.), RECARBONIZATION OF THE BIOSPHERE (2012). Another observer reports an even higher estimate (about 78 gigatons) of carbon lost from soil organic matter since 1850. R. Lal, *Soil Carbon Sequestration Impacts on Global Climate Change and Food Security*, 304 SCIENCE 1623–1627 (2012). See also R. Lal, *Climate Change Mitigation by Managing the Terrestrial Biosphere*, appearing in R. Lal, K. Korenz, *et al.* (eds.), RECARBONIZATION OF THE BIOSPHERE (2012). The salient point is this: a conversion of crop land from annuals to perennials is highly likely to cause a re-capturing (sequestration) of these amounts of carbon – in the range of 25 to 78 gigatons – released earlier as a result of the loss of soil organic matter through modern extractive agriculture. I am grateful to Tim Crews of The Land Institute for bringing these points to my attention, and also for offering this additional observation:

> The development and planting of perennial polycultures will recapture much of the carbon that has been lost from the soil to the atmosphere as a consequence of farming annual grains. Annual grains occupy 70% of the croplands on Earth. If planted to perennial grains, carbon sequestration could offset close to 10% of current annual emissions from fossil fuel combustion, and this could continue for close to a century. This recapture, or sequestration, will help reduce atmospheric CO_2 as we undertake the critical task of backing society out of fossil fuel dependence.

Personal correspondence with Tim Crews, December 2015.

9 For further details on soil carbon sequestration and agriculture, see notes 82–84 and accompanying text in Chapter 3, *supra*.

greenhouse gas than carbon dioxide.[10] This reduction in nitrous oxide emissions would come from the fact that nitrogen can be provided in polycultures by including legumes, which fix nitrogen, instead of by relying on synthetic nitrogen as is done now with annual monocultures.

- In some settings, perennial polycultures can also reduce emissions of methane, another potent greenhouse gas. This has already been made possible by the development and use, particularly in China, of perennial rice, which relies on its perennial character as an alternative weed control strategy to traditional flooding.

- Perennial polycultures can reduce groundwater contamination that results from nitrate leaching in annual monocultures.

- Perennial polycultures can, more generally, better maintain the health and fertility of a landscape over longer periods of time. This is attributable to several factors, such as these:

  - Perennial roots have years to grow much deeper into soil and thereby to gain access to nutrients and moisture that annuals cannot reach;

  - The initial canopy development of perennials in early spring is fast compared to that of annuals – these will have just been planted and therefore will be tiny in early spring – so that perennials are able to intercept and use more light early in the season, and thereby to suppress the establishment of weeds;

  - At the other end of the life cycle of annuals – typically this comes in the late fall – they are harvested by a form of "clear-cutting" that kills them, rendering them incapable of continued photosynthesis, whereas perennials can continue photosynthesis after harvest; and

  - Microhabitats that may be present in perennial crops for some organisms, such as nitrogen fixers, might be absent or much less robust in annual crops.

  - The below-ground soil ecosystem as a whole, of course, maintains a more beneficial community of soil organisms when soil disturbance ceases and organic inputs via roots increase.

IA2. Economic issues

Although most of the points made on the preceding three pages are presented mainly from an *ecological* standpoint – emphasizing, that is, the ways in which perennial polycultures can arrest or reduce (and often reverse) damage caused by modern extractive agriculture to the local or global ecosystem – many of those points also have important *economic* components to them. Recall from the discussion in Chapter 2 that the costs involved in farming operations

---

10 As noted earlier, nitrous oxide is said to be about 300 times more potent, molecule for molecule, than carbon dioxide (the much more prevalent greenhouse gas) in terms of causing climate change. See note 78 and accompanying text in Chapter 3, *supra*.

today are very high. They include the prices of massive farm implements (and their fuel) for planting, maintaining, and harvesting crops, as well as the costs of the seeds themselves and of an array of chemicals – herbicides, insecticides, rodenticides, fungicides, and fertilizers – some of which are heavily dependent on oil prices.

A goal of the research into perennial grains, including legumes,[11] grown in polycultures is to reduce the costs of many of those inputs. Such a reduction could reverse the trend of recent years, in which farmers who actually rely on farm income for their financial prosperity need to achieve economies of scale by farming vast tracts of land, usually including much land that they rent, not own. Looked at from another perspective, a goal of developing and implementing perennial-polyculture farm operations is to allow for a re-entry of small-scale farmers who have been largely elbowed out of competition because of the high costs of farm operations.

The prospects for this "re-population" of the farming communities (or what used to be farming communities), and for economic improvement generally for all farmers, are very good under perennial-polyculture opera-tions. In such operations there would be no annual purchase or production of new seed for sowing. Less mechanical energy would be expended in field operations, thus reducing fuel costs. Irrigation needs could be reduced. The need for pesticides, and therefore the cost of purchasing them, would drop.

Some of these same ecological and economic benefits that result from a perennial-polyculture approach recently caught the attention of The National Geographic Society. A 2011 article highlights the work of The Land Institute and then enumerates some of the key advantages of perennials:

> Today [some scientists are] trying to breed perennial wheat, rice, and other grains. Wes Jackson, co-founder and president of the Land Institute in Salina, Kansas, has promoted the idea for decades . . . [and] plant breeders in Salina and elsewhere are now crossing modern grains with wild perennial relatives; they're also trying to domesticate the wild plants directly. Either way the goal is crops that would tap the main advantage of perennials – the deep, dense root systems that fuel the plants' rebirth each spring and that make them so resilient and resource efficient – without sacrificing too much of the grain yield that millennia of selection have bred into annuals.
>
> We pay a steep price for our reliance on high yields and shallow roots . . . . Because annual root crops mostly tap into only the top foot or

---

11 Definitions, examples, and descriptions of grains and legumes are offered in Box 1.1, in Chapter 1, *supra*. As noted there, some well-known legumes that are widely used in modern agricultural production are alfalfa and clover – these two are "forage" legumes – and soybeans (soya beans) and lentils, which are grain legumes. Legume plants are notable for their ability to fix atmospheric nitrogen, thanks to a symbiotic relationship with bacteria (rhizobia) found in root nodules of these plants.

so of soil, that layer gets depleted, forcing farmers to rely on large amounts of fertilizers to maintain high yields. Often less than half the fertilizer in the Midwest gets taken up by crops; much of it washes into the Gulf of Mexico, where it fertilizes algae blooms that cause a vast dead zone around the mouth of the Mississippi. Annuals also promote heavy use of pesticides or tillage because they leave the ground bare much of the year. That allows weeds to invade.

Above all, leaving the ground bare after harvest and plowing it in planting season erodes the soil. No-till farming and other conservation practices have reduced the rate of soil loss in the US by more than 40 percent since the 1980s, but it's still around 1.7 billion tons a year. Worldwide, one estimate put the rate of soil erosion from plowed fields at ten to a hundred times the rate of soil production . . .

Perennial grains would help with all these problems. They would keep the ground covered, reducing erosion and the need for pesticides, and their deep roots would stabilize the soil and make the grains more suitable for [undertaking farming operations on] marginal lands . . . . The deep roots and ground cover would also hold on to fertilizer – reducing the cost to the farmer as well as to the environment.[12]

IA3. Health benefits

Recall that in Chapters 2 and 3, I catalogued not only ecological and economic disadvantages inherent in modern extractive agriculture; I also emphasized human-health concerns associated with it. What, if any, human-health *benefits* could come from agroecological husbandry – with perennial-polyculture operations at its core – if it can be satisfactorily developed?

Consider for a moment the principal sources of health problems that were highlighted in section II of Chapter 3. Most prominent among those were certain dangers posed by agricultural chemicals. The perennial-polycultures component of agroecological husbandry would directly address these dangers by reducing the use of such agricultural chemicals.[13] It could also contribute to reforms in biotechnological manipulation, since it would emphasize an appreciation for the complexity of nature.[14]

In sum, agroecological husbandry offers grounds for hope that the ills directly associated with modern extractive agriculture can be largely overcome. From an economic standpoint, an ecological standpoint, and a human-health standpoint, agroecological husbandry – assuming it can in fact be achieved – is preferable to modern extractive agriculture on these grounds.

---

12 Robert Kunzig, *Perennial Solution*, NATIONAL GEOGRAPHIC (April 2011).
13 In this respect, see Alan R. Townsend *et al.*, *Human Health Effects of a Changing Global Nitrogen Cycle*, 1 FRONTIERS IN ECOLOGY AND ENVIRONMENT 240 (2003).
14 See text accompanying notes 100–103 in Chapter 3, *supra*.

IB. Bringing food production closer to nature

The assertions made by Wes Jackson, and by enthusiasts of the work of The Land Institute more generally, include another element beyond the hard-and-fast scientific and physical elements. That is, in addition to asserting that the replacement of modern extractive agriculture with a perennial-polycultures form of farming will bring tangible benefits that can help address deep economic, environmental, and health problems, these advocates of reform state that such a transformation of agriculture would bring it – and all those associated with it – *closer to nature*, and that this is a good thing. Such a claim is evident in the titles to many of Jackson's books: *Nature as Measure, Consulting the Genius of the Place*, and *Becoming Native to This Place*, to name three.

Is this a valid and important claim? In order to address that question, let us consider it from two angles that emerge from a study of Jackson's work. The first angle is a narrow one: Does "bringing food production closer to nature" increase our *ability* to feed ourselves as a society over the long term? The second angle is wider: Does "bringing food production closer to nature" offer some broader form of nourishment, beyond merely keeping people fed? After briefly addressing these questions, I present an even more fundamental question about the "closeness" of food production to nature: What *is* "nature", for purposes of the effort of agroecological husbandry to "bring food production closer to nature"? Specifically, what balance if any is to be sought between a "pristine landscape" and a "manipulated landscape" as our model of the natural world? I explain these two terms below on pages 193–200.

IB1. Keeping people fed over the long term

Wes Jackson, Fred Kirschenmann, and several other observers claim that having a form of food production (along with other aspects of agrarian and rural life) that is "closer to nature" has great value in terms of our *ability* to feed ourselves over the long term, and therefore in terms of our survival as a species. In this regard, Jackson argues that as human beings we simply cannot know enough, or learn fast enough, to rely on our own knowledge to design effective systems of food production. Instead, we must try to "mimic" nature – and specifically, when it comes to grains and other major sources of human food, we must try to "mimic" the prairie ecology.

Jackson explains his argument in various ways. Here is an assortment of his observations and assertions as to how a mimicking of prairie ecosystems can provide solutions:

*   Prairies do not erode but instead accumulate soil; they do not require commercial pesticides, fertilizer, or as much fossil fuel; and they can avoid "epidemics of insects and pathogens" because of diversity.[15]

---

15 Wes Jackson, Becoming Native To this Place 43 (1994). On this point of avoiding epidemics, see *supra* note 2.

- "Learning from nature's wisdom means that we look to natural ecosystems precisely because they have featured recycling of essentially all materials and have run on contemporary sunlight", not on fossilized carbon.[16]
- "[N]one of our technologies can do better than nature's renewability powers."[17]
- "The future of agriculture . . . will depend on knowledge gained from our ecosphere's wild ecosystems."[18]
- "[E]ven though humans may learn faster than nature, natural plant and animal communities have been shaped by climate and evolutionary histories beyond complete human comprehension."[19]

Wes Jackson is of course not the only advocate for fashioning a new "natural-systems" form of agriculture that would mimic nature. Fred Kirschenmann has offered these observations about the "biomimicry movement":

The biomimicry movement . . . [was popularized] by physicist Janine Benyus, [and its] governing principle is that nature, by virtue of its long evolutionary journey, has already solved most of the problems we are grappling with. Consequently, biomimicry suggests that we explore nature as a reservoir of solutions to be discovered, rather than a series of defects to be corrected.

Since life's emergence on this planet some 4 billion years ago, nature has solved a lot of problems by extracting the "actual from the possible." As [Harold] Morowitz reminds us, "nature yields at every level novel structures and behaviors selected from the huge domain of the possible by pruning." Since nature has done this pruning, solutions to many of the problems we want to solve are already available.

Forests serve as a permaculture design of edible landscapes using many-storied cropping systems. What potential might that have for designing agriculture systems that depend less on exogenous inputs and disrupt natural systems less? Marshes serve as models for constructing wetlands that function as waste-treatment facilities while providing refuges for wildlife. Providing habitat for wildlife can provide free ecosystem services to agriculture; more pollinators come to mind. As the Land Institute has demonstrated, the prairie serves as a model for perennial cropping systems that are more resilient, ecologically restorative, and economically profitable than annual monocropping systems that hardly mimic nature at all.

---

16 Wes Jackson, Consulting the Genius of the Place: An Ecological Approach to a New Agriculture 43 (2010) [hereinafter Genius of the Place].

17 *Id.* at 13–14.

18 *Id.*

19 Wes Jackson and Jon Piper, *The Necessary Marriage Between Ecology and Agriculture*, 70 Ecology 1591 (No. 6, Dec. 1989).

These are all ways of redesigning systems to potentially *eliminate* problems rather than introducing external counterforces into systems in an effort to *solve* problems.[20]

The point being emphasized here by Kirschenmann, similar to the observations offered by Jackson on the importance of mimicking natural systems, is this: by being "closer to nature" in our agriculture – that is, by learning from nature's wisdom – we can build on the science of ecology and evolutionary biology in order to feed the world without depleting our natural ecosystems.[21]

IB2. A different nourishment destination

But there is more than this to the arguments of those who urge a move "closer to nature" in our agriculture and more generally. Beyond the claim examined just above – that having a form of food production (and other aspects of agrarian and rural life) that is "closer to nature" has great value in terms of our *ability* to feed ourselves as a global society over the long term, and therefore in terms of our survival as a species – there is also a claim that making agriculture "closer to nature" is personally, culturally, and communally *enriching*. In his 2010 book, *Consulting the Genius of the Place*, Wes Jackson

---

20 Frederick L. Kirschenmann, CULTIVATING AN ECOLOGICAL CONSCIENCE: ESSAYS FROM A FARMER PHILOSOPHER 210–211 (2010) (emphasis in original), citing Janine Benyus's 1977 book *Biomimicry: Innovation Inspired by Nature* and Harold J. Morowitz's *Emergence of Everything*. Kirschenmann offers some related observations on improving agricultural production by making it "closer to nature". For instance, he explains how researchers at the University of Missouri have found that instead of using Bt corn – that is, a variety of corn that has been genetically modified to produce the insecticide Bt in order to control rootworm – it is possible to "insert" into some lines of corn "native-plant resistance to corn rootworms, a resistance based on multiple proteins". Kirschenmann quotes a university official as saying that the researchers "aren't necessarily trying to eradicate corn rootworms completely" but rather to simply hold "rootworm damage below the economic threshold". This approach, Kirschenmann explains, represents "an effort to understand why the rootworm is a pest and find ways to alter the system so that it will no longer be a pest, rather than using an external counterforce to eradicate it". Kirschenmann, *supra*, at 173–174.

21 Kirschenmann explains just what "closer to nature" means in this context by citing Paul Hawken's book *The Ecology of Commerce* for the proposition that there are guidelines "embodying nature's principles [and] that serve as indicators of sustainability". The first principle is that "waste equals food" – that is, sustainability requires that production systems be designed so that waste from one part of the system returns as food for another part of the system. Kirscshenmann notes that in agriculture, "this principle has long been known as the law of return". The second sustainability principle is that operations must, on an ongoing basis, rely predominantly on current resources, not borrowed resources. This, Kirschenmann says, "suggests that we transition from a carbon-based economy to an economy based on hydrogen and sunshine". The third sustainability principle "requires that we pay heed to ecological *restoration*, inasmuch as "[m]aintaining the capacity for renewal is essential to sustainability." Kirschenmann, *supra* note 20, at 179.

explores this claim partly by recounting how he became acquainted with the works of such agrarian-values writer-philosophers as Aldo Leopold and Wendell Berry. For instance, Jackson quotes the closing paragraph of a letter that Wendell Berry sent to Jackson in November 1980, following Berry's reading of Jackson's *New Roots for Agriculture*:

> As one who has farmed with both tractors and teams [of horses or mules, presumably], I would insist . . . that with the use of a tractor certain vital excitements, pleasures, and sensitivities are lost. How much numb metal can we put between ourselves and our land and still know where we are and what we are doing? Working with a tractor is damned dulling and boring. It is like making love in boxing gloves.[22]

A similar view appears in Aldo Leopold's 1949 masterpiece, *A Sand County Almanac*. Leopold urges in that book the adoption of a "land ethic", which he says "enlarges the boundaries of the community to include soils, waters, plants, and animals" and thereby "changes the role of *Homo sapiens* from conqueror of the land community to plain member and citizen of it."[23]

In the minds of Aldo Leopold and Wendell Berry and Wes Jackson and many other enthusiasts of what I refer to as agroecological husbandry, *closeness to the Earth* is an essential ingredient of an enriched life and an enriched culture. If that is true, then it seems obvious that agroecological husbandry as I have described it in this chapter will contribute to it. The use of perennial polycultures, which is a centerpiece of agroecological husbandry, requires close attention to the ecology of a particular place (hence the title of Jackson's book *Consulting the Genius of the Place*). It will require understanding and expertise regarding that place's specific soils, species, schedules, rotations, natural pests, and rainfall. It will both require and facilitate small-scale operations, with the integration not only of various foodcrops but often of livestock as well. Further, agroecological husbandry will feature heavy reliance on more localized energy production and use. Taken together, these and other features of agroecological husbandry will indeed bring a great

---

22 GENIUS OF THE PLACE, *supra* note 16, at 35 (quoting from letter of November 11, 1980 from Wendell Berry to Wes Jackson).

23 Aldo Leopold, A SAND COUNTY ALMANAC 239–240 (1966, 1949). John Ehrenfeld uses similarly sweeping terms to urge a change in the way humans perceive of themselves in nature. Ehrenfeld, who was honored in 1999 with a lifetime achievement award from the World Resources Institute, is quoted by his biographer as saying that "[i]f we are going to address sustainability fully and meaningfully, . . . we must make fundamental shifts in the way we think and the way we organize our society. What's needed is a deep shift in values that is on a par with the Reformation, the Renaissance, the Enlightenment, or the Industrial Revolution." Such a "paradigm shift" as this "takes a movement to reexamine who we are, why we are here, and how we are connected to everything around us." John R. Ehrenfeld and Andrew J. Hoffman, FLOURISHING: A FRANK CONVERSATION ABOUT SUSTAINABILITY (2013).

closeness to nature, at least for those persons engaged in it, and will both produce and require an enriched wisdom.

I offered an odd title as the heading of this subsection – "A Different Nourishment Destination" – in order to draw further on the Metaphor A that I introduced Chapter 4. In that metaphor, a tribe or band of travelers has journeyed for many years toward a destination that they think holds great promise. However, the further they proceed along the route they have chosen, the clearer it becomes that this route poses serious dangers and that in fact the destination on which they had set their sights is one to be avoided and that some *other* destination should be sought instead. Therefore, the group sets about a "bushwhacking" effort to blaze a new trail to a new destination that has come into view.

Likewise, humankind as a whole has journeyed for about the past 10,000 years in search of food security. The "extractive agriculture" route that the human species has followed – emphasizing the development of ever-higher-yielding annual grains grown in monocultures – has seemed during most of the journey thus far to show great promise as an effective route to the desired destination. Recently, though, it has become clear that the destination itself is a dangerous one, because it rests on the supposition that humans can create a system of food production that runs against the grain of nature: instead of relying on the form of plant life that predominates in the most fertile ecosystems of the Earth's surface – perennial plants flourishing in the polycultures that characterize the world's grassland areas – the destination sought by humans has relied on an entirely different model, one that emphasizes annual plants grown in monocultures. Because that destination now appears dangerous for economic, ecological, and other reasons, a different path must be blazed to a different destination.

In order to explain more fully what I mean in referring to the need to seek "a different destination", I offer some observations by Fred Kirschenmann about how scientific advances require frequent, sometimes dispiriting, "midcourse corrections".

> I propose we examine Albert Einstein's analogy of how scientific work progresses. Progress in scientific thought, Einstein argued, was not like tearing down an old barn to build a new one. It was rather more like climbing a mountain in which each new plateau provides an increasingly clearer view of the landscape. *It is always easy, from the vantage point of a new plateau, to see that one might have taken a better path to climb the mountain. But it is pointless to curse the path one took to get there. On the other hand, one should reassess one's position from the new plateau and make midcourse corrections before continuing the climb.*[24]

---

24 Kirschenmann, *supra* note 20, at 124–125 (emphasis added).

I have italicized three sentences in the extract quoted from Kirschenmann in order to highlight just how revolutionary a shift to agroecological husbandry would be, and must be. In the first two of those three sentences, Kirschenmann suggests that even when a new vantage point reveals that a different path should have been taken in climbing the mountain, "it is pointless to curse the path one took to get there". In the third italicized sentence, Kirschenmann says that the climber should reassess his or her position from the new plateau and "make midcourse corrections before continuing the climb".

I believe Kirschenmann's observations *do not* in fact apply to agroecological husbandry. Because it requires a shift from modern extractive agriculture relying on annual monocultures to a fundamentally different system with perennial polycultures at its center, agroecological husbandry is most accurately seen as involving a different destination – a *different mountain* – and not merely a different road or a different path for climbing the same mountain. Whereas "bringing food production closer to nature" can be regarded in part as a more effective means of "keeping people fed over the long term" (the title I gave to the preceding heading), many commentators – Wes Jackson included – urge that a natural-systems agriculture has a different and more compelling goal. That goal is to provide a different form of nourishment: an enriched life and an enriched culture. It is a life and culture that in some ways harkens back to certain benefits and values that evolved in rural, pastoral, agrarian communities over many generations. At the same time, though, it is a life and culture that because of new developments can be free of much of the drudgery and risk of earlier centuries.

In short, perhaps it is *not* pointless to "curse the path one took" to get to the current "plateau" of scientific achievement in bringing high yields (albeit at enormous expense) in grain crops. If "cursing the path one took" thus far is part of a larger reorientation – one that involves selecting not only a new path but also a new *destination* – then I wholeheartedly support such a cursing. Maybe there should in fact be a lot more cursing going on, at least in this respect, than we hear now.

IB3. Beyond the "pristine continent" myth

Before concluding a look at these "closer-to-nature" aspects of agroecological husbandry and examining just what *progress* has been made in that direction, I wish to consider this question: just what *is* "nature" for these purposes? If a key appeal of perennial polycultures is that they mimic the grassland eco-systems that have been largely destroyed to create today's most productive "breadbasket" areas, such as those of the North American Great Plains, it makes sense to ask just how "natural" those grassland ecosystems are. Specifically, I wish to explore what balance if any is to be sought between a "pristine landscape" and a "manipulated landscape" as our model of the natural world.

A common perception of the vast prairies of the North American Great Plains in the years preceding the European invasion of the continent starting in the fifteenth century regards the region as being in pristine condition largely free of human interference or degradation. Under this view, the sheer scope of the grasslands made it possible, maybe even inevitable, that they would be safe from ruin and could quite easily support the relatively small numbers of indigenous peoples who shared them with other species, including of course the great herds of bison.

Recent scholarship has thrown this view into doubt. In particular, some authorities assert that the American continent as a whole, including the North American Great Plains, carried a much larger human population than has heretofore been assumed, and that in fact some portions of the landscape were subjected to an intense, long-term manipulation by those indigenous people. Because of the implications that these new perspectives might have for our consideration of creating a new form of agriculture, I summarize them in the following paragraphs.

Perhaps the most prominent authority espousing these new views on the state of the American landscape(s) before the European invasion is Charles C. Mann. Although most well-known for his book *1491* (recently followed by his book *1493*), Mann gave a synopsis of his views in an article published in the March 2002 issue of *The Atlantic Monthly*. He opens the article with an account of an airplane ride that he took into the Amazon region to conduct research with two anthropologists. As they flew over grassland areas, they saw below them "an archipelago of forest islands, many of them startlingly round and hundreds of acres across". Each of these islands, Mann reports, rose at least ten feet above the floodplain, allowing trees to grow and create small forests. Moreover, those forests "were linked by raised berms, as straight as a rifle shot and up to three miles long", supporting the belief by one of Mann's colleagues "that this entire landscape – 30,000 square miles of forest mounds surrounded by raised fields and linked by causeways – was constructed by a complex, populous society more than 2,000 years ago."[25]

Having given this illustration, Mann then widens the scope of his analysis and introduces his main thesis about the population and manipulation of the Americas in the centuries before Europeans arrived in great numbers:

> [A] cohort of scholars . . . has radically challenged conventional notions of what the Western Hemisphere was like before Columbus. When I went to high school, in the 1970s, I was taught that Indians came to the Americas across the Bering Strait about 12,000 years ago, that they lived for the most part in small, isolated groups, and that they had so little impact on their environment that even after millennia of habitation it remained mostly wilderness. My son picked up the same ideas at his schools. [But a contrary view asserts that] this picture

---

25 Charles C. Mann, *1491*, THE ATLANTIC MONTHLY, March 2002, at 1–2.

of Indian life is wrong in almost every aspect. Indians were here far longer than previously thought, [some] researchers believe, and in much greater numbers. And they were so successful at imposing their will on the landscape that in 1492 Columbus set foot in a hemisphere thoroughly dominated by humankind.[26]

What are some of the implications of that more recent, contrarian, view of indigenous American life that Mann presents? Three specific implications strike me as especially pertinent to the issue of bringing food production "closer to nature", especially in the region of the world most responsible for the rise of industrial agriculture. These implications relate to (i) population scales in that region before 1492, (ii) agricultural and environmental practices before 1492, and (iii) the so-called "pristine myth". I start with the last of these, which Mann describes (and critiques) as follows:

> Much of the environmental movement [in the USA] is animated, consciously or not, by what William Denevan, a geographer at the University of Wisconsin, calls, polemically, "the pristine myth" – the belief that the Americas in 1491 were an almost unmarked, even Edenic land, "untrammeled by man," in the words of the Wilderness Act of 1964, one of the nation's first and most important environmental laws. As the University of Wisconsin historian William Cronon has written, restoring this long-ago, putatively natural state is, in the view of environmentalists, a task that society is morally bound to undertake. Yet if the [more recent scholarship cited above] . . . is correct and the work of humankind [in the Americas before 1491] was pervasive, where does that leave efforts to restore nature?[27]

In order to explore that question about what it means to "restore nature", especially in the Americas, let us first examine just why recent scholarship has suggested that the demographic picture of the American hemisphere was in fact so much larger and more sophisticated than earlier believed. According to Mann, the first scholarly estimate of the indigenous population in North America as of 1491 was made in 1910 by James Mooney, who put the figure at 1.15 million.[28] That view changed, however, several decades later, and especially in 1966. In that year, Mann reports, Henry F. Dobyns drew on birth-and-death data from Mexico and Peru that had emerged since the 1940s – these data came particularly from research by Sherburne Cook and Woodrow Borah – and published a work concluding that terrible plagues

---

26 *Id*. at 2.
27 *Id*.
28 *Id*. at 4. Mann explains that "Mooney's glittering reputation ensured that most subsequent researchers accepted his figure uncritically".

had swept south from Mexico beginning in about 1525 and eliminated vast numbers of people. "So complete was the chaos" caused by this wave of death, Mann asserts, "that Francisco Pizarro was able to seize an empire the size of Spain and Italy combined with a force of 168 men."[29]

The obvious implication of Dobyns' (and other researchers') findings about this early-sixteenth-century wave of death in the Americas is that the Western Hemisphere's population in the years before Europeans came in 1492 must have been much higher than the figure of 1.15 million offered earlier by James Mooney. Indeed, as Mann reports, "Dobyns calculated [that] the Western Hemisphere held ninety to 112 million people. Another way of saying this is that in 1491 more people lived in the Americas than in Europe."[30]

In order to provide further support for this view, Mann offers another illustration of the massive die-off of indigenous Americans just following European contact:

> [Supporting this "large pre-1942 population" thesis is the work of] Charles Hudson, an anthropologist at the University of Georgia who spent fifteen years reconstructing the path of the expedition [of Hernando de Soto in the 1540s. When that expedition] crossed the Mississippi a few miles downstream from the present site of Memphis, . . . the Spaniards were watched by several thousand Indian warriors. [Then de Soto's men entered] into what is now eastern Arkansas, through thickly settled land – "very well peopled with large towns," one of his men later recalled, "two or three of which were to be seen from one town.". . .
>
> After Soto left, no Europeans visited this part of the Mississippi Valley for more than a century. Early in 1682 whites appeared again, this time Frenchmen in canoes. One of them was Réné-Robert Cavelier, Sieur de la Salle. The French passed through the area where Soto had found cities cheek by jowl. It was deserted – La Salle didn't see an Indian village for 200 miles . . . . [It seems, then, that] Soto "had a privileged glimpse" of an Indian world, Hudson says. "The window opened and slammed shut. When the French came in and the record opened up again, it was a transformed reality. A civilization crumbled. The question is, how did this happen?"
>
> [The answer probably involves pigs that were brought by de Soto, some of which escaped and then] were able to transmit their diseases to wildlife in the surrounding forest . . . . [After all, pigs can] disseminate anthrax, brucellosis, leptospirosis, taeniasis, trichinosis, and tuberculosis.

---

29 *Id.* (citing Dobyns' paper "Estimating Aboriginal American Population: An Appraisal of Techniques With a New Hemispheric Estimate", published in the journal *Current Anthropology*). Mann goes on to note that "[s]mallpox was only the first epidemic [to emerge; it was followed by others that] ravaged the remains of Incan culture." *Id.*

30 *Id.*

Pigs breed exuberantly and can transmit diseases to deer and turkeys. Only a few of Soto's pigs would have had to wander off to infect the forest. [This could serve then to infect the Indians.][31]

So much for population. What about agriculture? That is, if the Americas were home to such large populations as the more recent research suggests, how does this fact bear on the picture of pre-1492 agriculture in the Americas, and particularly on the notion that a "natural-systems" agriculture would be a return to some "pristine" form of nature?

Charles Mann suggests in this regard that human history can be seen as being "marked by two world-altering centers of invention". The first is the Middle East, where a Neolithic revolution started about 10,000 years ago and allowed the Sumerians to become "the world's first civilization". The second is central Mexico, where Mann says "Indian groups independently created nearly all of the Neolithic innovations, writing included".[32] Then came 1492.

When Columbus appeared in the Caribbean, the descendants of the world's two Neolithic civilizations collided, with overwhelming consequences for both. [It should be borne in mind that] in agriculture . . . [the American Neolithic civilization] handily outstripped the children of Sumeria. Every tomato in Italy, every potato in Ireland, and every hot pepper in Thailand came from this hemisphere. Worldwide, more than half the crops grown today were initially developed in the Americas. [Moreover,] Indian agriculture long sustained some of the world's largest cities. The Aztec capital of Tenochtitlán dazzled Hernán Cortés in 1519; it was bigger than Paris, Europe's greatest metropolis. The Spaniards gawped like hayseeds at the wide streets, ornately carved buildings, and markets bright with goods from hundreds of miles away . . . . [Moreover,] Central America was not the only locus of prosperity [built on agriculture]. Thousands of miles north, John Smith, of Pocahontas fame, visited Massachusetts in 1614, before it was emptied by disease, and declared that the land was "so planted with Gardens and Corne fields, and so well inhabited with a goodly, strong and well proportioned people . . . [that] I would rather live here than any where."

---

31 *Id.* at 7–8. Anticipating skepticism about the ability of a few pigs to wreak this much destruction, Mann explains that "Indians were fresh territory for many plagues, not just one", so that a compounding effect probably was felt; moreover, "Indians are characterized by unusual homogenous MHC types" that serve as defenses against bacteria and viruses, so that most of the natives were susceptible to the same diseases, which could therefore sweep quickly through vast numbers of people. *Id.* at 9. For an analysis of the interplay of epidemic disease and other aspects of European invasion of North America, see Russell R. Menard, Epidemics and Enslavement: Biological Catastrophe in the Native Southeast 1492–1715 (2009).

32 Mann, *supra* note 25, at 13.

Smith was promoting colonization, and so had reason to exaggerate. But he also knew the hunger, sickness, and oppression of European life. France . . . experienced seven nationwide famines in the Fifteenth century and thirteen in the Sixteenth. Disease was hunger's constant companion. During epidemics in London the dead were heaped onto carts . . . and trundled through the streets . . . .

The Earth Shall Weep, James Wilson's history of Indian America, puts the comparison bluntly: "the western hemisphere was larger, richer, and more populous than Europe."[33]

To what extent, if at all, did this (allegedly) large population of Native Americans manipulate their environment in order to produce food? Mann explains that they broke land areas up into small plots, as Europeans did, but that they also managed surprisingly large agricultural areas as well.

Like people everywhere, Indians survived by cleverly exploiting their environment . . . [and they] often worked on such a grand scale that the scope of their ambition can be hard to grasp. [They] reshaped entire landscapes to suit their purposes. A principal tool was fire, used to keep down underbrush and create the open, grassy conditions favorable for game. Rather than domesticating animals for meat, Indians retooled whole ecosystems to grow bumper crops of elk, deer, and bison. The first white settlers in Ohio found forests as open as English parks – they could drive carriages through the woods. Along the Hudson River the annual fall burning lit up the banks for miles on end; so flashy was the show that the Dutch in New Amsterdam boated upriver to goggle at the blaze like children at fireworks. In North America, Indian torches had their biggest impact on the Midwestern prairie, much or most of which was created and maintained by fire. Millennia of exuberant burning shaped the plains into vast buffalo farms . . . . Is it possible that the Indians changed the Americas more than the invading Europeans did? "The answer is probably yes for most regions for the next 250 years or so" after Columbus, William Denevan [a University of Wisconsin geographer] wrote, "and for some regions right up to the present time."[34]

---

33 *Id.* at 13–14. Mann hastens to add that it would hardly be claimed, at least on reflection, that America was "a disease-free paradise", since Indians "had ailments of their own, notably parasites, tuberculosis, and anemia" and faced difficult lives that made life-spans "only as long as or a little longer than those in Europe, if the evidence of indigenous graveyards is to be believed." *Id.* at 14.

34 *Id.* at 15. Thus, according to Mann and those he cites, it might be that "about 12 percent of the nonflooded Amazon forest was of anthropogenic origin – directly or indirectly created by human beings". *Id.* at 18. Indeed, some researchers "think it's all human-created", Mann reports. *Id.*

In sum, recent studies suggest that before 1492, both in the northern and in the southern hemispheres of the Americas, populations were much larger than earlier thought, and that *before* they suffered such a massive die-off from European diseases, for which they had inadequate defenses, those people heavily manipulated the landscape, partly in ways that we can still discern or infer and surely also in ways that we will never know. Mann concludes his article with these observations about the supposed "pristine" character of pre-European America:

> Crediting Indians with the role of keystone species has implications for the way the current Euro-American members of that keystone species manage the forests, watersheds, and endangered species of America. Because a third of the United States is owned by the federal government, the issue inevitably has political ramifications. In Amazonia, fabled storehouse of biodiversity, the stakes are global.
>
> Guided by the pristine myth, mainstream environmentalists want to preserve as much of the world's land as possible in a putatively intact state. But "intact," if the new research is correct, means "run by human beings for human purposes." Environmentalists dislike this, because it seems to mean that anything goes. In a sense they are correct. Native Americans managed the continent as they saw fit. Modern nations must do the same. If they want to return as much of the landscape as possible to its 1491 state, they will have to find it within themselves to create the world's largest garden.[35]

In my view, this perspective on what *might* have been the state of the ecosystems in the Americas – including the current "breadbasket" areas of North America where modern extractive agriculture exists in its most robust form – prompts several observations. First, the assertions offered by Charles Mann have yet to be universally accepted;[36] further studies will no doubt

---

35 *Id.* at 21–22.
36 For an example of some disagreement with elements of Mann's findings, see *Paper challenges 1491 Amazonian population theories*, available on the website of the American Association for the Advancement of Science at: www.eurekalert.org/pub_releases/2007-03/fiot-pc1030607.php. According to this source, Mark Bush of the Florida Institute of Technology has argued that although there is indeed evidence of large human populations in certain specific areas in North and South America, "when you start to look away from known settlements, you may see very long-term local use" in small pockets rather than a broad management of ecosystems. According to Bush, "[t]hese people didn't stray very far from home, or from local bodies of water for several thousands of years." On this basis, Bush rejects Mann's suggestion that there were vast tracts of human-manufactured landscapes in the Western Hemisphere. The source explains that Mann's suggestion has been relied on in an attempt to influence conservation policy in the Americas. Specifically, the proposition that millions of people once populated the Americas – and that in Amazonia, at least, the rainforest is the product of long-term human use – has been used as farmers and loggers as justification for clearcutting rainforests. Their argument, that the ecosystem already experienced

sharpen our understanding of the role humans played in various epochs in various parts of the world in shaping (and often degrading) ecological conditions – and the significance this had had for agriculture.

Second, regardless of these uncertainties, it is surely true that humans have managed ecosystems to a very large extent, especially through agriculture, and that in many cases these efforts at management have failed miserably. This is a point that has been highlighted in David Montgomery's book on "dirt" and the erosion of civilizations.[37]

Third, the fact that human efforts at managing ecosystems for food production have often ended in disaster does not necessarily mean that such efforts are *intrinsically* bound to fail. A key assertion of the work done by Wes Jackson and others is that ecosystems management *can* be conducted successfully by humans *if* carried out in a way that mimics natural systems. I have explored that assertion at some length earlier. Indeed, the proposal to shift away from modern extractive agriculture to agroecological husbandry rests on this assertion.

Fourth, a central truth of ecosystems management must be that of *limits*. Limits on the appropriate level of human population – a topic I touched on briefly in projecting a future based on agroecological husbandry (in section III of Chapter 4) – imply that humans need not and should not exercise their full power as "keystone species", to use Charles Mann's term, in the entire ecosphere. Instead, as I also touched on briefly in discussing an agroecological future, some areas of the world *should not*, in my view, be "run by human beings for human purposes", as Mann urges. To the contrary: large areas should be returned carefully to wilderness and kept that way with as little human interference as possible.[38]

However, for those areas of the world that *are* to be devoted to human settlement and use, it *is* appropriate that humans create an improved managed environment to produce food and to secure the values that I have described earlier by citing Aldo Leopold's views of a "land ethic". The process I am proposing in this book for doing so is the widespread adoption of agroecological husbandry.

---

vast landscape disturbance and proved resilient, relies on the ubiquitous influence of Pre-Columbian people, the suggestion that Bush's work rejects. More recent scholarship, however, has rejected Bush's conclusions on various grounds, thus providing support for Mann's findings. Jeremy Hance, *Experts Dispute Recent Study that Claims Little Impact by Pre-Columbian Tribes in Amazon* (2012), http://news.mongabay.com/2012/07/experts-dispute-recent-study-that-claims-little-impact-by-pre-columbian-tribes-in-amazon/.

37 David R. Montgomery, Dirt: The Erosion of Civilizations (2007).

38 Such an attention to *limits* is consistent with what Frederick Kirschenmann has called "the agriculture of restraint". See text accompanying note 118 in Chapter 3, *supra*. Another way of expressing the idea that humans need not and should not exercise their full powers to affect the rest of the natural world is to think in terms not just of an agriculture of restraint but also of a consumption of restraint, an economics of restraint, and a technology of restraint. In Chapters 7 and 8, I urge a "sovereignty of restraint" as well.

An obvious question is whether this is possible. I turn next to that question: How realistic is it, as a practical matter, to count on developing a form of grain production – agroecological husbandry – that can supplant modern extractive agriculture? Do we have the science?

II. Prospects for success in moving toward agroecological husbandry

I have tried to establish thus far in this chapter that agroecological husbandry *should* be embraced in order to avoid a range of deeply damaging consequences of modern extractive agriculture. For example, the economic unsustainability and ecological degradation that characterize modern extractive agriculture could be reversed through widespread practice of agroecological husbandry – and particularly through a perennial-polyculture approach to food production. But is such a thing *possible?* In the following paragraphs I explore the scientific progress that has been made thus far in this direction and how this progress is accelerating.

IIA. Progress thus far – general acceptance of perennial polyculture science

Progress on the "science side" of agroecological husbandry, and specifically in developing herbaceous perennial seed-producing polycultures, can be seen both (i) in the general acceptance of the scientific underpinnings of the effort and (ii) in the actual outcomes of field research. I will address those two points in order.

The cohort of scientists involved in the research to develop perennial herbaceous seed-producing polycultures has expanded steadily. Over the years since The Land Institute started its work, several books and articles have emerged that adopt its goals. These include the following:[39]

- A 1992 book, *Farming in Nature's Image*, is described as a work "expanding on an idea conceived at The Land Institute" in order to "outline an innovate agricultural model in which farms mimic native ecosystems".[40]
- A 2001 Australian study, *Polyculture Production*, emphasizes – based on an extensive review of the literature then available – both (i) the benefits of polycultures in crop production and (ii) the expectation that perennials grown in polycultures will prove ultimately the most viable approach to producing food, feed, and fiber.[41] The report cites several reasons for

---

39 This account draws in part on GRASSLANDS, *supra* note 1, at 220–221. Some of its content also appears in *Next Synthesis, supra* note 2, at 207–210.

40 See Judith D. Soule and Jon K. Piper, FARMING IN NATURE'S IMAGE [back cover] (1992).

41 See generally Larry Geno and Barbara Geno, POLYCULTURE PRODUCTION: PRINCIPLES, BENEFITS AND RISKS OF MULTIPLE CROPPING LAND MANAGEMENT SYSTEMS FOR AUSTRALIA (2001). As for polycultures, the report offers this summary:

this, including the fact that "[p]olycultures, especially perennial, utilise far less soil disturbance, use canopy closure to shade weeds and use physical and chemical competition among root systems to reduce weed loads",[42] as well as the fact that "the bulk of a perennial polyculture biomass is what sustains the 'protective structure of the ecosystem' ".[43]

- Numerous articles in international scientific journal have reported on research regarding perennials grains and/or the use of polycultures in agriculture.[44]
- A two-day workshop on perennial agriculture took place in New York in February 2012, sponsored by the Carbon Farming organization and Gaia Northeast LLC.[45]
- A special conference was convened at FAO headquarters in Rome in the summer of 2013, and chaired by researchers from The Land Institute, to discuss developments in perennialization of grains (see pages 206–225 for more details of that conference and its proceedings).
- A group of researchers assembled by the Royal Academy (UK) has endorsed perennial crops, citing these reasons:

Extensive individual case studies and broader review papers provided evidence that polycultures yield more from smaller areas and as an approach to agricultural intensification they suffer less than energy, gene, or expertise intensive strategies to increase production. Their yield is more stable over space and time than monocultures in terms of income level, stability and risk. Polycultures generally showed advantages while highlighting methodological problems in assessing subsistence versus market values and the role of full cost accounting in addressing benefits and costs.

Id. at ix. As for perennials, the report notes the work done at The Land Institute and elsewhere and identifies as one of the conclusions emerging from other studies that "[p]erennial plants in polyculture will likely be the base of successful mimics". *Id.* at 55.

42 *Id.* at 53. As Tim Crews of The Land Institute has explained to me, a more complete description of this "competition" would draw attention not just to the root system but more generally to the system of roots *and shoots*.

43 *Id.* at 56.

44 See, e.g., J. K. Piper, *Neighborhood Effects on Growth, Seed Yield and Weed Biomass for Three Perennial Grains in Polyculture*, 4 JOURNAL OF SUSTAINABLE AGRICULTURE (1993), no. 2, at 11–31 (as cited in Geno and Geno, *supra* note 41, at 96); K. Wilson, 1993, *Perennial Wheat is Achievable*, AUSTRALIAN FARM JOURNAL, Feb. 1993, at. 75 (also as cited in Geno and Geno, *supra* note 41, at 99); Jerry D. Glover *et al.*, *Increased Food and Ecosystem Security via Perennial Grains*, 328 SCIENCE 1638–1639 (2010); N.G. Jordan, *et al.*, *Sustainable Development of the Agricultural Bioeconomy*, 316 SCIENCE 1570–1571 (2007); T. S. Cox, *Breeding Perennial Grain Crops*, 21 CRITICAL REVIEWS IN PLANT SCIENCES 59–91 (2002); Christopher B. Field, *Sharing the Garden*, 294 SCIENCE 2490–2491 (2001) (noting that in general, "nature's ecosystems have greater net primary production than the human-managed 'modern' systems that have been imposed on the areas wrested from the control of those natural systems"). See also Crews *et al.*, *infra* note 126, and Cox *et al.*, *infra* note 64.

45 For details, see the Carbon Farming website at http://carbonfarmingcourse.com/workshops/perennial-agriculture#5.

Perennial crops would store more carbon, maintain better soil and water quality and would be consistent with minimum till practice. These crops would also manage nutrients more conservatively than conventional annual crops, and they would have greater biomass and resource management capacity. [46]

An overall endorsement of perennials has also been given by the National Research Council of the National Academy of Sciences. A 2010 report titled *Toward Sustainable Agricultural Systems in the 21st Century* includes the following observations:[47]

- "Perennial crops have deeper root systems than annuals, providing access to more water and nutrients."
- "Perennials have greater ability to maintain the health and fertility of a landscape over longer periods of time."
- Perennials also have a longer growing season, allowing more sunlight to be captured by the crop.
- "Perennial plants reduce erosion risks, sequester more carbon, and require less fuel, fertilizer, and pesticides to grow than their annual counterparts."
- "[P]erennial production systems [have been found to have] resulted in lower net greenhouse-gas emissions [– at least of carbon dioxide –] than annual cropping systems."[48]
- Some researchers have "found perennial systems to be effective at reducing the potential for ground water contamination by nitrate leaching."

---

46 The Royal Society, Reaping the Benefits: Science and the Sustainable Intensification of Global Agriculture 25 (2009) (asserting that for these reasons, "[t]he conversion of annual crops into perennial plants could help sustain the health of cultivated soils").

47 See Committee on Twenty-First Century Systems Agriculture, National Research Council, Toward Sustainable Agricultural Systems in the 21st Century 250–251 (2010). Several of the quoted passages cite articles and studies authored by scientists associated with The Land Institute.

48 As Tim Crews of The Land Institute has explained to me, although this observation about greenhouse gas emissions is accurate in respect of carbon dioxide, it is not yet clear that perennial cropping systems *per se* result in lower emissions of nitrous oxide and methane than result from annual cropping systems. However, perennial crops grown *in polycultures* do hold this potential. See note 10, *supra*. For a summary of the main ways in which agricultural and livestock production release greenhouse gases into the atmosphere, and how (in general) this release of greenhouse gases bears on climate change, see subsection IB8 of Chapter 3, *supra*. As indicated there, one set of calculations suggests that 13.5% of world greenhouse gas emissions are currently attributable to agricultural activities, including livestock-related activities, and that nearly all of those emissions take the form of the more high-impact greenhouse gases – namely, nitrous oxide and methane. For more detailed discussion of methane emissions from livestock operations, see notes 79–80 in Chapter 3, *supra*.

- Overall, "[r]eplacing some of the single-season crops with perennials would create large root systems capable of preserving the soil and would allow [grain production] in areas currently considered marginal."
- "Recent advances in plant breeding, such as the use of marker-assisted breeding, genomic in situ hybridization, transgenic technologies, and embryo rescue, provide new opportunities for plant breeders to select for desired characteristics."
- "In [recent] years, plant breeders in the United States, Argentina, Australia, China, India, and Sweden have initiated plant genetic research and breeding programs to develop wheat, rice, corn, sorghum, sunflower, intermediate wheatgrass, and other species as perennial grain crops."

In short, the effort to develop herbaceous perennial seed-producing poly-cultures has gathered considerable momentum in recent years. It has been recognized as both *worthwhile*, because of the benefits that such an approach would bring, and *feasible*, in part because of recent developments and tech-niques that promise to increase the efficiency of the exercise, especially the plant-breeding aspects of it. As the title to this subsection indicates, there is "general acceptance of perennial polyculture science" and a commitment to ensuring that the research goals are in fact met.

IIB. Progress thus far – specific development of perennial polyculture crops

IIB1. Challenges and solutions

Let us turn now to specifics: what progress has been made in actually developing the science that would make it possible to supplant, or even replace, modern extractive agriculture with agroecological husbandry? To address that question, consider first this basic scientific reality that bears on developing *perennials* that would take the place of *annuals* for use in producing food crops: at least as they appear on the Earth today, perennials concentrate the great majority of their effort in growing *roots*, not *seeds*. That is, perennial populations depend for their survival not so much on the distribution of seeds that will germinate and grow in a later season but rather on the building of below-ground root mass that will endure severe weather, such as cold and drought. Consequently, the current "inventory" of perennial plants on our planet can produce only a small fraction of the total calories required for direct consumption by a growing human population. In addition, wild *annual* species of plants (that is, those annuals that have not been domesticated) can produce only a small fraction of the volume needed for human consumption.

Given this reality, a central challenge of natural-systems-agriculture researchers is to develop new strains of perennials, in order to enlarge the current "inventory", so that the broad array of *annuals* used as human food crops can be matched, at least in part, by an adequately broad array of *perennials* used as human food crops. Beyond that, however, the underlying philosophy of natural-systems agriculture, especially as posited by Wes Jackson and like-minded researchers, insists that because such a new form of agriculture should "mimic the prairie" (that is, the ecology of grasslands), which primarily involves *polycultures*, not *monocultures*, then it is necessary not only (i) to develop perennials, and the modified ecosystems that perennials facilitate by increasing soil organic matter and the soil organisms that come with it, but also (ii) to ascertain how these perennials can be grown in polycultures, in order to give adequate yields while reaping the other benefits enumerated earlier.

On the basis of this multiple-step reasoning, four related questions have been seen as central to the work of researchers at The Land Institute:

- Can perennialism and high grain yield go together?
- If so, can the performance of a polyculture of perennials compare favorably with the performance of a monoculture of perennials using minimum inputs?
- Under what conditions might such an ecosystem be free from fertility inputs?
- Can we manage such complexity well enough to keep pests from out-competing us?

Researchers at The Land Institute and elsewhere began to work on the "hardware" – that is, on actually breeding perennials – around 2001. How? As a very general matter, two principal approaches can be used to develop "new" perennial crops: (i) perennializing domestic annuals and (ii) domesticating wild perennials. A more complete answer, however, was provided recently by David Van Tassel of The Land Institute:

> There are currently four main routes to combining the yield of an annual crop with the longevity of a wild perennial. The first three require wide hybridization between an annual species and a perennial species. Under the "triticale option", a full or partial amphiploid with whole chromosomes of both species is created. If the chromosome constitution remains stable in successive generations, a new species has been created. Option 2 utilizes introgression breeding to move "domestication genes" from the cultigen into the wild perennial background. Option 3 is similar, but aims to move "perennialism genes" from the wild species into the genetic background of the cultivated species. The fourth option – domestication – attempts to find rare alleles in the primary gene pool of a wild perennial that confer favourable "domestication syndrome" traits such as reduced shattering, increased threshability, etc.[49]

---

49 David L. Van Tassel, *et al.*, *Evaluating Perennial Candidates for Domestication: Lessons from Wild Sunflower Relatives*, appearing as Chapter 9 in Perennial Crops Proceedings, *infra* note 50, at 114.

As noted in the following text, very significant progress has now been made toward developing *perennial* versions of several types of food crops, with special attention being given to sorghum, wheat, intermediate wheatgrass ("Kernza"®), sunflowers, and rice. The other part of the scientific project – developing *polycultures* – involves ecology and evolutionary biology, and a new phase of this work was initiated recently at The Land Institute. This polyculture research is taking place with more developed protocrops – something that was not possible before the "hardware" (in the form of new perennials) was developed.

In order to gauge the progress made thus far toward developing perennial crops, the FAO convened a special workshop in the summer of 2013 at its headquarters in Rome. The workshop was co-chaired by researchers from The Land Institute, and its main aim was to discuss work underway on the perennialization of grains and legumes (and some other crops). The report of that workshop[50] began by explaining why intense study of perennials is now particularly crucial:

> Sustainable production systems have always relied on the flexibility, efficiency, and multiple functions of perennial trees and forages grown in combination with annual cereals, legumes, and oil species. But over the last 50 years, research, technologies and markets have focused mainly on a limited number of *annual* species to meet the increased demand for food. Furthermore, the primary focus was on increasing grain yields with reduced attention given to the social, environmental and market consequences of these food systems.
>
> However, food security and agriculture are now entering an era characterized by scarce and depleted resources, climate change, price volatility and job losses. To adapt to this new era, agricultural technologies, science and markets have to be transformed to ensure sufficient food is produced for a growing population, while meeting simultaneously the economic, social and environmental challenges of the twenty-first century.
>
> *Perennial cereals, legumes and oil species* represent a paradigm shift in agriculture and hold great potential to move towards sustainable production systems. Today, most agronomic practices used to grow annual crops require excessive water consumption, significant amounts of synthetic mineral fertilizers, labour, emissions of CO_2[,] and [they] disrupt natural biological processes. Perennial crops instead are more rustic, improve

---

50   See FAO, Perennial Crops for Food Security: Proceedings of the FAO Expert Workshop (2014), available on the website of The Land Institute at: https://landinstitute.org/wp-content/uploads/2014/11/PF FAO14_intro.pdf [hereinafter Perennial Crops Proceedings]. As explained by one of the principal conveners, "[f]orty-one people from ten nations participated in the meeting. The goals were to aggregate and put in context all research done on perennial grains up to now, begin forming a researchers' network, and plan for more extensive, well-coordinated and better-supported research in coming years." *Id.* at 2.

soil structure and water retention capacity[,] and [they] contribute to increase climate change adaptation and mitigation practices and promote biodiversity and ecosystem functions.[51]

A principal organizer of the 2013 FAO workshop – Stan Cox of The Land Institute – drew special attention to the benefits of perennial grains for soil protection:

> Perhaps the most important benefit of perennial agriculture will be the protection and development of healthy soil ecosystems that can ensure food security over the long term. That would achieve an important reversal of what is now an alarming trend. In 2011, the Food and Agriculture Organization (FAO) released its report *The State of the World's Land and Water Resources for Food and Agriculture*, concluding that 25 percent of the world's food-producing soils are highly degraded or are rapidly being degraded and that if moderately degraded soils are included, one-third of Earth's entire endowment of cropland is under threat. Loss of productive soil is most severe in the Himalayan and Andean regions; semi-arid tropical regions of Africa and India; rice-growing lands of Southeast Asia; and areas of intensive and industrialised farming throughout the world. *Eighteen countries* – nine of them in sub-Saharan Africa and four in Southeast Asia – *now see more than half of their entire land area degrading rapidly*. And while past production increases have received much of their impetus from irrigation, future freshwater resources are in at least as much trouble as the world's soils.[52]

In Chapter 1, I characterized the Earth's soil as a "thin skin of life stretched over a rock", and I emphasized both (i) the fragility of that "skin of life" and (ii) its profound significance to the existence of life, including human life, on Earth. What Stan Cox is highlighting in the passage excerpted immediately above is that even if there were no other benefits that would flow from the use of perennials rather than annuals for purposes of producing the bulk of the world's food, the dangerous state of the world's soils would give urgent reason for conducting intense research now on perennials.

The report of the FAO workshop proceeds then to demonstrate that such research has in fact been proceeding briskly and producing important results. A range of crops – rice, wheat, sorghum, maize (corn), sunflowers, and others – have been the subject of extensive field studies and laboratory research.[53] In the following paragraphs I provide a rough progress report on

---

51 *Id.* at iii (emphasis added).

52 *Id.* at 1 (emphasis added).

53 As Stan Cox also writes in the report of the FAO workshop, the researchers who gathered in Rome in 2013 offered "a tour of the ecological and human landscape where perennial grains currently grow and are being developed" – with a shared ultimate goal of "[beginning to chart]

those efforts, drawing mainly from two sources: (i) the FAO workshop proceedings and (ii) research updates issued by The Land Institute and its staff members. For the most part, this "progress report" is current as of late 2015. It does not, however, purport to be *comprehensive*. Instead, I intend for it to be *indicative* of the momentum already achieved.

IIB2. Wheatgrass and wheat

Researchers working first at Rodale Institute[54] and then at The Land Institute have created a domesticated grain-producing perennial – referred to as intermediate wheatgrass – that is now being sold under the trademark Kernza®. Wholegrain Kernza® flour has been used to make breads, pancakes, and tortillas[55] – and, more recently, beer.[56] The success of this form of intermediate wheatgrass is attributable to the fact that researchers at The Land Institute have, over the course of two cycles of selection, been able to increase seed yield by about 77 percent and seed mass by about 23 percent[57] – an increase that, if extrapolated, could, in roughly 20 years, "result in yields of approximately 2,500 kg/ha^{-1}, similar to annual wheat [yields] in Kansas".[58]

---

a course that will take perennial-grain research – which now consists of geographically and scientifically diverse, conceptually bold, but largely autonomous and independent projects – and weave them into a global network that can make this new agricultural concept a reality." *Id.* at 3.

54 For information about the Rodale Institute, located in Kutztown, Pennsylvania, see http://rodaleinstitute.org/. See also *infra* note 57 and accompanying text.

55 See *The Land Institute: What We Do and Why* (undated). See also THE LAND INSTITUTE ANNUAL REPORT 2011 (June 2010–July 2011), at 4. Each bag of Kernza® grain comes with recipes and an explanatory note indicating that initial analyses show higher levels of certain vitamins and minerals than annual whole wheat.

56 For information on the use of Kernza® in beer production, see *Kernza in Food and Drink: Brewing, Baking, and the Search for Heavy Molecules to Make Light Bread*, LAND REPORT, Spring 2013, at 19–22, https://landinstitute.org/wp-content/uploads/2014/11/LR-105.pdf. For information about the use of Kernza® in making whiskey, see *infra* note 60 and accompanying text.

57 Lee R. DeHaan *et al.*, *Current Efforts to Develop Perennial Wheat and Domesticate* Thinopyrum intermedium *as a Perennial Grain*, appearing as Chapter 6 in PERENNIAL CROPS PROCEEDINGS, *supra* note 50, at 72, 73. As noted there, scientists at The Land Institute achieved these gains in studies that selected for yield per head, increased seed mass, free threshing ability, reduced height, and early maturity. The 77 percent and 23 percent gains referred to here are for plants grown in a solid stand. In offering a historical explanation of the 30-year development of what became Kernza®, the researchers explain that (1) intermediate wheatgrass was selected in 1983 for domestication by the Rodale Institute following studies of nearly a hundred species of perennial grasses, (2) the Rodale Institute performed two cycles of selection beginning in 1988, and then (3) breeding work began at The Land Institute in 2002. *Id.* at 72.

58 *Id.* at 78. The researchers point out, however, that the improved intermediate wheatgrass (Kernza®) produced thus far "is currently inferior to wheat for most potential uses" due to "a small grain size and . . . low gluten quality", the latter of which limits the utility of Kernza® in baking raised breads. Both of these limitations are being addressed by further research at The

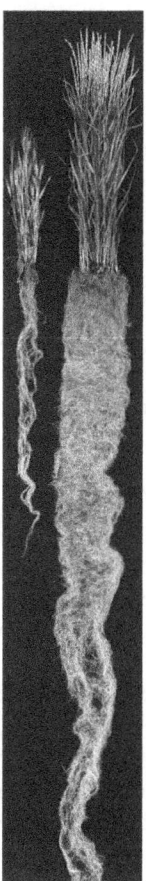

Figure 5.1 Annual wheat versus perennial intermediate
wheatgrass[59]

Figure 5.1 illustrates the difference between annual wheat
and perennial intermediate wheatgrass as developed at The
Land Institute. The most prominent feature of the image,
which has become something of an icon for The Land
Institute, is the latter's enormous root system.

The significance of this progress was underscored recently
when a 90-acre production plot of Kernza® perennial grain
was planted at the University of Minnesota, a long-time
collaborator with The Land Institute. The Minnesota field is
intended to supply grain to two California companies –
Patagonia Provisions (an environmentally conscious food
company) and Ventura Spirits Company, which will use the
Kernza® to make whiskey.[60]

In sum, the research at The Land Institute in developing
grain-producing perennial wheat-grass has shown excellent
results.[61] Although the production of Kernza® has been
primarily carried out in monocultures thus far, the ultimate
aim of The Land Institute scientists and others is to

Land Institute. *Id.* at 84. A recent update from researchers there indicates that three additional
selection cycles were recently performed but that "populations developed from these later selection cycles have not been evaluated in a replicated multi-location trial, as the results of the first
two cycles were" – so in that respect, the breeding work is ongoing. Personal correspondence
with Lee DeHaan and Tim Crews of The Land Institute, December 2015.

59 Printed with permission from Steve Renich and The Land Institute. A color version of this image
can be seen as an illustration in a 2013 article by two researchers at The Land Institute. See David
Van Tassel and Lee DeHaan, *Wild Plants to the Rescue*, 101 AMERICAN SCIENTIST 218, 219 (2013).

60 See Lee DeHaan, *Two Firms Eyeing Kernza Food Products*, in THE LAND INSTITUTE ANNUAL
REPORT FALL 2013, at 8–9. For a description of the Kernza®-based whiskey that Ventura Spirits
Company plans to market, see http://venturaspirits.com/home/. The website highlights the fact
that although whiskey production typically uses annual grains that "are often destructive to local
landscapes", Kernza Whiskey will be made from "a perennial wheat grass developed by The
Land Institute to help build and maintain healthy soil with its incredibly deep root system." *Id.*

61 A late-2014 update from The Land Institute reported that studies at the University of Minnesota
showed Kernza® grain yields to be twice as great as regular intermediate wheatgrass – and that
"besides producing grain that can be used as food for humans, Kernza produced as much non-
grain biomass [for use as forage] as switchgrass" produced. *In Search of Elusive Haploids*, in THE
LAND INSTITUTE ANNUAL REPORT FALL 2014, at 12.

incorporate it into polycultures – and, to this end, research is currently underway to develop Kernza®-legume bicultures.[62]

Another aspect of wheat and wheatgrass research at The Land Institute involves wide-hybrid cross-breeding[63] of wheatgrass with wheat to create a perennial version of wheat itself. This type of breeding work started around 2001 after testing many old perennial wheat lines with frustrating results. Achieving perenniality, though, is complicated. Some perennial wheat lines, produced from bread wheat and wheatgrass hybridization, can be made to perform in ways that are similar to those of annual wheat, with 50 percent to 70 percent of the grain yield found in annual wheat. However, the perenniality level is highly influenced by environmental factors, and this hinders the adoption of these wheat lines in production.[64]

Several steps have been taken in order to make progress toward robust perennial wheat. Some such steps, and their results, include the following:

- Researchers have developed DNA markers to assist the research and selection of perennial wheat. A few years ago, 56 new DNA markers were developed that can be used to identify all 21 wheat chromosomes in the "wheat × wheatgrass" hybrid plant populations. The loss of more wheat chromosomes has been seen to be associated with stronger perenniality.[65]

- Researchers have found that wheat flowering time genes were related to insufficient perenniality in perennial wheat. They noted specifically that nearly all hybrid plants flowered multiple times a year starting from the second year. This characteristic makes plants die quickly under stressed conditions such as heat and cold. Natural and induced variations are now being exploited in order for hybrid plants to flower only once a year, as wheatgrass does.[66]

- By evaluating a broad range of perennial wheat lines, researchers have found that an old line called "MT-2" could live in the field for years without special management. Distinct from regular perennial wheat lines, MT-2 was developed by means of a cross between spring durum wheat and intermediate wheatgrass. More than 2,500 individual "wheat × wheatgrass" hybrid plants have been investigated for seed

---

62 For details, see text accompanying note 115, *infra*.

63 For an explanation of wide-hybrid cross-breeding, see *infra* note 69.

64 For background on efforts at achieving perennialization in wheat (and several other crops), see Thomas S. Cox, Jerry D. Glover, David L. Van Tassel, Cindy M. Cox, and Lee R. DeHaan, *Prospects for Developing Perennial Grain Crops*, 56 BIOSCIENCE 649–659 (2006).

65 THE LAND INSTITUTE ANNUAL REPORT FALL 2011 (JUNE 2010–JULY 2011), at 3, 4. For further details about efforts by scientists at The Land Institute to develop perennial wheat, see *Faster Track to Perennial Wheat*, LAND REPORT, Spring 2011, at 5–7. For information about the perennial-wheat-breeding work underway at Washington State University, see its website at: http://plantbreeding.wsu.edu/research.html.

66 See *Too-Eager Hybrids Face Danger*, in THE LAND INSTITUTE ANNUAL REPORT FALL 2015, at 7.

fertility, seed size, perenniality, vigor, flowering time, disease resistance, and other traits. A plant called "12F620" has been found to survive for four years, with a regrowth habit similar to that of wheatgrass. Researchers at The Land Institute have concluded that it is possible to obtain long-lived perennial wheat through wheat and wheatgrass hybridization.

• Thousands of crosses have also been made between a variety of other wheat and wheatgrass species. Among these, the new crosses between winter durum and intermediate wheatgrass have shown excellent combination of seed weight, perenniality, and seed fertility. Researchers at The Land Institute are participating in a global experiment that involves testing several especially promising breeding lines at 21 locations across nearly a dozen countries.[67]

Research on perennial wheat and related species also has produced promising results for scientists at institutions other than The Land Institute,[68] including on other continents. In Australia,[69] for instance, germplasm samples imported from the USA, Russia, and China (along with germplasm from the Australian

---

67 Shuwen Wang, *Work In Lab, The World Aid Wheat Research*, in THE LAND INSTITUTE ANNUAL REPORT FALL 2014, at 10. See also *Too-Eager Hybrids Face Danger*, *supra* note 66.

68 For a description of research into perennial wheat and perennial wheatgrass in Michigan, see Sieglinde Snapp, *Agriculture Redesign through Perennial Grains: Case Studies*, appearing as Chapter 11 in PERENNIAL CROPS PROCEEDINGS, *supra* note 50, at 148, 149–151. For further information regarding similar efforts underway at Michigan State University, see its website at: www.kbs. msu.edu/people/faculty/snapp/perennial-wheat.

69 See Len J. Wade, *Perennial Crops: Needs, Perceptions, Essentials*, appearing as Chapter 1 in PERENNIAL CROPS PROCEEDINGS, *supra* note 50, at 6 et seq. See also Philip J. Larkin and Matthew T. Newell, *Perennial Wheat Breeding: Current Germplasm and a Way Forward for Breeding and Global Cooperation*, appearing as Chapter 4 in PERENNIAL CROPS PROCEEDINGS, *supra* note 50, at 39, 42. These authors suggest that of the various methods for developing perennial wheat-like plants, "[h]ybridisation between annual grain crops and perennial relatives offers [a better] avenue to combine the traits of perenniality and grain quality in a new crop species" than do the efforts to domesticate perennial wheat-like plants, which is the approach that has been followed by The Land Institute thus far in developing its perennial wheatgrass called Kernza®. Another set of researchers participating in the FAO conference offer this abbreviated comparison of these two alternative approaches:

> Two major approaches are being used to develop perennial small grains: wide hybridization and domestication. The two approaches present unique strengths and challenges. Wide hybridization involves crossing an annual grain such as wheat with related perennial species. Wide crosses will in theory make available genes controlling traits such as yield, seed size, free threshing ability, and quality, which have been accumulated in current grain crops. With wheat the challenge has been to obtain cytogenetic stability in wide hybrids while preserving perenniality and domestication traits. Direct domestication of wild perennials has the potential benefit of working with populations that are vigorous perennials. However, the necessary genetic variation for domestication may be lacking in perennial species, or substantial time may be required for selection to achieve adequate seed size, yield, or other domestic traits.

DeHaan *et al.*, *supra* note 57, at 73.

Wheat Collection) have been studied in an effort to find "putative perennial wheat derivatives" that could be developed into varieties that would have the capacity to regrow after harvest and to yield grain over successive years. Some of the "wheat × wheatgrass" derivatives studied in these efforts did in fact show the ability to produce grain over three successive years, and the regrowth of the plants "was associated with the presence of at least one whole genome equivalent (14 chromosomes) from the perennial donor species", raising prospects that the research can yield a perennial wheat with commercially and ecologically appropriate traits.[70]

A significant element of the perennial-wheat research in Australia is its concentration on developing a "dual-purpose" crop – that is, perennial wheat that not only produces grain but that also provides additional forage for livestock during summer and autumn. Such a "dual-purpose" approach "could increase farm profitability substantially" – for farming operations that include livestock, that is – and could, because of its forage value, reduce the relative grain yield required from the perennial wheat to just 40 percent of the yield that would be required if annual wheat without the forage potential were produced instead.[71]

This "dual-purpose" approach[72] is only one way in which researchers working to develop perennial wheat and wheat-related crops (such as intermediate wheatgrass) seem to be taking a broad and flexible perspective on their efforts, exploring a range of ways in which those efforts should be undertaken and evaluated. One of the research teams participating in the summer 2013 FAO workshop emphasized the need for flexibility in explaining the alternative approaches to developing perennial wheat and other small grains:

> "Perennial wheat" hybrids and intermediate wheatgrass are currently at far ends of a spectrum [in terms of the efforts to create high-yielding, high-quality perennial small-grain crops. Ultimately, these efforts will be

---

70 Wade, *supra* note 69, at 10. Wade explains that the research involves evaluating "forage biomass production under serial defoliation" (that is, how much material is produced for livestock to eat over several grazings and seasons), on changes in the development of root bundles, and under varying levels of water availability, including dehydration tolerance. *Id.* at 11.

71 *Id.* at 10. Wade goes on to explain that the results of the work thus far have been encouraging enough to warrant extending the tests to include previously untested germplasm and to study outcomes from an additional fourth year of those derivatives that survived the first three years – and, more generally, to propose a specific "breeding approach for developing adapted perennial wheat for Australian farmers". *Id.* at 11.

72 Tim Crews of The Land Institute has explained to me that in addition to the dual-use concentration on grain production and forage for livestock, another form of dual-use research underway in a collaborative effort by The Land Institute and the University of Minnesota combines grain production with bio-fuel production. Moreover, research is underway to investigate grazing and grain systems in Iowa, Minnesota, Wisconsin, Ohio, New York, Kansas, and Colorado. Personal correspondence with Lee DeHaan and Tim Crews of The Land Institute, December 2015.

most effective if they involve] simultaneously working from both ends of the spectrum . . . . Whether the perennial grain of the future will be wheat with grass-like traits added or a perennial grass with the addition of wheat-like traits is a question that we need not dwell on. *What is clear is that progress is being made at an accelerating pace toward the day when useful, high-yield, long-lived small grains are a reality.*[73]

IIB3. Rice

Researchers are pursuing two principal approaches in order to develop perennial rice: (i) introducing genes from a perennial form of rice – in particular, the species *Oryza longistaminata*, which has rhizomes – into the domesticated Asian rice, *Oryza sativa*, which is an annual (but was derived from perennial ancestors); (ii) actually domesticating *Oryza longistaminata* and/or *Oriza australiensis*, another perennial species also with rhizomes.[74] Promising results have emerged from the first of these techniques. As reported by one participant in the 2013 FAO workshop, "diligent breeding work over many years at the Yunnan Academy of Agricultural Sciences [in China, with Dr. Fengyi Hu as program leader,] has resulted in the production of interspecific progenies that have both long rhizomes and high fertility".[75]

Another set of researchers represented at the FAO workshop provided specific details of the successes realized thus far in the work at Yunnan Academy:

[A particular cultivar] was chosen as the *O. sativa* parent, as this cultivar was widely grown in lowland or upland [settings and provided high yields with] good grain quality . . . and with disease resistance to rust etc. It was crossed with *O. longistaminata*, and the F1 was intermediate in characteristics between the parents . . . . [P]rogress was made in

---

73 DeHaan *et al.*, *supra* note 57, at 87 (emphasis added).

74 Eric J. Sacks, *Perennial Rice: Challenges and Opportunities*, appearing as Chapter 2 in PERENNIAL CROPS PROCEEDINGS, *supra* note 50, at 16, 17. As Sacks explains, there are two species of domesticated rice – *Oryza sativa* from Asia and *Oryza glaberrima* from Africa – and both of them "are derived from perennial ancestors, either directly or via an annual intermediary." *Id.* at 18. In the case of Asian rice, the perennial progenitor was *Oryza rufipogon*, which gains its perenniality from stolons, not from rhizomes. *Id.* Stolons, which are horizontal-creeping plant stems or runners that take root at points along their length to form new plants, are "not well-suited to surviving drought because they may be exposed to sun and dry air on the surface of the soil if grown under upland conditions, or under rainfed production during the dry season. In contrast, the rhizomes of *O. longistaminata* are protected from desiccation by insulating soil. Additionally, *O. longistaminata*, which can form large monocultures in the wild, is more vigorous than *O. rufipogon*." *Id.* at 23. Accordingly, most attention for perennialization of rice by introducing perennial genes into annual rice has focused on *Oryza longistaminata*.

75 *Id.* at 23.

developing perennial rice [through a process of intercrossing and backcrossing,] and selection for desired traits.[76]

As an overall conclusion, the researchers report, "perennial rice breeding is on the way and [so far there] are five perennial rice (PR) lines, namely PR23, PR57, PR129, PR137 and PR139, that have been bred.[77] On the strength of the progress they report, those researchers offer this more general set of observations:

A successful perennial rice breeding program has been established at Yunnan Academy of Agricultural Sciences, with the line PR23 now in pre-release testing in Yunnan Province. Development of perennial rice is consequently at the forefront of perennial grain development, and will hopefully act as an incentive to success in other species. The time is ripe to build on this success by establishing a consortium of perennial crop researchers, supported by a suite of donors to ensure the continuity of efforts needed for success in this challenging but important endeavour.[78]

Prospects are also favorable for the second technique – that is, domesticating a species of rice that is naturally perennial. This is particularly true in respect of the need to withstand drought conditions.[79] Researchers already know about some specific challenges to such a domestication program. For instance, both *Oryza longistaminata* and *Oryza australiensis* have long, vigorous rhizomes, which makes them potentially invasive; accordingly, the domestication efforts need to keep rhizome lengths moderate in order to reduce the risk of such invasiveness. Still, the sequence of many of the genes that control for certain key traits in annual rice, such as those for non-shattering and semi-dwarfism, are already known, and this "should allow for relatively rapid domestication of wild rice species via targeted screening of germplasm and selection."[80]

---

76 Shila Zhang *et al.*, *The Progression of Perennial Rice Breeding and Genetics Research in China*, appearing as Chapter 3 in PERENNIAL CROPS PROCEEDINGS, *supra* note 50, at 27, 30–31.

77 *Id.* at 28.

78 *Id.* at 37. A recent report from the field research in China, where 660 hectares is planted in the PR23 perennial rice, indicates that the plants have now undergone their third harvest with no decline in stands or in yield. As currently developed, PR23 is not rhizomatous and will probably work well only in continuously flooded conditions. However, ongoing research there does involve efforts to select for hardier rhizomatous rice that would be suitable for other conditions – namely, intermittently flooded, rainfed, and dryland conditions. Personal correspondence with Tim Crews of The Land Institute, December 2015.

79 Sacks, *supra* note 74, at 24 (explaining that "adaptation of rice to perennial growth while surviving seasonal drought may be a case in which domestication is the best option [because both] *O. longistaminata* and *O. australiensis* have rhizomes which may enable the plants to survive in a dormant state during drought").

80 *Id.*

In addition to these two techniques for developing perennial rice, some attention has been given to the encouragement of rice "ratooning" – that is, the process by which rice, once harvested, can be managed to create a viable re-growth crop without re-planting. Although not a true perennialization of rice, some observers suggest that rice ratooning could provide useful increases in yield.[81]

IIB4. Corn (maize)

Perennialization of maize (corn) has proven particularly challenging. Two of the participants in the 2013 FAO workshop outlined some reasons for the difficulty – reasons that also apply more generally to the other perennial crops described here:

> For a maize plant to act as a perennial a number of conditions must be met: the plant must not senesce at the end of the season; the plant must accumulate energy into structures that can overwinter; the plant must be able to prevent its overwintering structures from both freezing and degradation; and finally the plant must remobilize energy from the overwintering structures into new regrowth in the spring. For perennial maize to actually be grown by farmers this must all be accomplished in a plant that can produce adequate grain yields and can switch back and forth between reproductive and vegetative growth – in addition to the other suites of traits that farmers desire such as disease resistance, nutrient and water use efficiency. [82]

Based on these general observations, the same researchers explained that in the case of maize, "[t]he breeding progress for selection on these many different quantitative traits is certain to be slow, but [we] feel all must be selected simultaneously to avoid breeding into a corner."[83] They also offered this observation about maize-perennialization efforts:

> [I]t has taken over 100 years and billions of research dollars to get maize [as an annual crop] to reach its current productivity. Developing high-yielding perennial maize is likely to take an additional 10 to 40 years, at which point hypotheses of yield and eco-system service comparisons can be formally tested.[84]

---

81 Ronald D. Hill, *Back to the Future! Thoughts on Ratoon Rice in Southeast and East Asia*, appearing as Chapter 26 in PERENNIAL CROPS PROCEEDINGS, *supra* note 50, at 362 et seq.

82 Seth C. Murray and Russell W. Jessup, *Breeding and Genetics of Perennial Maize: Progress, Opportunities, and Challenges*, appearing as Chapter 8 in PERENNIAL CROPS PROCEEDINGS, *supra* note 50, at 103, 107.

83 *Id.* at 107.

84 *Id.* at 109. Tim Crews of The Land Institute has indicated that some research underway in Texas focuses on developing perennial, *non-grain-producing* biofuel corn for the far-southern part of the USA.

It is worth noting that there might well be some specific reasons for giving less attention to perennialization of maize than to the perennialization of other crops: maize plays a larger role than any other crop does in modern extractive agriculture, especially because of its use in producing livestock feed, high-fructose corn syrup, animal food products (such as dog food), heating fuel, ethanol fuel, medicines, plastics, fabrics, adhesives, and more.[85] At least some of these uses – livestock feed in large cattle-feedlot operations, for instance, and the production of ethanol – would not be favored in an agricultural system aiming for lower greenhouse gas emissions and at a reintegration of livestock and cropping operations.

Thus far, researchers at The Land Institute have not concentrated much effort on creating a perennial form of corn. They have pointed out, though, that recent years have seen a "great advance in mapping genetic code and in computer statistical modeling for researchers".[86] Some of that new technology has been put to use by scientists to press forward on the efforts referred to previously to develop a strain of perennial wheat, and will help also in perennial corn research.

IIB5. Sorghum

Despite the evident significance of wheat, rice, and corn as staple food crops – a significance that I highlighted in Chapter 1 – perhaps it is sorghum that holds the most surprising promise, the most "headroom", for contributing to global food security. Consider these facts[87]:

- *Significance for food security.* Sorghum is native to the Sahelian region of Africa, where it is estimated to provide about one-third of calories in the human diet. This makes it essential to some of the poorest and most food-insecure regions of the world – regions that for various reasons have recently experienced dramatic degradation of soils.
- *Drought resistance.* Sorghum has a strong and inherent drought tolerance – also important to much of Africa and to certain parts of the southwestern

---

85 See Box 1.2 in Chapter 1, *supra*, for further details about the uses of maize.

86 Scott Bontz, *To Make a Perennial Corn*, LAND REPORT, Spring 2011, at 13–15 (explaining the possibilities of crossing *Zea diploperennis*, a rare perennial species, with annual corn familiar to most US farmers in order to make a suitable perennial strain that can stand up to hot summer temperatures and that will not have a "photoperiod problem" under which it will not flower during certain days with long hours of sunlight).

87 Most of the details, and all of the quoted passages, in these bullet-points are drawn from Andrew H. Paterson *et al.*, *Viewpoint: Multiple-Harvest Sorghums Toward Improved Food Security*, appearing as Chapter 7 in PERENNIAL CROPS PROCEEDINGS, *supra* note 50, at 90–95. For further discussion of sorghum perennialization efforts, see Stan Cox, Timothy Crews, and Wes Jackson, *From Genetics and Breeding to Agronomy to Ecology*, appearing as Chapter 12 in PERENNIAL CROPS PROCEEDINGS, *supra* note 50, at 158, 162–166.

USA, particularly with climate change. In fact, sorghum "is the most drought-resistant of the world's top five cereals".

- *Potential for perennialization.* The annual form of sorghum that is currently used as a food crop – *Sorghum bicolor* – is fairly closely related to Johnsongrass (*Sorghum halapense*), and therefore it is likely that crosses between certain forms of the two relatives could successfully yield plants that express desirable qualities of both. (For instance, Johnsongrass reproduces in part by rhizomes, which not only allows the species to be a hardy perennial but also, recent research suggests, may host nitrogen-fixing endophytic bacteria; by contrast, no members of the cultivated species, *Sorghum bicolor*, are rhizomatous.) There is some irony, or at least curiosity, in the promise shown by a cross between sorghum bicolor and Johnsongrass, since the latter of these is widely known among farmers as a highly aggressive weed that can choke out desirable crops.

- *Multiple uses.* Sorghum production can be directed not only at yielding grain but at three other benefits as well: sweet sorghum can be used for livestock feed; sweet sorghum can also be used for sugar production (with sugar yields comparable to those of sugarcane); and bioenergy sorghums are a promising source of cellulosic ethanol. Moreover, in most areas of sub-Saharan Africa, after grain harvest, sorghum stalks are used for fodder or in construction.

- *Room for improvement.* Despite the attractive attributes of sorghum as a crop, and the promise it holds for improvement through perennialization, the effort to improve sorghum in its annual form "lags [behind] that of maize, wheat and rice, each of which have more than doubled in world-wide average yield in the last 38 years [whereas] [s]orghum yields only gained 51 percent."

Researchers at The Land Institute have focused their attention on (i) exploring the traits of conventional sorghum (especially *Sorghum bicolor* and one of its relatives, *Sorghum propinquum* (a perennial), and on (ii) crossing *Sorghum bicolor* with *Sorghum halapense* in order to combine desirable qualities of both – including of course perenniality. With financial support from two grants totaling more than $650,000,[88] The Land Institute researchers continue their selection and backcrossing work, with yield and kernel weight gradually increasing. Although the yields for the perennial sorghum are not yet at the levels of the yields for annual commercial grain sorghum,[89] the chief sorghum

---

88 Stan Cox, *Grants Spur Sorghum Work*, in The Land Institute Annual Report Fall 2013, at 7. Both grants – one from the US Agency for International Development and the other from the USDA – are part of larger grants awarded to the University of Georgia, a long-time collaborator with The Land Institute. One focuses on whether perennial sorghum strains developed in Kansas and Georgia would be suitable for Africa and South Asia; the other funds genomic analysis of sorghum populations. *Id.*

89 *Id.*

researcher at The Land Institute reported in 2014 that "[a]fter more than 10 years of careful selective breeding, we're having success" in diverting above-ground biomass into grain by reducing the number of stems and increasing the size of the heads.[90]

A side benefit of the sorghum-related research summarized here is that "information about *rhizomatousness from a few models (that are also major crops) may extrapolate broadly to a wide range of taxa.* Successes in clarifying the genetics of perenniality in this genus, may accelerate progress in implementing new production systems in other genera."[91] In other words, lessons learned in the context of the sorghum-related research might facilitate perennialization of other food crops.

IIB6. Oilseeds

Researchers at The Land Institute have given considerable attention to the development of perennial oilseeds – especially Maximilian sunflower (*Helianthus maximiliani*)[92] and *Silphium integrifolium* (rosinweed, or simply silphium). That work has shown good progress thus far, in that stronger stalks and resistance to diseases are being cross-bred into domesticated versions of these strains.[93]

Before offering further details about *what* has been done in this area of research, it is worth considering *why* research into oilseed perennialization is important. A team of researchers participating in the 2013 FAO workshop offers this explanation:

> Perhaps the most compelling argument in favour of perennial oilseed research is simply that oil crops currently occupy 281 million hectares [which is] about 26 percent of the world's acreage devoted to staple crops (cereals, pulses, oil crops and roots/tubers) and about 18 percent

---

90 *Our Sorghum Is Hitting the Road*, in THE LAND INSTITUTE ANNUAL REPORT FALL 2014, at 6 (quoting Stan Cox, and also noting that the success thus far warrants expanding the research to several other climates, including those in Mali, Ethiopia, Uganda, South Africa, China, and elsewhere in the USA). For an account of how expanded facilities at The Land Institute created a "population explosion" in sorghum specimens – thus providing genetic raw material for breeding more productive perennials and gathering fundamental gene-mapping data for potential perennial sorghum that will give high yields and resiliency – see THE LAND INSTITUTE ANNUAL REPORT FALL 2011 (JUNE 2010–JULY 2011), at 3. For another explanation on progress at The Land Institute in developing a high-yield perennial sorghum, see *Overcoming Negative Relationships*, LAND REPORT, Spring 2011, at 7–15.

91 Paterson *et al.*, *supra* note 87, at 98 (emphasis in original).

92 The Maximilian sunflower is a principal North American perennial species of sunflower. The annual sunflower, *Helianthus annuus* L., is a major grain crop in North America.

93 See THE LAND INSTITUTE ANNUAL REPORT FALL 2011 (JUNE 2010–JULY 2011), at 4. A self-pollinating version of sunflowers has also been discovered, but it is rare and has not received priority in the research at The Land Institute.

of all land classified as arable or under permanent crops .... For whatever reasons, humanity has chosen to devote a large proportion of the planet's land area to oil crops and providing more perennial oil crop options could reduce the ecological disservices associated with annual oil crops. Encouragingly, 41 percent of all vegetable oil already comes from perennials such as olive, oil palm, and coconut ....

However, these perennial oil crops are all tropical or subtropical tree-like species and virtually all the oil crops in temperate regions are annuals such as soybean, sunflower, flaxseed and canola. Furthermore, tree-like crops are slow to establish and expensive to harvest and immobilize moisture and nutrients from the environment for extended periods.[94] Perennial oilseed forbs could provide additional options even in the tropics, supplying some of the ecosystem services of the tree crops, as well as new services, such as fodder, and new harvesting and management (mowing, grazing and burning) options.

In both tropical and temperate regions, the development of perennial oilseeds, particularly those with diverse end-uses and capable of producing on marginal lands where annual cropping is at a high risk of failure, has the potential to increase the productivity of global agricultural systems.[95]

The same researchers then explain the four main alternative approaches to developing perennial food crops. These four approaches, as noted on page 205,[96] involve various ways of either (i) perennializing domesticated annuals or (ii) domesticating wild perennials. Researchers focusing their efforts on sunflowers have attempted all of the four approaches at various research centers. The conclusion drawn thus far is that wide hybridization (the first of the four approaches) is unlikely to succeed. By contrast, they characterize as "slow but steady" the successes achieved recently in domesticating perennial varieties of sunflowers. Moreover, they emphasize that "domestication can now be accelerated" because of technological breakthroughs:

Advances in sequencing technology and statistical modelling are starting to eliminate barriers to the genetic dissection of complex quantitative and plastic ecological traits even in species which lack genomic resources or known pedigree .... [In addition,] [w]hole genome GS [gene sequencing] using markers development techniques such as

---

94 Moreover, some such crops, such as those found in palm oil plantations, have contributed to deforestation and the ills that accompany it. These include habitat degradation, species extinction, climate change, and abuses of the rights of indigenous peoples. For details, see the relevant page on the website of the World Wildlife Fund, www.worldwildlife.org/industries/palm-oil. See also, www.saynotopalmoil.com.

95 Van Tassel *et al.*, *supra* note 49, at 112, 114.

96 The four approaches are described in the text accompanying note 49, *supra*.

genotype-by-sequencing and RAD-seq have been effective in predicting selection candidates for complex traits such as grain yield in wheat . . . and are becoming more common in both plant breeding and ecological genomics.[97]

In sum, there seems to be substantial yet-untapped potential for domesticating wild versions of sunflower with attributes that are most important for agricultural use. Just what are those attributes? The researchers participating in the FAO workshop identified several such attributes and offered the following suggestions about which ones should be regarded as most important in future research:

> We suggest, from both experience and deduction, that seed yield and the classic domestication traits such as shattering may often be less important in selecting [sunflower] perennial species for domestication than genetically complex traits such as growth rate, phenology [effects of seasonal and cyclical changes] or branching pattern.[98]

Researchers at The Land Institute have concentrated their attention to some extent on the last of those traits – branching pattern. In 2007, they observed a single unusual individual Maximilian sunflower (*Helianthus maximiliani*) in their nursery with a "strongly reduced branching and a single, larger than average head on each main branch." Within a few years, large numbers of single-headed plants were being evaluated "and some individuals had yields per head exceeding anything in the normally branching population".[99] Further work on this line and on other lines has continued in Kansas and Manitoba,[100] and research on other sunflowers has also been undertaken elsewhere.[101]

---

97 Van Tassel *et al.*, *supra* note 49, at 117.

98 *Id.* at 113.

99 *Id.* at 122. As for other research efforts with Maximilian sunflowers – that, is selecting for other attributes – the data gathered as of early 2014 "suggest that yield has increased" over several cycles spanning about a decade. *Id.*

100 The participants in the FAO Workshop explained that "Maximilian sunflower is being domesticated simultaneously in Kansas and Manitoba. While genetic materials have been exchanged between these programs, the day length and climate differences make it probable that two distinct domesticated races are likely to emerge." *Id.* at 118.

101 Earlier research, now discontinued, in Wisconsin concentrated on *Helianthus pauciflorus*, which (along with the related *Helianthus tuberosus*) is a tuberous perennial sunflower species (or, by some accounts, subspecies) known to hybridize naturally with annual sunflowers – and, in some cases to have larger seeds in heads that are positioned above the leafy portion of the plant, making mechanical harvesting easier. Although research began a few years ago in Kansas on *Helianthus pauciflorus*, its sensitivity to drought led The Land Institute to phase out its evaluation of this species in Kansas. *Id.* at 123–124.

In addition to studying sunflowers, The Land Institute is also studying another oilseed called wholeleaf rosinweed, or *Silphium integrifolium*, which has fairly large seeds that are very similar in flavor and composition to sunflowers.[102] Although researchers at The Land Institute had not given special emphasis to working on breeding perennial rosinweed (silphium) for a number of years, the exceptionally hot, dry summers that Kansas experienced in 2012 and 2013 put on display the high drought tolerance of this species. This fact, along with the observation that some selected plants in The Land Institute test plots were producing three times the number of seeds per head as those found in typical wild plants, prompted researchers to concentrate more effort on rosinweed. They have found thus far that the plants have two particularly distinct sets of root systems. One set of roots spreads horizontally just below the surface of the soil, capturing the moisture produced by light rains; another set of "thicker roots plunge nearly straight down to capture water six or more feet in the ground."[103] A depiction of rosinweed (silphium) appears in Figure 5.2 as the fifth plant from the right end of the drawing. Also depicted there are numerous other prairie plants of North America – and, at the far left end of the drawing, the relatively short-rooted Kentucky bluegrass.

Some of the continuing research at The Land Institute on rosinweed is focusing on improved screening techniques (that is, in looking for favorable characteristics in the plants) by testing for correlations between (i) seedling growth rates immediately after germination and (ii) the vigor shown in adult plants in terms of the number of leaves, stem thickness, time of flowering, and biomass produced. Developing these screening techniques might facilitate dramatically more efficient test cycles.[104]

Silphium research is underway also at the University of Minnesota and at USDA offices in Fargo, North Dakota. The researchers in Minnesota have these areas of emphasis:

- understanding the genetics basis of important domestication-related and yield-related traits in silphium;
- examining the agronomy of silphium – and in particular, comparing the yields of silphium when grown with different level of plant nutrients, or when planted in different densities (plants per hectare) – as well as other ways of improving planting techniques; and

---

102 David Van Tassell, *Silphium: The Heat Is Turned Up*, in THE LAND INSTITUTE ANNUAL REPORT FALL 2013, at 11.

103 *Id.* As a consequence of these and other factors, The Land Institute's breeding nursery "yielded more than 300 pounds [of rosinweed] per acre in 2012, a drought year in which many annual sunflower plots were not worth harvesting." *Id.*

104 *Taking Measure of* Silphium, in THE LAND INSTITUTE ANNUAL REPORT FALL 2014, at 11. Research on silphium (rosinweed) is being conducted in test plots in Kansas, Wisconsin, North Dakota, Vermont, and Minnesota, as well as in South America. Personal correspondence with David Van Tassell of The Land Institute, September 2015 and December 2015.

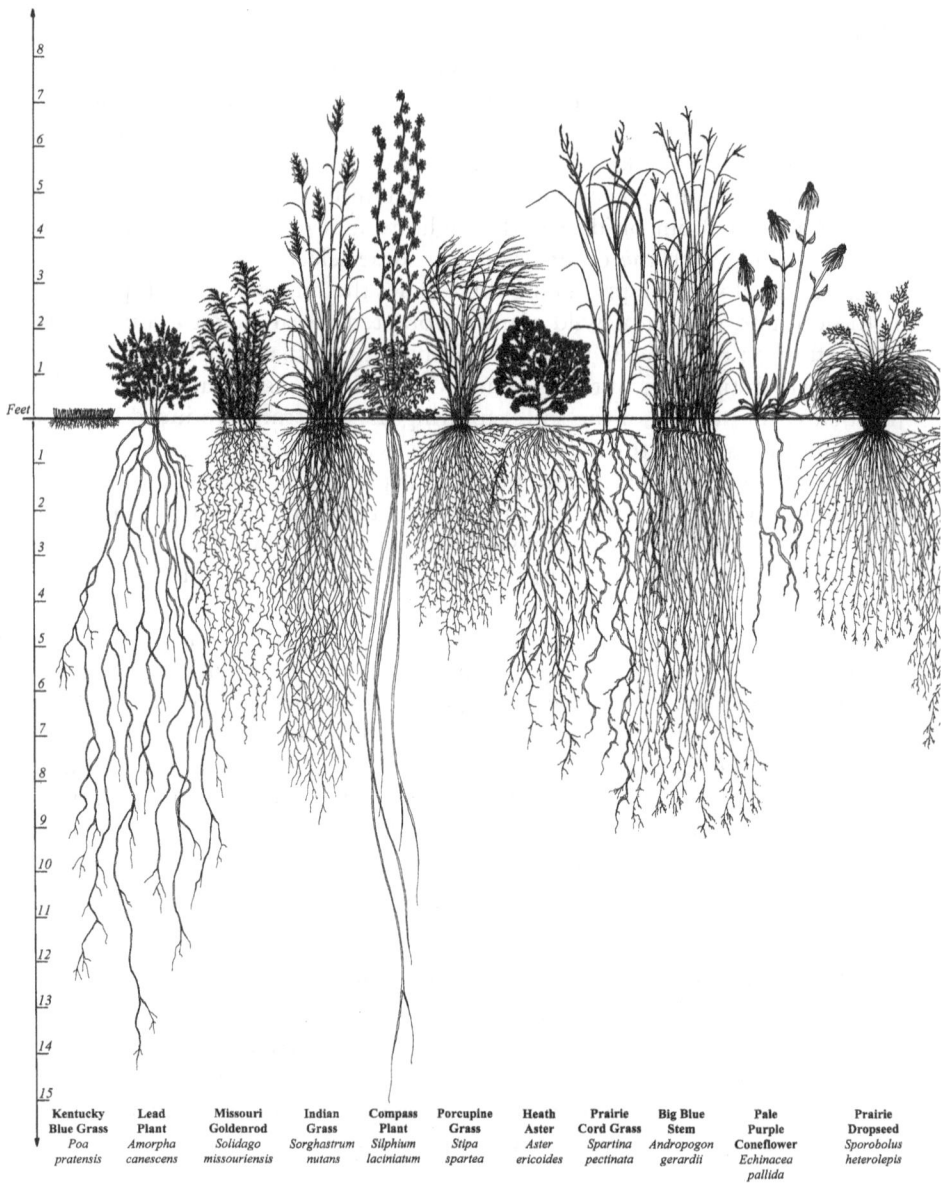

Kentucky Blue Grass	Lead Plant	Missouri Goldenrod	Indian Grass	Compass Plant	Porcupine Grass	Heath Aster	Prairie Cord Grass	Big Blue Stem	Pale Purple Coneflower	Prairie Dropseed
Poa pratensis	*Amorpha canescens*	*Solidago missouriensis*	*Sorghastrum nutans*	*Silphium laciniatum*	*Stipa spartea*	*Aster ericoides*	*Spartina pectinata*	*Andropogon gerardii*	*Echinacea pallida*	*Sporobolus heterolepis*

Root Systems of Prairie Plants

Figure 5.2 Rosinweed (*silphium integrifolium*) and other North American prairie plants[105]

105 Root System Prairie Plants, copyright 1995, Heidi Natura and Conservation Research Institute – reprinted with permission from Conservation Research Institute (Cedarburg, Wisconsin), www.conservationresearchinstitute.org.

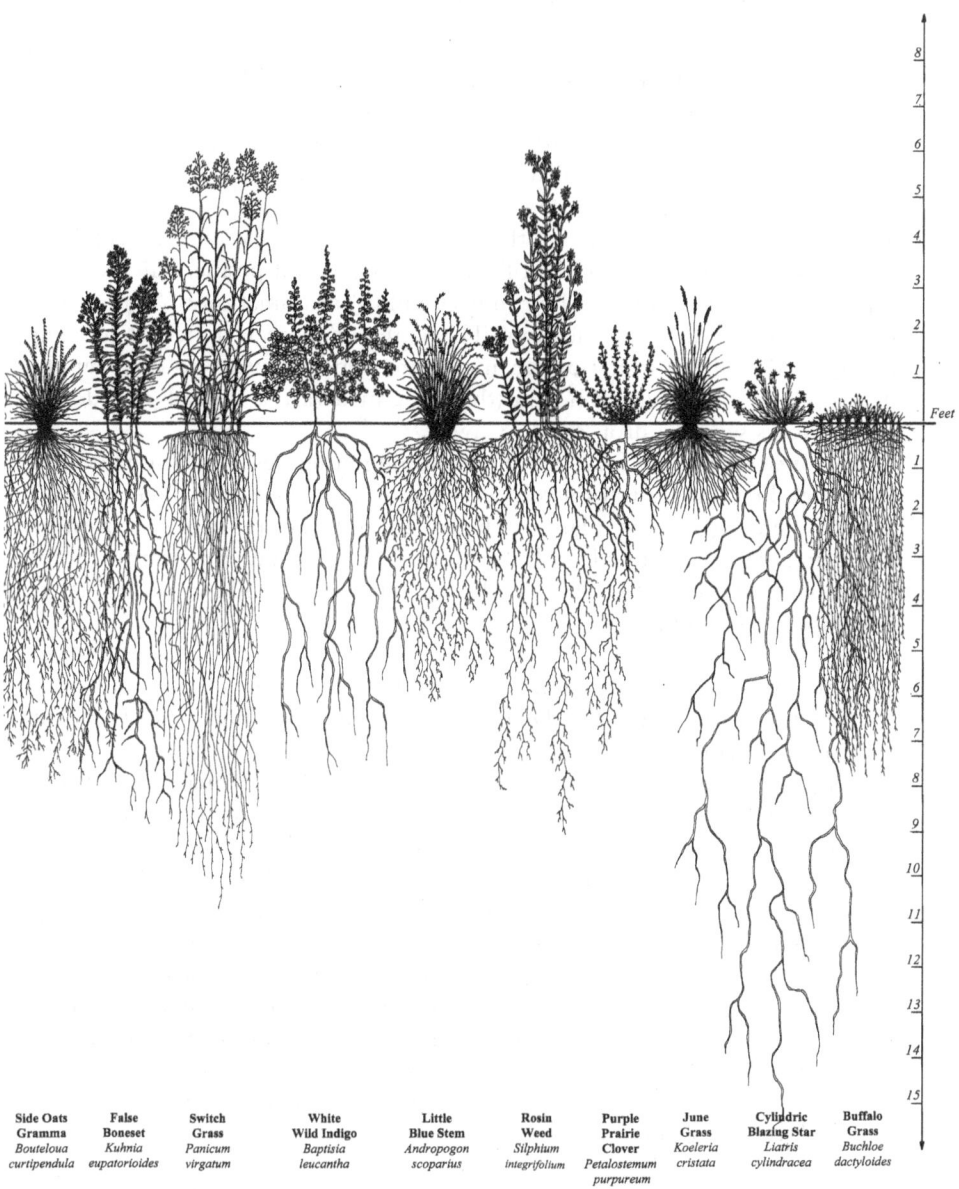

Side Oats Gramma	False Boneset	Switch Grass	White Wild Indigo	Little Blue Stem	Rosin Weed	Purple Prairie Clover	June Grass	Cylindric Blazing Star	Buffalo Grass
Bouteloua curtipendula	*Kuhnia eupatorioides*	*Panicum virgatum*	*Baptisia leucantha*	*Andropogon scoparius*	*Silphium integrifolium*	*Petalostemum purpureum*	*Koeleria cristata*	*Liatris cylindracea*	*Buchloe dactyloides*

Feet

8
7
6
5
4
3
2
1

1
2
3
4
5
6
7
8
9
10
11
12
13
14
15

Conservation Research Institute

Heidi Natura 1995
©

- studying the population genetics/genomics and the evolution of silphium
  – particularly the pattern of relatedness and genetic diversity within and
  between populations (both within the species and between different
  species of silphium), in order to determine how most efficiently to use
  the genetic resources available in silphium.[106]

The silphium-related work at the USDA office in Fargo involves both
(i) entomology research to identify the beneficial and pestilential insects of
silphium and understanding their impact and (ii) examination of "large effect
genes" in silphium domestication. The latter of these is aimed at locating
natural and artificial mutations in important genes in order to help rapidly
develop lines with reduced height (that is, semi-dwarf) and heads that do not
easily allows seeds to drop or blow away.[107]

Some researchers are focusing on developing other perennial oilseed crops.
One such line of research relates to field cress (*Lepidium campestre*), which is
a high-yielding oilseed plant with an upright stature and synchronous
flowering. It is a biennial with a potential to be a perennial crop. Some
participants at the 2013 FAO workshop reported as follows on efforts to
develop that potential.

> Inter-specific hybridizations were made between field cress and other
> species in the genus *Lepidium* to transfer desirable traits from the latter
> to the former. The most successful inter-specific hybridization was
> between field cress and *Lepidium heterophyllum*, a perennial close relative
> of field cress. The F1 hybrids produced from these species were perennial
> and showed very strong hybrid vigour with a significantly higher seed
> yield and a root system superior to those of both parents. Interesting
> lines have been selected from the F2 populations for further breeding.
> Interspecific hybridizations were also made between field cress and
> *Lepidium draba* to transfer shatter-proof genes from *L. draba* to field cress;
> and between field cress and *Lepidium graminifolium* to increase the oil
> content in field cress. Overall, a significant increase in oil content and
> seed yield, a significant decrease in pod shattering, and improvement in
> root systems are all highly promising developments, as is the progress
> in developing perennial field cress.[108]

In short, early results show that field cress, with potential both for cover and
oilseed production, might be perennialized successfully.

---

106 Personal correspondence with David Van Tassell and Tim Crews of The Land Institute,
    December 2015.
107 *Id.*
108 Mulatu Gelata *et al.*, *Domestication of* Lepidium campestre *as Part of Mistra Biotech, a Research
    Programme Focused on Agro-Biotechnology for Sustainable Food*, appearing as Chapter 10 in
    PERENNIAL CROPS PROCEEDINGS, *supra* note 50, at 142, 144.

IIB7. Other perennialization efforts

In addition to the work and progress summarized above, researchers at The Land Institute and elsewhere have also given considerable attention to the development of some other perennial crops. These include the following:

- Illinois bundleflower (*Desmanthus illinoensis*) is a productive native member of the pea family. Its nitrogen–fixing character can add functional diversity to future perennial grain systems. Researchers at The Land Institute have begun domesticating this perennial species as a grain legume by selection to reduce shattering of the seeds and to increase seed size and yield, but have not yet improved the flavor of the seeds.[109]
- Initial work has also been done at The Land Institute toward developing domesticated versions of other crops. This has included evaluations of germplasm of eastern gamagrass, wild rye and blue flax, and perennial versions of chickpea, maize, and millets.[110] The Land Institute is in the process of hiring a legume breeder to undertake further research, and it is also collaborating with researchers at St. Louis University and the Missouri Botanical Garden to begin inventorying perennials that may be worthy of domestication.[111]
- Researchers in Malawi are working to develop semi-perennial pigeon pea (*Cajanus cajan*).[112]

IIB8. Developing the production of perennials in polycultures

The foregoing summaries relate to developing *perennial* crops. In recent years, research work has accelerated on the other key element of herbaceous perennial seed-producing polycultures – that is, growing perennials in mixed

---

109 Personal correspondence with Lee DeHaan and Tim Crews of The Land Institute, December 2015. The primary objective of the research focusing on Illinois bundleflower is to develop a grain-producing legume that might also have dual use roles as a forage and a nitrogen fixer. See also The Land Institute Annual Report Fall 2011 (June 2010–July 2011), at 4. See also *Illinois Bundleflower: A Perennial Multiple Purpose Third Crop for Iowa*, available on the website of the USDA's Sustainable Agriculture Research and Education program, at: http://mysare.sare.org/MySare/ProjectReport.aspx?do=viewRept&pn=GNC05-055&y=2007&t=1.

110 Personal correspondence with David Van Tassell and Tim Crews of The Land Institute, December 2015. See also *Roots: The Land Institute*, May 2008, at 3.

111 Personal correspondence with Tim Crews of The Land Institute, December 2015.

112 See Snapp, *supra* note 68, at 152–154. The significance of semi-perennial pigeon pea in this context is that it is often intercropped with annual corn or sorghum – but it may be intercropped also with a perennial sorghum once one is developed. Like some other legumes, pigeon pea has an important ability to solubilize recalcitrant forms of phosphorus in soils, a characteristic that can be particularly helpful in more highly weathered African soils where fertilizers are not available or not affordable. Personal correspondence with Tim Crews of The Land Institute, December 2015.

populations following the model of native grasslands and prairies. It is worth noting that the study of polycultures *per se* is not new. Tim Crews, the Director of Research at The Land Institute, has offered this perspective:

> A great deal of research has been undertaken on polycultures in annual or annual/non-herbaceous perennial hybrid agroecosystems *and* on the roles of diversity in native ecosystems. So the polyculture work [relating to perennial grains in particular, although novel in itself, is nevertheless] starting from a deep foundation.[113]

Building on that foundation, researchers at The Land Institute involved in studying the operation of polycultures in the context of perennial grains have focused their attention most intensely in recent years on nutrient cycling. Tim Crews offers this explanation:

> I joined the research staff at The Land Institute[114] to investigate a range of ecological questions relevant to the design and functioning of perennial polycultures . . . . I had a good idea of where I wanted to begin: To understand the unique nature of the nutrient economy under perennial species, and work on increasing biological sources of nitrogen (such as intercropping with legumes) as well as managing organically bound phosphorus. [So at The Land Institute] we established two sets of legume-Kernza[115] polycultures. In these plantings we are measuring how much nitrogen is fixed by the legume and transferred to the Kernza, and how things like crop spacing, soil fertility and legume type change the nitrogen equation. We also established a two-crop polyculture experiment involving perennial sorghum and Silphium (one of our oilseed crops)[116] to see if competition for soil water resources can be relaxed in a multi-species planting.[117]

The overall aim of these experiments is to understand better the operation of the ecosystem in which plants grow and flourish and yield food. The

---

113 Personal correspondence with Tim Crews, August 2015.
114 Crews joined the staff at The Land Institute in 2012. Like the other key members of the research staff there, Crews has extensive scientific training and background. After receiving his doctorate degree from Cornell and carrying out a post-doctoral fellowship at Stanford, Crews developed an agroecology program at Prescott College in northern Arizona.
115 As noted in subsection IIB2 of this chapter, Kernza® is the perennial intermediate wheatgrass developed at The Land Institute.
116 For a summary of work at The Land Institute on wholeleaf rosinweed (*Silphium integrifolium*), see text accompanying notes 102–105, *supra*.
117 Tim Crews, *Ecology Study, Polycultures are Taking Root*, in THE LAND INSTITUTE ANNUAL REPORT FALL 2013, at 4. As noted above, The Land Institute is, in keeping with its renewed emphasis on the polyculture-specific research, currently in the process of hiring a legume breeder as a new member of the research staff.

subject is complicated because so much of that ecosystem is below-ground and composed of very small structures – including, for instance, the mycorrhizal structures, mucilage, and rhizosphere around the roots of the plants.[118] The complexity is greater, not surprisingly, in a polyculture than it is in a monoculture. However, some understanding of how the ecosystem works in, say, a natural prairie is essential in order to develop mixtures of new perennial crops that will have adequate yield and hardiness.

It is with this aim in mind that studies are underway in West Africa to study the use of certain crops – these include pigeon pea, sorghum, and rice – to enhance the production of organic matter in the soil, thereby improving the ecosystem in which agricultural crops are grown in that part of Africa. Tim Crews has offered this explanation:

> [M]ost of these fragile soils [in West Africa] have lost enough of their organic matter that they have become essentially unresponsive. In other words, if you amend them with fertilizers, you get very little response, because *the ecosystem as a whole is broken* – and loss of soil organic matter has been identified as the primary culprit. Soil organic matter helps maintain soil structure so that oxygen and rainfall can get in and CO_2 can get out, and organic matter provides the necessary food to feed the billions of microbes that we now know help plants grow. How do you build soil organic matter [in that setting]?[119]

Some answers to that question are coming in the form of research recently discussed at a conference in Mali. The research investigates how both perennial sorghum and pigeon pea can be used in the West African region to fix nitrogen, and how pigeon pea can also be used to "go after phosphorus bound to iron, which is an extremely useful attribute in highly weathered soils".[120]

The preliminary results of these and other polyculture-specific research efforts suggest that perennial legume-grain bi-cultures can be stable for years, and that nitrogen originating from nitrogen fixation by the legume component of such a bi-culture can supply most if not all of the main crop requirements. Further research is taking place to fine-tune the legume–alfalfa intercrop in particular, and work is also being carried out to see if water is partitioned between the two crops in such a setting. In addition, the researchers at The Land Institute are evaluating whether legume–wheatgrass intercrops maintain

---

118 For a relatively elementary introduction to mycorrhizal associations, particularly around roots, see M. C. Brundrett, *Mycorrhizal Associations* (section 3, Roots) (2008), at: http://mycorrhizas.info/root.html.

119 Tim Crews, *Research Round-up* (remarks at Prairie Festival, September 2015), at 2, 6 (on file with author).

120 *Id.*

lower fluxes of nitrous oxide (a potent greenhouse gas)[121] compared to wheatgrass fertilized with synthetic nitrogen fertilizers.

An increasingly important area of ecological research at The Land Institute is how agro-ecosystems change as they undergo succession from a highly disturbed early successional state under annual crops to a mid-successional state under perennials. The successional changes are predicted to include significant carbon and nitrogen sequestration – recapturing, that is, what was lost when the prairie was first plowed – as well as changes in the microorganismal community or microbiome.

This issue of "succession" warrants further explanation, because it focuses on a key question regarding the *scientific* or *technical* feasibility of transitioning from modern extractive agriculture to agroecological husbandry centered on perennial polycultures. Imagine a field of corn in northeast Missouri (this comes easily to me because I grew up surrounded by such fields). If the field is farmed using the methods that prevail in today's world, it will be (1) planted in the spring, (2) managed through the growing season in some fashion, either by mechanical tilling or by chemical treatment, in order to suppress weeds that would otherwise overcome the planted crop, (3) harvested in the fall by a "clear-cutting" operation, and then (4) left alone over the winter and thus "exposed as a flayed skin laid open", as one observer expressed it in another context.[122]

Having imagined that field, now consider the concept of agroecological "succession" – that is, the changes in maturity and diversity occurring in an agro-ecosystem *after* it has experienced severe disturbance of some sort.[123] In the case of a farmer's field, the soil is largely sterilized, at least from the perspective of its living components, by the combination of weed-suppression operations (step 2 in the preceding paragraph) and the "clear-cutting" form of harvesting (step 3 in the preceding paragraph). That set of steps leaves the field in what can be regarded as an "early stage of secondary succession", during which the soil – which I referred to earlier as the "thin skin of life" that all organisms depend on and participate in[124] – is in recovery mode. The rich diversity of its endemic species of living organisms constantly try to return to it, to enliven it, with only the short interval between one crop's harvest and the next crop's planting available for such recovery to occur.

---

121 As noted earlier, nitrous oxide is said to be about 300 times more potent, molecule for molecule, than carbon dioxide (the much more prevalent greenhouse gas) in terms of causing climate change. See note 78 and accompanying text in Chapter 3, *supra*.

122 See note 2 in Chapter 4, *supra*.

123 In a simple definition, the concept of "secondary succession" refers to the process of an ecosystem reviving itself after all or a portion of it has been destroyed. More detailed explanations of the concept can be found in Crews *et al.*, *Mid-Succession*, *infra* note 126 (explaining that secondary succession "is distinguished from primary succession as it begins following a disturbance to an already established soil and ecosystem by fire, flood, drought, or any other impact that drastically upsets the dominance of the established community").

124 See text accompanying note 27 in Chapter 1, *supra*.

Often that interval occurs over winter months when conditions for renewal are inhospitable anyway.

By contrast to that "early stage of secondary succession" of a modern agro-ecosystem, consider next a "mid-successional" agro-ecosystem. That sort of agro-ecosystem is one that lies further down the "successional gradient" than the field planted annually in corn — just as the ecosystem in a 30-year-old secondary-growth forest would lie further down the "successional gradient" than an area that was logged just last year by means of a "clear-cutting" operation. Even further down the "successional gradient", to pursue that same illustration, would be the old-growth forest that has stood undisturbed by humans or serious other forms of destruction for several hundred years.

Consider finally the question of nitrogen, which is an essential input for plants generally and especially important for fast growth of corn plants and the cobs and seeds they produce. In modern extractive agriculture, a cornfield gets its nitrogen largely by means of synthetic nitrogen fertilizer — a matter I have discussed at some length earlier.[125] Researchers at The Land Institute and elsewhere have directed their attention to this question: Can adequate nitrogen be secured and made available to *perennial* crops (not planted anew each year) through some natural internal means — that is, *other* than through application of synthetic nitrogen fertilizer produced from exogenous inputs — and still achieve respectable yield?

A recent research paper reported the outcome of such studies. The authors of the paper offer this synopsis, which begins with a reminder of key advantages that perennial grains would have over annual grains *if* such perennial grains were really viable commercially:

> Annual-based arable agroecosystems experience among the greatest frequency, extent and magnitude of disturbance regimes of all terrestrial ecosystems. In order to control non-crop vegetation, farmers implement tillage practices and/or utilize herbicides. These practices effectively shift the farmed ecosystems to early stages of secondary succession where they remain as long as annual crops are grown. Humanity's long-standing dependence on a disturbance-based food and fiber producing ecosystem has resulted in degraded soil structure, unsustainable levels of soil erosion, losses of soil organic matter, low nutrient and water retention, severe weed challenges, and a less-diverse or functional soil microbiome. While no-till cropping systems have reduced some hazards like soil erosion, they remain compromised with respect to ecosystem functions like water and nutrient uptake, and carbon sequestration compared to many later successional ecosystems. Recent advances in the development of perennial grain crop species invite consideration of *the ecological implications of farming grains further down the successional gradient than ever before possible.*

---

125 See, for example, note 20 in Chapter 1, *supra*, and accompanying text.

In this review, we specifically explore how the nitrogen (N) economy of a mid-successional agroecosystem might differ from early-successional annual grain ecosystems as well as native mid-successional grassland ecosystems. We present a conceptual model that compares changes in soil organic matter, net ecosystem productivity, N availability, and N retention through ecosystem succession. Research from the agronomic and ecological literatures suggest that mid-successional grain agriculture should feature several ecological functions that could greatly improve synchrony between soil N supply and crop demands . . . . Compared to native mid-successional grasslands that cycle the majority of N required to maintain productivity within the ecosystem, a mid-successional agriculture would require greater external N inputs to balance N exports in food. Synthetic N fertilizer could make up this deficit, but in the interest of maximizing ecological intensification in order to minimize inputs and associated environmental consequences, we explore making up the N deficit with biological N_2 fixation. The dominant approach to addressing problems in agriculture is to target specific shortcomings such as nutrient retention or weed invasion. *Moving agriculture down the successional gradient promises to change the nature of the ecosystem itself, shifting attention from symptom to cause, such that ecological intensification and provision of a broader suite of ecosystem services happen not in spite of, but as a consequence of agriculture.*[126]

The paper concludes affirmatively: It *does* seem possible that adequate nitrogen can be secured and made available to *perennial* crops (not planted anew each year) through some natural internal means – that is, *other* than through application of synthetic nitrogen fertilizer produced from exogenous inputs – and still achieve respectable yield. In particular, a "legume-grain intercrop" polyculture can be designed, put in place, and managed in order to achieve adequate nitrogen availability.[127]

IIB9. A Summing-up on the state of development of perennial polycultures

Several general themes emerge from the detailed results presented at the 2013 FAO workshop and the other accounts offered above starting on page 206. One theme is this: developing perennial crops will prove challenging as a technical matter. As one of the workshop participants explained,

---

126 Timothy E. Crews *et al.*, *Going Where No Grains Have Gone Before: From Early to Mid-Succession*, 223 Agriculture, Ecosystems and Environment 223–238 (2016) (hereinafter *Mid-Succession*).

127 *Id.* In this respect, the report illustrates its findings by citing research in Kansas and elsewhere showing that "native perennial grasslands . . . had been harvested without fertilization for 75 years, yet still remained productive and maintained much greater levels of soil C [carbon], N and aggregation than adjacent wheat fields". The report emphasizes the central role that legumes play in the operation of such grassland ecosystems.

"[p]erenniality seems to be under the control of multiple genes responsible for specific biological structures such as bulbs, rhizomes and meristems, as well as for physiological traits such as resistance to cold, drought and biotic stresses."[128] Because of this complexity in genetic instructions, the "transformation of annual crops into perennial crops with high grain yield, coupled with superior technological and nutritional quality[,] could turn out to be a very demanding and challenging goal."[129]

However, another theme also emerges from the foregoing account: despite the demands and challenges presented, researchers have made substantial progress to date on the development of perennial food grains (and legumes) to be grown in polycultures. Not surprisingly, this progress is mixed. As one participant in the 2013 FAO workshop expressed it, the "development of perennial crops ranges from its infancy (e.g. perennial maize and others, . . .) to intermediate (e.g. perennial wheat . . .), to approaching reality (e.g. perennial sorghum [and] . . . perennial rice [and] . . . perennial legumes)."[130] Moreover, the related but distinct project of finding how such new perennials can be grown in polycultures rather than monocultures remains at a very early stage of development. Still, the momentum is strong and increasing on the *perennial* element, and generally promising on the *polyculture* element, of creating herbaceous perennial seed-producing polycultures of the sort that Wes Jackson identified over 30 years ago as a "missing item" among the categories of plants from which food might be produced for human consumption.[131] Indeed, it is the strength of that momentum that was on display at the FAO conference of experts in the summer of 2013. The proceedings from that conference constitute a firm endorsement of perennial crops as a necessary part of agricultural development.

An even more recent initiative to bring researchers together for a discussion of developing perennial polycultures was undertaken in late 2014 in Colorado. The Land Institute, which co-sponsored a four-day conference at Estes Park in October of that year, offers this summary:

Four dozen researchers from five continents converged on a historic hotel in the Rockies and brainstormed about helping The Land Institute achieve its aim – perennial grains grown in mixtures.

The four-day conference, titled New Roots for Ecological Intensification, was staged in late October at the Stanley Hotel in Estes

---

128   Norberto E. Pogna *et al.*, *Evaluation of Nine Perennial Wheat Derivatives Grown in Italy*, appearing as Chapter 5 in Perennial Crops Proceedings, *supra* note 50, at 54, 55.

129   *Id.*

130   Wade, *supra* note 69, at 9. Tim Crews of The Land Institute has suggested that perennial sorghum might better be classified as "intermediate" in its development (rather than "approaching reality") and that Kernza® (intermediate wheatgrass) should also be listed as "approaching reality" in its development. Personal correspondence with Tim Crews, August 2015.

131   See text accompanying Figure 4.1 and Figure 4.2 in Chapter 4, *supra*.

Park, Colo. The sessions, which lasted deep into the evening, were focused on the question of how to speed development of perennial grains. Included were many small, crop-specific breakout discussions where researchers shared what they've learned and challenges they've encountered.[132]

Researchers who gathered for the Estes Park conference are expected to develop new collaborations that are likely to build upon the momentum described in the foregoing paragraphs. One result of such collaborations is the research summarized at pages 229–230 about nitrogen flows in perennial polycultures;[133] another result is a research paper examining research priorities for ecological intensification with perennial grain cropping systems.[134] The Estes Park conference also yielded other initiatives and collaborations. These included joint field operations, "the first ever sequencing of Silphium RNA", joint grant proposals for perennial grain research, and team studies of such matters as insect pollinators and pests, controlling seed dormancy, water use efficiency, and photosynthetic efficiency in various plants being subjected to perennialization research.[135]

IIC. Frequently asked questions

In a brochure prepared in 2009, The Land Institute provided a four-page set of questions and answers about perennial grains – the likelihood that they can in fact be developed, the benefit they could provide in addressing issues of world hunger, the ecological need they might meet, their ability to withstand attacks by pests and disease, how many of them could (or must) be grown together in order to "mimic" the ecological architecture of a natural prairie, and even how such a polyculture of perennial grains could be harvested as a practical matter.[136] As a form of summing-up on perennial polycultures, I provide in these last pages of Chapter 5 a slightly modified and updated version of that set of questions and answers. My modifications offer some additional explanation to a few points and provide some new information.

---

132 *Ecological Intensification at Estes Park*, available on the website of The Land Institute, at: https://landinstitute.org/news-post/ecological-intensification-at-estes-park.

133 See Crews *et al.*, *Mid-Succession, supra* note 126.

134 See Matthew R. Ryan, Matthew G. Bakker, Jacob M. Jungers, Timothy E. Crews, Steven W. Culman, Sivaramakrishna Damaraju, Lee R. DeHaan, Richard C. Hayes, and Maged Nosshi, *Management Considerations and Research Priorities for Ecological Intensification with Perennial Grain Cropping Systems* (forthcoming 2016; on file with author).

135 See *A Synopsis of Outcomes One Year Out from the Meeting "New Roots for Ecological Intensification"* (2016, on file with author).

136 The Land Institute, *A 50-Year Farm Bill* (June 2009), at 10 (on file with author and available from The Land Institute). For further details about the proposed 50-year farm bill, see text accompanying notes 36–39 in Chapter 6, *infra*.

Q1: *Too little too late?* The development of perennial grains is still at very early stages – in fact, it is expected to take at least 25 years to achieve more than two or three profitable, productive perennial grain crops. Isn't that too late to address the problems facing the world today?

A: We hope not, but we do need to move as fast as possible. New strategies are needed that emphasize efficient nutrient use in order to lower production costs and minimize negative environmental impacts. The sooner that successful alternatives are available, the more land we can save from degradation. It is likely that global agricultural acreage will expand over the next two to three decades especially if the human population increases to 8 to 10 billion people. Recent projections predict an 18 percent or more increase in agricultural land between 2009 and 2020. This is very troubling because the best soils on the best landscapes are already being used for agriculture. Much of the future expansion of agriculture will be onto marginal lands, where the risk of irreversible degradation under annual grain production is high. As these areas become degraded, expensive chemical, energy and equipment inputs will become less effective and much less affordable.

So yes, the need to move ahead quickly with perennials is urgent. Bear in mind that in regions of the world where high inputs of fertilizers, chemicals and fuels are not an option, agricultural systems that are highly efficient, productive, and conservative of natural resources are needed – and will be needed even more 25 years from now. In other words, the need for a natural-systems agriculture, featuring perennial grains, is both urgent and growing.

Q2: *Low yields and eco-damage?* Can we expect perennial grain crops to be as productive as annual grain crops and, if not, won't they actually worsen environmental problems by requiring more land for agricultural production?

A: There is sufficient evidence that "reasonable reference yields" of annual crops can be matched on high-quality lands and exceeded on poor-quality lands by diverse perennial systems with fewer negative impacts. That evidence is very strong. But it is important to be careful about which annual yields are used as a standard. For example, the world record wheat yield was harvested in the Palouse region of Eastern Washington State where wheat yields can top 100 bushels per acre. Annual wheat production in that region, though, has resulted in extensive erosion. All of the topsoil has been lost from over 10 percent of the region's landscapes. On eroded sites Palouse wheat yields may be less than 25–30 bushels per acre. Crop yields that come at such a high cost to the soil resource – or that depend on an extravagant use of chemical fertilizers – should not be used as a standard of comparison. If we use reasonable, sustainable, ecologically responsible yields as our standard, perennials can perform well.

Q3:   *Inevitability of low yields on perennial grains?* Won't the seed yield of perennials always be limited by the fact that a perennial plant needs to save some energy for overwintering – energy that could have been used to produce seed?

A: The short answer is no. The theoretical limitations to seed yield in perennials are no more serious than in annuals. In annuals, yield is limited by shorter growing seasons, water shortage due to short roots and poor seedling establishment. In perennials, yield can be constrained by the need to overwinter, but rapid spring growth of perennials, combined with season-long access to water deep in the soil profile, means that perennials such as alfalfa are overall more productive than related annuals like soybeans. Much of the journey-work of plant breeders has been to shift the allocation of resources away from leaves, stems, and crowns, and toward seed in the development of perennial grain crops.

Q4:   *Are perennials really ecologically necessary?* With advances in no-till production of annual grain crops, do we need perennial grains to mitigate the environmental problems associated with agriculture?

A: Unfortunately, yes. Although no-till technology has reduced erosion in many areas, some problems remain due to the biological limitations of annual plants. Chief among the problems associated with no-till is water quality. Annual crops, even in no-till situations, are relatively inefficient in capturing nutrients and water. Because annual crop plants are often either absent or too small to use and manage water during times of rainfall, as much as 45 percent of precipitation may drain below root zones of annual crops, whether produced with no-tillage or conventional tillage practices. These rates of water loss under annual crops can be five times greater than under perennials.

This is a problem because water flowing through the soil profile carries soil nutrients and agrichemicals that pollute rivers, lakes and coastal waters. This problem can be compounded under no-till production, which often requires greater inputs of agrichemicals and fertilizers. (That is, heavy chemical use puts more chemicals in the soil, and these are carried down into streams and rivers as pollution.) In sum, the problem is getting worse, despite increased adoption of no-till and minimum-till systems.

Crop seeds need warm, well-drained seedbeds in order to properly germinate. No-till operations limit this. This is why tillage remains an attractive practice in some regions; warming and drying of the seedbed can be hastened. Advances in plant breeding may eventually allow for optimal germination in cooler, wetter conditions, but in many agricultural regions – including the Midwestern USA – seedlings will still be small when the rains come.

Q5:   *How much diversity and what kind?* If the perennial-polyculture farming systems that The Land Institute (and others) will hope to develop are supposed to "mimic" natural ecosystems to some degree, just what level and kind of plant diversity is needed, and how will it be deployed?

A: The answer to both parts of the question is, "It depends." It depends on the resilience and fertility of the soil, climate, disease pressures, and types of crops. Nearly all of nature's land-based ecosystems feature perennial plants grown in diverse mixtures.

Natural ecosystems, in general, use and manage water and nutrients most efficiently and build and maintain soils. For that reason, nature is our standard. The level and spread of diversity varies. The characteristics of the region in which they are to be grown will have to be assessed.

Diversity is of two kinds: multiple species and genetic diversity within species. Current grain production practices commonly involve planting a single genotype (near-zero genetic diversity) across a field often larger than 100 acres. Furthermore, that single genotype and other genetically similar plants are being grown on millions of acres in a region. Increases in genetic diversity at the species, field, and landscape levels are needed. The final ordering of the components of the diversity will be determined by what is useful and can be practically achieved by local farmers.

Q6:   *Already tried and failed?* Several serious attempts have been made in the past to perennialize grain crops, and yet we don't have any such crops today that have emerged from those earlier efforts. What has changed that offers promise of success now?

A.   History need not be a source of discouragement. In the case of wheat, most involvement with perennials had to do with bringing desirable genes – for resistance, say – from a wild perennial relative into the annual crop. The perennialization effort, in most cases, was carried on, more or less as a hobby, by an interested researcher but with no institutional commitment for a sustained program to guarantee continuity. When the researcher retired, the effort ended. The Soviets had the most ambitious perennial wheat program, but political decisions halted these efforts in the late 1950s or early 1960s.

Besides, we are now in a new era in two ways:

- *Our economic perspective is more sophisticated and realistic.* In recent years, the costs (both ecological costs and economic costs) that annual cropping imposes on our soils and waters are increasingly weighed against bushels per acre. As a result of this different, more all-encompassing accounting (balancing real costs with benefits), we now see that some reduction in yields (bushels per acre) is acceptable, and indeed preferable, if that reduction is accompanied by an even bigger reduction in costs. In short, we have a broader (and smarter) perspective on "efficiency" and "yield".

- *Our technology is profoundly improved.* With recent advances in plant breeding, more knowledge of the genome, and greatly increased computational power, thinking about breeding limits has changed. In other words, much more is possible now than ever before in developing perennial grains.

Q7:   *Will pests destroy perennial grain crops?* Since mechanical tillage and annual rotations are largely eliminated in perennial systems, don't the perennial plants become "sitting ducks" for pests and disease?

A: Of course not. Here the proof is in the pudding: perennials dominate most native landscapes. They constitute, for instance, roughly 80 percent of North America's native flora. Perennials have thrived throughout evolutionary history despite the pressures of pests and disease.

In some fields or some regions, some perennial crops will prove to be more problematic than others, and breeding for complex traits such as yield and perenniality might unintentionally purge (displace) genes involved in resistance responses. There will, in other words, undoubtedly be pest and disease problems. But these problems also afflict our most productive annual crops. And there are many examples of herbaceous perennial plants − alfalfa, switchgrass, brome − that remain highly productive for many years despite exposure to pests or disease. So we will continue focusing on strengthening the perennial grains we develop to withstand pests and disease. In this regard, diversity (whether at the field or landscape scale or over time), field burning, and selecting for resistance in a plant breeding program are essential elements of our work.

Q8:   *Why not work on other aspects of agricultural reform?* How do alternative methods of production such as permaculture, biointensive, or organic fit in with perennial grain crops? For example, what about vegetables and fruits? And how do community-supported agriculture farms fit in?

A: We focus on the crops that occupy 68 percent of global cropland and provide about the same percentage of food calories: annual grain crops that are today grown primarily in monocultures. Any number of approaches, alternative or conventional, could be used in managing perennial crops and distributing the harvest.

This is not to say that efforts aimed at reducing the scale of industrial agriculture and increasing local food security are misguided. They are not. They are necessary to transform our food system over the long term. While promoting local small-scale, organic agriculture we must also assess how and where the bulk of our calories can best be produced. Think about this: if all or even a large portion of the calories consumed by New Yorkers came from New York State, there would be few trees left there, and the state's thin, poor soils would be quickly degraded. The bulk of the calories consumed by New Yorkers must instead, as a practical matter, come directly or indirectly from grain crops which grow well in the Midwest and Great Plains states. Similar examples apply in many parts of the world. Grain crops are essential, and that is why great attention must be paid to improving them.

It is important to note that the emphasis on perennial polycultures is consistent with – and might even be regarded as part of – certain other agricultural-reform movements and initiatives that have (fortunately) gained considerable traction in recent years. One of these is "climate smart agriculture" ("CSA"). The FAO has carried out especially important work in developing CSA, as explained in its *Climate-Smart Agriculture Sourcebook* of 2013.[137] The CSA initiative, however, has apparently given little attention thus far to the prospects for perennial grains to transform modern extractive agriculture – as evidenced by the fact that only a very few references to perennials appear in that *Sourcebook* (and then mainly in connection with pastures and groundcover).[138] Another initiative is that of "sustainable intensification" of agriculture.[139] For reasons that have been highlighted

---

137 Food and Agriculture Organization, CLIMATE-SMART AGRICULTURE SOURCEBOOK (2013), www.fao.org/docrep/018/i3325e/i3325e.pdf. As noted in the *Sourcebook*, climate-smart agriculture "integrates the three dimensions of sustainable development (economic, social and environmental) by jointly addressing food security and climate challenges", and it has three main pillars: "sustainably increasing agricultural productivity and incomes", "adapting and building resilience to climate change", and "reducing and/or removing greenhouse gases emissions, where possible." *Id.* at ix. The *Handbook* goes on to assert that CSA "is not a single specific agricultural technology or practice that can be universally applied". *Id.* at 10. Perhaps this is a feature that distinguishes CSA from the sort of agroecological husbandry at issue in this book, which *does* in fact concentrate on a single specific agricultural technology – perennial grains and legumes grown in polycultures – as a revolutionary new approach to producing the food that constitutes nearly 70% of the global human diet.

138 In a table listing several "changes in agricultural systems", the *Sourcebook* has a line for "Perennials/Agro-Forestry", but the information there seems to anticipate the use of perennials only as species in areas "adjacent" to traditional croplands featuring annual species. See *id.* at 386.

139 See Tara Garnett and Charles Godfray, *Sustainable Intensification in Agriculture: Navigating a Course through Competing Food System Priorities* (2012), available on the website of the Oxford Martin Programme on the Future of Food, at www.futureoffood.ox.ac.uk/sustainable-intensification. That report emphasizes the need for research that will lead to "[g]reater understanding of how

earlier, perennial polycultures epitomize *sustainability* of agricultural production because they would mimic the natural ecosystems of prairies that have been converted over the centuries into croplands for (unsustainable) annual crop production.

Q9:   *Will the public eat perennial grains?*
A: People like the taste of Kernza®, which is a perennial wheat that is already produced by The Land Institute, and there is little reason for people to find significant or undesirable taste differences in perennial grains generally. Greatest short-term success in developing suitable perennial crops will come with perennializing current grain crops with which the public are already familiar. Indeed, one of the strongest arguments for perennializing those grains is that it does not require large dietary shifts.

Q10:   *Finally, how are you going to harvest a perennial grain polyculture?*
A: This question arose so frequently over the years that The Land Institute researchers decided to plant a polyculture of four annual crop species: corn, soybean, sorghum, and sunflower. The seed mixture was planted with an air drill. At harvest time, we made slight adjustments to our harvesting equipment. Progress through the field was slow, but not prohibitively so. Seeds were separated with a seed cleaner. The point is that mechanical equipment that is already in existence can do the job with just a little fine tuning. The larger problems are agronomic, not engineering.

As the answers to these "frequently-asked questions" suggest, perennial polycultures have a strong likelihood (i) of being developed successfully using scientific methods largely unavailable until quite recently and (ii) of providing key answers in addressing issues of world hunger. This new form of agriculture, by mimicking the ecological architecture of natural prairies, has the proven potential to produce yields that compare favorably – and on a global scale – with conventional annual-monoculture crops, while mitigating much of the ecological degradation associated with modern extractive agriculture. While it offers by no means the only alternative model for food production, agroecological husbandry's emphasis on perennial polycultures holds great promise.

---

the various elements of complex systems interact . . . , both at fine grained and broader spatial and temporal scales. This understanding needs to encompass not just environmental interactions but also the relationship between the environment, human health, ethics and livelihoods." For an explanation of the relationship between "sustainable intensification" and "climate-smart agriculture", see Bruce M. Campbell *et al.*, *Sustainable Intensification: What Is its Role in Climate Smart Agriculture?*, 8 CURRENT OPINION IN ENVIRONMENTAL SUSTAINABILITY 39–43 (October 2014).

In the Preface to this book, I quoted from Henry David Thoreau's advice about building castles in the air and putting foundations under them. Perhaps we can consider one such castle to be an ecologically robust and resilient future in which humans live sustainably and in concert with the rest of the natural world. Agroecological husbandry offers a key design component for such a castle. Now how do we put the legal, policy, and institutional foundations under it?

Part IV

Building new legal foundations for agroecological husbandry

6 Necessary changes in substantive national law

In Chapters 2 and 3 of this book, I highlighted several fundamental flaws that are inherent in modern extractive agriculture – that is, in that form of food production, especially grain and legume production, that dominates the world today. In particular, I concentrated on its economic unsustainability, its ecological unsustainability, the risks it poses to human health, and the ways in which it goes "against the grain", both historically and socially, of human development.

In contrast to modern extractive agriculture, a "natural-systems" agriculture that I described in Chapters 4 and 5 offers the promise of overcoming many of the disadvantages and dangers of modern extractive agriculture. Centered on mimicking nature in its diversity, its resilience, and its ecoregion specificity, this natural-systems agriculture would feature perennial grains grown in polycultures. It would rely not on fossil carbon but on current solar energy, and for this reason it would not present the climate-change risks that modern extractive agriculture does. This new agriculture, which I have labeled "agroecological husbandry" for reasons explored especially in Chapter 4, would also give close attention to agrarian values, intergenerational equity, and multi-species interests.

The closing segments of Chapter 5 emphasize the momentum that has been achieved thus far on what we might consider "the science part" of agroecological husbandry. As summarized there, (i) agroecological husbandry has already gained broad acceptance in principle, (ii) recent field research and crop developments have demonstrated the practical feasibility and potential of perennial grains in terms of yield, performance, and value, and (iii) steady progress is also being made in understanding the interactions of chemical processes and nutrient cycles in order to blend various species of those perennials into robust polycultures.

Now I shift attention from "the science part" to "the legal part". Here comes some heavy lifting. It would be comforting, of course, to assume that the strong momentum being achieved in the science of developing agroecological husbandry as a viable alternative to modern extractive agriculture will somehow usher in a smooth transition from the latter to the former. It would also be naïve. In my view, it is highly unlikely that modern extractive agriculture, with roots reaching back 10,000 years, will give way to agroecological husbandry automatically and without other forces at work to facilitate such a transition.

One particular force that must be at work for a transition to occur from modern extractive agriculture to agroecological husbandry is *legal reform*. Changes in law, based of course on changes in policy and values, are necessary in order for the promise of agroecological husbandry to be realized. In the pages that follow, I will trace the broad contours of the legal reforms that I consider most important in order to achieve the transformation of agriculture.

In doing so, I will first (in section I of this chapter) identify some legal changes that the analysis in Chapters 2 and 3 show to be needed. Hence this first question: "In what ways can the specific ills and risks summarized in those two chapters – relating to economics, ecology, human health, and social issues – be reversed or overcome through legal changes?" Some answers to this question will be obvious and rather simple. For instance, legal changes can be made to drastically reorient the system of agricultural subsidies that so heavily favor a small handful of annual crops and the forces of industrial agriculture by which they are produced. Other legal initiatives can be taken to ensure that all externalities of modern extractive agriculture, including ecological degradation, are fully accounted for, so that the true long-term costs of this system can be seen and taken into account. Some other answers will not be so obvious. In short, the discussion in section I will enumerate and briefly explain roughly a dozen categories of legal changes responding to the problems highlighted in Chapters 2 and 3.

Second, in section II, I will survey some proposals for legal reforms in the USA, where modern extractive agriculture began and has taken shape in its most extreme form. My aim is to highlight and elaborate on the rich literature regarding substantive[1] legal and policy change that observers closely familiar

---

1 Throughout this chapter, as reflected even in the title of the chapter, my focus will stay on *substantive* legal changes as distinct from procedural, structural, or institutional changes. The same

with US farm-related laws have proposed. For instance, one highly valuable contribution to that literature is the proposal for a "50-year farm bill" inspired by Wes Jackson and Wendell Berry and expanded on by The Land Institute. Numerous other US-centered legislative and policy proposals also warrant attention.

Section II will close by insisting that the sort of drastic reformulation of law and policy that I and others call for in the USA today parallels the drastic reformulation that occurred about 70 years ago, when the US federal government put in motion the initiatives that resulted in the Green Revolution. My point in those closing pages is that the legal reforms I recommend at the national level are not at all unprecedented in character. Those observations will then set the stage for the discussion in Chapter 7, which shifts attention to the global stage. It is in that chapter that I outline a new collaborative treaty system by which global support can be mustered and mobilized for a transition from modern extractive agriculture to agroecological husbandry. That effort too, while extremely ambitious, is not at all unprecedented in character.

A central theme runs through both this chapter and Chapter 7: *extensive legal and regulatory changes must be made in order to facilitate a transition from modern extractive agriculture to agroecological husbandry.* As Mary Angelo of the University of Florida has expressed it in the US context, legal changes are needed that will amount *not* to a "mere tinkering with the existing regulatory regime" but rather a "complete overhaul of existing agricultural policy."[2]

I. Counteracting the economic, ecological, health, and social unsustainability of modern extractive agriculture

IA. Legal changes to counter economic unsustainability

At the end of Chapter 2, I offered a "summing-up" on the economic unsustainability of modern extractive agriculture.[3] I provide below an abbreviated version of that "summing-up":

• Although agriculture has always involved risk, the most contemporary "industrial" form of agriculture as it has developed in the last few decades forces today's farmers – and especially small farmers and would-be farmers – to face extremely high entry costs and operational costs.

---

will be true in most of Chapter 7, where I turn my attention to substantive legal initiatives to be made at the *global* level. For my views on structural and institutional changes – including a new form of international institutional cooperation that would emerge from a reorientation of the concept of sovereignty, which serves as a backbone for international relations – see section I of Chapter 8, as well as my forthcoming book A GLOBAL CORPORATE TRUST FOR AGROECOLOGICAL INTEGRITY: MANAGING A NEW AGRICULTURE IN A WORLD OF ECO-STATES.

2 See *infra* note 74 and accompanying text.

3 See subsection IV of Chapter 2, *supra*.

- This explains in part both (i) why there has been such a dramatic decline in the past few decades in the proportion of people engaging in farming as an occupation – particularly in the developed countries where the most high-cost resource-intensive techniques of agriculture predominate – and (ii) why the bulk of agricultural production has come to be concentrated in a remarkably narrow assortment of crops.

- As a result, the actual control and operation today of agricultural production is concentrated largely in a few giant agribusiness enterprises that have accumulated the political power to strengthen even further the trend away from diversification – despite the fact that smaller farms with diversified ownership and crops are widely thought to have greater potential for poverty reduction and efficient productivity if given adequate and equitable support.

- Even though large farm operations and giant agribusiness enterprises seem at first glance to be succeeding, those operations do not reflect an agricultural model that is economically sustainable. Without heavy direct and indirect financial subsidization – that is, without generous public resources from society as a whole directed to prop up the relatively few giant agribusiness enterprises – their economic viability would crumble.

- Moreover, what some would see as the ultimate justification of modern extractive agriculture – that it has succeeded in creating an ever-increasing world food supply – is itself deeply suspect: there is no sign that the system of modern extractive agriculture is meeting the world's needs now or can do so in the future.

- In this and other ways, the Green Revolution might be seen as a "flash in the pan" that has, while providing a short-term boost in productivity, brought economic injury to farmers world-wide – a consequence that is particularly unfortunate given the fact (emphasized at pages 51–52) that smaller, more environmentally-oriented food production is better in important ways than production through high-cost, fossil-carbon-dependent extractive agriculture.

- This combination of factors should make us question the appropriateness of providing such massive subsidies to the enterprises in whose hands agricultural production is currently concentrated. More fundamentally, this combination of factors suggests that the model itself – that is, modern extractive agriculture as described earlier in this book – is intrinsically flawed.

What legal reforms might be made in response to these various ways in which modern extractive agriculture is economically unsustainable? Four obvious reforms come immediately to mind.

A first reform would be to provide legal and regulatory support and financial incentives that would *reduce the high entry costs and other hurdles to small farmers and beginning farmers*. In the USA, some support of this type has appeared in the form of recent versions of the so-called "farm bill". The Food,

Conservation, and Energy Act of 2008, for instance, included various initiatives for attracting and retaining beginning and socially disadvantaged farmers and ranchers.[4] Similarly, the Agricultural Act of 2014 – several key features of which are summarized in Box 6.1 – included some funding for beginning-farmer development and for facilitating farmland transition to the next generation of farmers.[5] Important reforms and initiatives have also been proposed by Neil Hamilton of Iowa State University, and by others, for supporting new small farmers.[6]

Box 6.1 Selected key features of the US Agricultural Act of 2014[7]

The USA addresses agricultural and food policy through a variety of programs, including commodity support, nutrition assistance, and conservation. The primary legal framework for agricultural policy is set through a legislative process that occurs approximately every five years.

A new farm law, the Agricultural Act of 2014 ("2014 Farm Act"), was signed on February 7, 2014, and will remain in force through 2018 – and in the case of some provisions, beyond 2018. The 2014 Farm Act makes major changes in commodity programs, adds new crop insurance options, streamlines conservation programs, modifies some provisions of the Supplemental Nutrition Assistance Program (to which by far the most funding under the legislation is devoted), and expands programs for specialty crops, organic farmers, bioenergy, rural development, and beginning farmers and ranchers.

---

4 *2008 Farm Bill Side-By-Side*, USDA Economic Research Service [hereinafter *Side-by-Side*], http://webarchives.cdlib.org/sw1vh5dg3r/http://ers.usda.gov/FarmBill/2008/Overview.htm

5 For details, see the third-to-last paragraph in Box 6.1, highlighting certain provisions of the 2014 US farm bill. Related provisions, also referred to there, give support for veterans who enter into small farming.

6 See Neil Hamilton, *America's New Agrarians: Policy Opportunities and Legal Innovations to Support New Farmers*, 22 FORDHAM ENVIRONMENTAL LAW REVIEW 523 (2011). See also National Young Farmers Coalition, "Building a Future with Farmers: Challenges Faced by Young, American Farmers and a National Strategy to Help them Succeed," 2011, www.youngfarmers.org/newsroom/building-a-future-with-farmers-october-2011/; Laurie Ristino, *Back to the New: Millennials and the Sustainable Food Movement*, 15 VERMONT JOURNAL OF ENVIRONMENTAL LAW 1 (2013); Edward Cox, *Helping Landowners Help New Farmers: Incentive Programs and Other Legal Tools for Transitioning Land to the Next Generation of Farmers*, 17 DRAKE JOURNAL OF AGRICULTURAL LAW 37 (2012).

7 Details in Box 6.1 are drawn largely from a summary prepared by the USDA's Economic Research Service ("ERS") regarding the 2014 Farm Act – with special attention to those provisions that relate directly to certain topics addressed in this book, including agricultural subsidies, conservation, beginning farmers, and selected other topics. The summary is available at the ERS website, specifically at www.ers.usda.gov/agricultural-act-of-2014-highlights-and-implications.aspx.

The *crop commodity programs* provide benefits based on price or revenue targets for producers of corn and other feed grains, wheat, rice, soybeans and other oilseeds, peanuts, and pulses – all of which are referred to below as "covered commodities". They also provide for the continuation of benefits through marketing assistance loans for covered commodities. Several earlier programs, however, have been repealed, including the Direct Payments, Countercyclical Payments, and Average Crop Revenue Election programs. In their place, the legislation creates two new programs: Price Loss Coverage program and the Agriculture Risk Coverage program. In very general terms, this represents a shift away from direct payments to crop insurance.

In the Price Loss Coverage program, payments are provided to producers with base acres of wheat, feed grains, rice, oilseeds, peanuts, and pulses on a commodity-by-commodity basis when market prices fall below the reference price. A table of reference prices shows reference prices for wheat, corn, sorghum, and soybeans as $5.50, $3.70, $3.95, and $8.40 per bushel, respectively. The payment rate is the difference between the reference price and the annual national-average market price – or the marketing assistance loan rate, if that rate is higher). For each covered commodity enrolled on the farm, the payment amount is the payment rate, times 85 percent of base acres of the commodity, times payment yield.

Under the Agriculture Risk Coverage ("ARC") program, producers may choose county-based or individual coverage. For producers choosing county-based ARC, payments are provided to producers with base acres of covered commodities on a commodity-by-commodity basis when county crop revenue (actual average county yield times national farm price) drops below 86 percent of the county benchmark revenue (five-year average county yield times five-year average of national price or the reference price – whichever is higher for each year), calculated separately for irrigated and non-irrigated crops. For each covered commodity enrolled on the farm, the county ARC payment amount is the difference between the per-acre guarantee (as calculated above) and actual per-acre revenue (but no greater than 10 percent of the commodity's benchmark revenue), times 85 percent of base acres of the commodity.

Producers may choose to participate in ARC using individual farm revenue instead of county revenue. In that case, similarly complicated calculations apply.

Payment limitations apply: payments are limited to $125,000 for each individual actively engaged in farming, without specific limits for individual programs. A spouse may receive an additional $125,000. The limitation is applied to the total of payments for covered commodities from the PLC and ARC programs, and marketing loan gains and loan deficiency payments under the marketing assistance loan program. Adjusted gross income (AGI) limitations also apply. Any individual with an annual AGI above $900,000 (including nonfarm income) is ineligible to receive farm program payments under commodity or conservation programs.

The *conservation* provisions of the farm bill aim to provide assistance to producers and landowners to adopt conservation activities on agricultural and forest lands to protect and improve water quality and quantity, soil health, wildlife habitat and air quality. Program practices range from conservation activities that address natural resource issues and benefit productivity of agricultural working lands, forestlands, and grasslands to wetlands restoration and temporary or permanent land retirement.

However, the farm bill gradually reduces the Conservation Reserve Program cap from 32 million acres to 24 million acres by 2017. It also consolidates many conservation programs into new programs or merges them into existing programs, reducing the number of USDA conservation programs from 23 to 13. It also reestablishes, for the first time since 1996, the link between crop insurance premium subsidies and the conservation of highly erodible land and wetlands.

A principal new program is the Agricultural Conservation Easement program, providing funding for long-term easements for restoration and protection of on-farm wetlands and protection of eligible agricultural land from conversion to nonagricultural uses. The program consolidates the functions of the Wetlands Reserve Program, the Grassland Reserve Program (easement portion), and the Farmland Protection Program. Moreover, annual funding is significantly less than that provided for those predecessor programs. Other new or revised programs include the Regional Conservation Partnership Program, the Conservation Stewardship Program, and the Crop Production on Native Sod Program.

The current version of the *crop insurance* provisions provide new and continuing insurance products to protect producers against losses resulting from price and yield risks. Under the Federal crop insurance program, private-sector insurance companies sell and service the policies, and USDA's Risk Management Agency develops and/or approves the premium rate, administers premium and expense subsidies, approves and supports products, and reinsures the companies.

For instance, the new Supplemental Coverage Option creates a new insurance product for crop producers that provides area-based coverage in combination with coverage offered by traditional crop insurance policies.

Provisions in *support to beginning farmers* include increases in funding for beginning farmer development, facilitating farmland transition to the next generation of farmers, and improving outreach and communication to military veterans about farming and ranching opportunities. For instance, $33 million is available during 2014–2018 for the Conservation Reserve Program Transition Incentives Program to assist retired or retiring farmers when they transfer land to certain farmers, including beginning farmers. Moreover, $100 million is available during the same period for the Beginning Farmer and Rancher Development Program.

Other provisions of the 2014 Farm Act deal (i) with *horticulture* by supporting marketing and promotion of horticulture crops, such as through a pilot program on Procurement of Unprocessed Fruits and Vegetables (to facilitate procurement by schools of such foodstuffs), (ii) with *organic agriculture* by supporting organic certification, organic agricultural research, and organic crop insurance – partly in recognition of the fact that organic demand continues to outstrip supply, (iii) with *local and regional foods* by supporting increased consumer access to and marketing of locally and regionally produced food, via both farmer direct-to-consumer outlets and intermediated outlets, and (iv) with *rural development* by facilitating rural electrification, distance learning and telemedicine, water treatment, waste treatment, rural broadband availability, and rural micro-entrepreneurship.

Even with all these provisions and programs, 80 percent of the outlays under the 2014 Farm Act will go to funding nutrition programs, of which the Supplemental Nutrition Assistance Program is the centerpiece. About 8 percent will fund crop insurance, 6 percent conservation programs, and 5 percent commodity programs. The remaining 1 percent will fund all other programs mentioned previously.

Second and more generally, legislative and regulatory action should be taken that aims at *increasing the size and diversity of farm and rural populations* by improving economic and social conditions. In the USA, the 2008 farm bill gave support to rural development programs by providing some funding for enhanced infrastructure, health care, and communications systems.[8] As noted in Box 6.1, some support for rural development also appeared in the 2014 farm bill, but with generally reduced funding authorization levels. The trend of financial support should be up, not down.

Third, legal and regulatory steps are needed to provide support also for the *diversification of crops*. As noted in subsection IIB of Chapter 1

---

8 *Side-by-Side, supra* note 4.

(see especially Box 1.2), corn (maize), wheat, and rice together accounted for 89 percent of all cereal production worldwide in 2012, despite the fact that there are thousands of edible cereal grains that have been developed over the centuries.[9] A principal legal mechanism for re-introducing diversity into food crop production would be a *wholesale reorientation of agricultural subsidies*. Instead of continuing and expanding subsidy support for corn and other currently favored crops, laws – in the USA and elsewhere, of course – could sharply reduce financial support for those crops and sharply increase financial support for other crops, particularly the grains and legumes currently emerging from research into perennial polycultures that lie at the heart of natural-systems agriculture.[10]

As a blend of all of the points made immediately above, legislative and regulatory measures should ensure that a drastically revised system of agricultural subsidies places *effective "anti-concentration" subsidy ceilings* – that is, ceilings on the aggregate subsidy revenues that can be received (i) by any one farming entity, (ii) by large farming entities, and (iii) in respect of any particular crop. Such ceilings are not uncommon in US practice. For instance, some 1987 amendments to the 1985 farm bill set annual payment limitations at $50,000 per person for deficiency and "paid land diversion" payments and imposed a $250,000 aggregate payment limitation for certain types of subsidies.[11] Similarly, the 1973 farm bill reduced payment limitations from $55,000 (set in 1970) to $20,000 for all program crops.[12] As noted in Box 6.1, the ceiling established under the 2014 US farm bill is $125,000 for an individual actively engaged in farming; a spouse may receive another $125,000. Ceilings of this sort could usefully be kept lower and be applied in ways to encourage further diversification of crops and entities receiving the subsidy payments.

IB. Legal changes to counter ecological unsustainability

In the first section of Chapter 3, I offered a survey of reasons for concluding that modern extractive agriculture is ecologically unsustainable. At the outset of that survey, I quoted this passage from a 2000 journal article by J. B. Ruhl:

> Consider the typical farming process: first, remove all existing vegetation from the land and level it; second, deploy a single-species regime of crop

---

9 As noted above, several sources emphasize that 90% of the world's food comes from 30 crop species, even though about 7,000 crop species exist. See note 27 in Chapter 2, *supra*.

10 Such a reorientation of subsidies should also provide greatly expanded financial support for certain other types of food crops as well, particularly fruits and vegetables. For a reference to pertinent provisions in the 2014 US farm bill, see the next-to-last paragraph in Box 6.1.

11 *Agriculture: A Glossary of Terms, Programs, and Laws*, 2005 edition, at p. CRS-116, Congressional Research Service, www.cnie.org/NLE/CRSreports/05jun/97-905.pdf.

12 *Id.* at p. CRS-13 (see entry for P.L. 93–86, 1973 Agricultural and Consumer Protection Act).

or livestock; third, cultivate the crop or livestock with water and chemicals; finally, remove the crop or livestock and associated waste products from the land and start over. A number of environmental harms flow directly and necessarily from that basic reality of farming: (1) habitat loss and degradation; (2) soil erosion; (3) water resources depletion; (4) soil salinization; (5) chemical releases; (6) animal waste disposal; (7) water pollution; and (8) air pollution. In each of these categories, farms are a significant source of environmental harm.[13]

What should be arresting about this passage, especially against the backdrop of the description I have given of agroecological husbandry (in Chapters 4 and 5) is that Ruhl's litany of ecological ills applies to modern extractive agriculture but would *not* apply to agroecological husbandry. The "typical farming process" that Ruhl refers to is typical only to modern extractive agriculture. That form of agriculture does, for the most part, "remove all vegetation" each year (sometimes multiple times each year) during harvest.[14] Modern extractive agriculture does "deploy a single-species regime" of crops – that is, it employs monoculture. Modern extractive agriculture does rely heavily on chemicals. On the other hand, agroecological husbandry typically *does not* follow this pattern.

In considering, then, what legal changes should be made to counter the ecological unsustainability of modern extractive agriculture, we can focus not only on reducing the negatives – what Ruhl refers to as "environmental harms" – but also on enhancing the positives by facilitating a transformation from modern extractive agriculture to agroecological husbandry. Here are some specifics:

First, legislation should be adopted (and adequate funding provided) to *expand dramatically the ongoing scientific research into perennial species* of food grains and legumes that can gradually supplant the annual crops that dominate today's agriculture. The same sort of public (that is, government-funded) research support that has yielded high-performing annual grains – research that has been carried out in the USA, for instance, in land-grant colleges and universities under the Morrill Act of 1862[15] and the Hatch

---

13 J. B. Ruhl, *Farms, Their Environmental Harms, and Environmental Law*, 27 ECOLOGY LAW QUARTERLY 263, 274 (2000).

14 This practice of removing all vegetation is modified, of course, with no-till farming. However, the vegetation that remains following harvest in no-till farming consists largely of the dead remains of annual plants – and then any live vegetation remaining in the field is killed with chemical poisons such as glyphosate ("Roundup") just before the field is planted anew.

15 Officially titled "An Act Donating Public Lands to the Several States and Territories which may provide Colleges for the Benefit of Agriculture and the Mechanic Arts," the Morrill Act provided each state with sizeable acreages of Federal land that resulted in the establishment of state universities. In Kansas, for instance, the "land-grant" institution is Kansas State University, originally named Kansas State Agricultural College. Indeed, according to sources cited in the pertinent Wikipedia account, Kansas State University was the very first land-grant institution

Act of 1887[16] – should now be devoted instead to the development of natural-systems agriculture of all sorts, including perennial varieties of grains and other foodcrops.

Likewise, legislation should be adopted (and adequate funding provided) to *expand dramatically the ongoing scientific research into foodcrop polycultures.* Recall that a central feature of the natural-systems agriculture that I have described earlier is perennials grown not in monocultures but in polycultures. As described in Chapter 5, the development of perennial grains is further along thus far than the designing of polycultures. It is not too much of an oversimplification to say that the former – developing perennial grains – is a matter predominantly of plant-breeding research, whereas the latter requires new and intense study of ecology and evolutionary biology. In a sense, designing polycultures of perennials requires understanding the intricate chemical processes of prairie systems, since prairies (grasslands) provide the primary model for natural-systems agriculture. Accordingly, a massive increase in public funding should be directed toward research into foodcrop polycultures.

Legislative and regulatory steps should be taken also to *reorient agricultural subsidies.* Just as the "wholesale reorientation of agricultural subsidies" that I called for in subsection IA can help address the *economic* unsustainability of modern extractive agriculture, likewise such a reorientation can help address the *ecological* unsustainability of modern extractive agriculture. The reasoning is parallel: in both cases, reducing the current high subsidies given to a handful of specially-favored annual crops can facilitate a transition to a new form of agriculture – particularly if the public funds currently allocated for those existing subsidies are redirected in favor of the development and expansion of perennial grains and other forms of more sustainable food production. One means of reorienting subsidies is to tie them largely to a farmer's ecological performance (conditioning payments, for instance, on demonstrated improvements that the farmer has brought in soil quality over time), as opposed to the farmer's crop yields.

Legislative and regulatory steps should be taken also to *remove fossil-carbon subsidies.* Several of the "environmental harms" that Ruhl highlights result from the heavy use of fossil-carbon inputs on which modern extractive agriculture relies. Recall that I have used the term "extractive" not only to reflect the fact that modern agriculture extracts nutrients from the soil but also to reflect the fact that modern agriculture has, particularly in the past half-century or so, depended very heavily on extracted fossil carbon. This fossil carbon takes the form not only of fuel but also of feedstock (that is,

---

actually created under the Morrill Act; it was established on February 16, 1863, and opened on September 2, 1863. My daughter's KSU diploma carries a seal that refers to the institution still as "Kansas State University of Agriculture and Applied Science".

16 This legislation was styled "An Act to establish agricultural experiment stations in connection with the colleges established in the several States".

petrochemical inputs) for the production of synthetic fertilizers and various biocides (herbicides, fungicides, insecticides, rodenticides, etc.) – hence my use of the term "fossil carbon" rather than merely "fossil fuel".[17] In Box 6.2 ("An explanatory essay on fossil-carbon subsidies"), my colleague Caleb Hall explains that the fossil-carbon industry enjoys a broad range of publicly-funded support.

Box 6.2   An explanatory essay on fossil-carbon subsidies – Caleb Hall[18]

Although the fossil-carbon industry is probably the most profitable industry in history, with its profits at the international level reaching into the trillions of dollars, the industry is still sustained by a wide range of subsidies of one sort or another. Although many of these are indirect in character, just the *direct* subsidies – which take the form mainly of tax benefits – are of immense value and expense.

Tax policy is the most direct, and possibly largest overall, concession to the fossil-carbon industry in the USA and some other developed economies. It not only subsidizes the price of the end product, thus encouraging more people to rely on cheap fossil fuels and other

---

17 For instance, when natural gas, such as methane, is used in the creation of synthetic ammonia (widely used as fertilizer in modern agriculture), it is not used as a *fuel* but instead as a component in the chemical process that results in ammonia. For a summary explanation of that chemical process, see note 20 in Chapter 1, *supra*.

18 As noted in the Preface and the Acknowledgements to this book, Caleb Hall is an attorney with background in environmental science and legal experience in environmental law and policy. Now serving at the Missouri House of Representatives' Research Division, he served as a lead research assistant throughout my preparation of this book, contributing heavily to the information, analysis, and recommendations that it offers. In preparing this "essay on fossil-carbon subsidies", Mr. Hall has relied on numerous sources, including: (i) OECD iLibrary (Jan. 28, 2013), www.oecd-ilibrary.org/environment/inventory-of-estimated-budgetary-support-and-tax-expenditures-for-fossil-fuels-2013_9789264187610-en, (especially pages 370–379 of that OECD report); (ii) Organisation for Economic Co-operation and Development, *Taxing Energy Use*, OECD iLibrary (Jan. 28, 2013), www.oecd-ilibrary.org/taxation/taxing-energy-use_9789264183933-en; (iii) John Broder, *Obama's Bid to End Oil Subsidies Revives Debate*, THE NEW YORK TIMES, Jan. 31, 2011 at A14; (iv) IEA, OPEC, OECD, and World Bank, *Analysis of the Scope of Energy Subsidies and Suggestions for the G-20 Initiative*, Jan. 15, 2013, www.oecd.org/env/45575666.pdf; (v) Francisco Javier Arze Del Granado, David Coady, and Robert Gillingham, *The Unequal Benefits of Fuel Subsidies: A Review of Evidence for Developing Countries*, 40 WORLD DEVELOPMENT 2234 (2012); and (vi) International Monetary Fund, *Energy Subsidy Reform: Lessons and Implications*, (Jan. 28, 2013), www.imf.org/external/np/pp/eng/2013/012813.pdf. The last of these sources (from the IMF) provides an overview of international fossil fuel subsidy payments as well as profitability and the scope of certain industries involved in fossil fuels. For another overview of recent energy use and profitability in the USA and abroad, see US International Trade Administration, *2010 Industry Energy Industry Assessment*, Apr. 8, 2010, www.ita.doc.gov/td/energy/2010%20Energy%20Industry%20Assessment%20JAN10%20FINAL.pdf.

products that rely on fossil-carbon inputs; it also reduces the expenses of extraction and transportation. This reduces the risk and cost of doing business, and it encourages further unsustainable activity while at the same time dissuading renewable development and use.

Tax subsidies, by reducing or negating the cost of externalities for fossil-carbon industry participants, relieve the entire economic sector of costs that other businesses are typically required to pay. For instance, a business will usually be required to pay for the removal of its trash, but fossil-carbon industries largely escape that requirement – in the sense that they have not been expected to remove their "trash," manifesting itself as in large part as greenhouse-gas emissions, producing global climate change.

This tax policy does not reflect what anyone would consider a "free market", and therefore we should expect massive so-called "conservative" political outrage over such policy. However, this is not the case because the individuals who most vocally promote so-called "conservative" values are often the same persons who stand to benefit the most from this tax policy (and other subsidies to the fossil-carbon industry).

A common "conservative" response to criticism on this front is that these subsides are not intended to benefit the fossil fuel industries themselves but are instead intended to benefit the common consumer. If that were true, a wide range of other market mechanisms and tax structures could be employed to directly benefit those consumers. Instead, the current system of subsidies, taking the form of favorable tax policy, reflects a political choice made to favor fossil-carbon products over sustainability.

Thankfully, the fact that massive payouts (via subsidies of one sort and another) are made to fossil-carbon industries does have a bright side: by reverse implication, these payout amounts can be made available to be spent proactively – that is, in a way that will shift support *away* from fossil carbon and toward more sustainable uses. In other words, tax policy could be changed in the USA and other developed economies in ways that would substantially support renewable energy practices and natural-systems agriculture. Such a change in tax policy would, of course, require strong political will.

One useful source of details on the various forms and amounts of fossil-fuel subsidies, particularly preferential tax treatment, present in the member countries of the Organisation for Economic Co-operation and Development ("OECD"), is the OECD's *Inventory of Estimated Budgetary Support and Tax Expenditures for Fossil Fuels 2013*. This report provides details on the many different tax structures that the 34 member countries of the OECD have adopted, and it gives estimates on fossil fuel subsidies and preferential tax treatment in those countries. The report explains that the term "subsidies" refers to the myriad of ways of supporting one economic activity over another such as price quotas and controls, transferring funds directly, public assumption of risk, rebates, friendly tax treatment, and undercharging.

That OECD report provides the following highlights about the estimated budgetary support and tax expenditures for fossil fuels in the USA:

- Fossil fuels make up 84 percent of the US primary energy supply. Oil accounts for 36 percent, followed by natural gas at 25 percent and coal at 23 percent. Nuclear energy supplies 10 percent, and finally renewables, mainly biomass, account for 6 percent.
- The US was one of the first countries to deregulate the upstream oil and gas sector in the 1980s.
- Oil and gas producers are allowed to expense a share of intangible exploration and production drilling costs rather than to amortize them over time. Non-integrated oil and gas producers can amortize geological and geophysical expenditures over a two-year period and integrated producers over seven years. Also, oil producers are granted a tax credit of 15 percent of the investment costs related to the use of enhanced oil recovery methods when the real price of crude falls below a set level. Some states provide their own tax incentives as well, but more well-known are federal kickbacks such as the 2005 Energy Policy Act. It allowed refiners to expense 50 percent of the costs of capital equipment used to increase refinery capacity.
- Coal mining operations receive favorable taxes in the form of royalty income, the partial expensing of advanced safety equipment, and concessions for thin-seamed coal in Appalachian states.
- Many municipally owned utilities issue low-cost, tax-exempt debt to finance power plant construction and other facilities. Federally, power generators can amortize certain pollution control facilities over a seven year period, and tax credits are available for investment in "clean coal" technologies.
- The Low Income Home Energy Assistance Program, started in 1981, provides federal grants to low income households to finance energy bills.
- Off-road users of gasoline and diesel fuels, including farming, fishing, forestry, and mining sectors, are not subject to federal excise taxes on fuel, and most states then grant exemptions or lower taxes on those same sectors' fuel use.

- In 1975, the Strategic Petroleum Reserve was created to ensure a secure petroleum reserve in case of major supply disruptions. It accounts for half of the US emergency reserves, and it is financed completely by the federal government.
- The 2009 American Recovery and Reinvestment Act also increased funding for the federal fossil-energy research and development program looking into research such as coal liquefaction or fuel conversion.
- Alternative Fuels Production Credit (1987–present): The Energy Policy Act of 2005 provided a temporary income-tax credit of three dollars per BTU oil-barrel equivalent for coke and coke gas domestically produced. Coke or coke gas produced during a four-year period beginning on the later of January 1, 2006, or when the facility was placed into service, received the credit. However, the amount of credit-eligible coke could not exceed 4,000 barrels of oil-equivalent a day.
- In 1951, individual owners of coal-mining rights began to be able to use capital-gains tax rates rather than regular income-tax rates. This was meant as a way to boost coal production.
- Excess of Percentage over Cost Depletion: Usually the output from wells and mines allows producers to recoup the expenses in mineral exploration, but under percentage depletion producers can recover costs by claiming them as depletion allowances with a fixed percentage of gross income from the property. This can only be applied on up to 1,000 barrels of average daily production of domestic crude oil or an equivalent amount of natural gas. Eventually, the sum of these deductions can exceed the original investment cost substantially. For oil and natural gas property owners, the percentage ranges from 15 percent to 25 percent, but in marginal wells the deduction may run up to 100 percent of the net income from the property.
- An enhanced oil recovery credit gives oil and natural gas producers a tax credit of 15 percent of the investment costs related to enhanced oil recovery methods.
- Originally enacted in the Energy Policy Act of 2005, there is now an investment tax credit for power plants using integrated gasification combined cycle ("IGCC") or other advanced coal-based electricity generating technologies. The credit amounts to 20 percent for investments, and 15 percent for otherwise qualifying investments that use other advanced coal-based electricity generating technologies. The Treasury Department may allocate $800 million to the former and $500 million to the latter. In 2008, the rate was increased to 30 percent for new IGCC investment, and the Treasury

can now allocate an additional $1.25 billion of credits to qualifying projects.

- A 20 percent tax credit was also available for qualifying gasification projects with a $350 million ceiling. In 2008, the tax credit increased to 30 percent, and the Treasury Department can now allocate an additional $250 million to qualifying projects that sequester 75 percent of total CO_2 emissions.

Some perspective on the significance of these various forms of tax-based subsidies – both in the USA and in other OECD countries – can be gained by considering the fact that the fossil fuel industry is the most profitable industry in the history of the planet. Notwithstanding this high profitability, the fossil-carbon industry receives heavy subsidies internationally just as it does domestically in the USA. As the IMF has pointed out, the international public paid for $480 billion of fossil fuel profits in 2011 in "pre-tax" subsidies – that is, those that do not take into account the costs of negative externalities and energy consumption itself. Post-tax subsidies have been estimated as reaching up to $1.9 trillion. That amounts to 2.5 percent of the world's GDP or 8 percent of total government revenues.

Another source draws also on IMF calculations to emphasize that the equivalent of $1.4 trillion is spent each year in "mispricing" – that is, in externalities that cannot be readily quantified, such as air pollution and climate change. Once those costs are taken into account, the countries that provide the most subsidization of fossil-carbon fuel are the USA ($502 billion per year), China ($279 billion equivalent per year), and Russia ($116 billion equivalent per year). Using an economic system that actually accounts for those externalities would reduce global greenhouse gas emissions by 13 percent.

Given the various forms of injury they cause, the subsidies enumerated in Box 6.2 should be *removed* as gradually as necessary to avoid chaos but as quickly as necessary to blunt the worst effects of global climate change. This can be accomplished through legal and regulatory means, and doing so will help address several of the "environmental harms" that Ruhl enumerates. These include most prominently soil erosion, chemical release, and water pollution, assuming that the removal of fossil-carbon subsidies helps facilitate a move to natural-systems agricultural practices that (i) do not strip fields bare every season through chemical neutralization and (ii) do not rely on large quantities of biocides to prevent pests from attacking vulnerable shallow-root annual foodcrops. The shift to agroecological husbandry would also tend to reduce habitat loss and degradation – at least to the extent that such loss and degradation results from chemical poisoning.

Legislative and regulatory steps should also be taken to *stiffen agriculture-specific anti-pollution protections* to impose much stronger penalties on point-source and non-point-source pollution from farms. As Ruhl himself emphasized, "the truly pernicious effects of farming on habitat today occur offsite", resulting in part from the fact that "gaseous and dissolved nitrogen oxide and ammonia emitted from agricultural ecosystems are transported to and deposited in downwind and downstream terrestrial and aquatic ecosystems."[19] Moreover, as I noted in section I of Chapter 3, research into the effects that pesticides have on water, air, soil, and wildlife suggests that the impacts include the destruction of beneficial species, increases in pest resistance, reduction in pollination, crop losses, and more.[20] Stiffening laws, regulations, enforcement, and fines on agricultural run-off and pesticide use would help counteract this ecological damage. Such legal measures would work in tandem with some of the measures I have mentioned previously in order to internalize the negative externalities of modern extractive farming and thereby help facilitate a shift to agroecological husbandry.

Moreover, steps should be taken to *impose a system of penalties and credits for greenhouse gas ("GHG") emissions* from agricultural operations. As explained in Chapter 3, William Ruddiman has emphasized that farming constitutes "the largest alteration of Earth's surface from its natural state that humans have yet achieved",[21] and this alteration has involved climate change through the release of massive quantities of greenhouse gases.[22] Given the intense necessity now to reduce GHG emissions, and thereby to offset the worst effects of global climate change, the negative externality of GHG emissions from agriculture should be internalized. This could be achieved, from a legal perspective, by implementing a system by which the annual greenhouse gas emissions attributable to a particular farm or field are monitored and estimated, along with the annual amount of carbon sequestration that is attributable also to that farm or field. The legal system would then impose penalties for GHG emissions and credits for carbon sequestration. An effect of such a regime would be to force farmers to find ways to reduce their agriculture-based GHG emissions; one way farmers could achieve such a reduction would be a shift toward natural-systems agriculture.

---

19 Ruhl, *supra* note 13, at 276–277.
20 See text accompanying note 61 in Chapter 3, *supra*.
21 William F. Ruddiman, PLOWS, PLAGUES, AND PETROLEUM: HOW HUMANS TOOK CONTROL OF CLIMATE 163 (2005).
22 *Id.* Ruddiman asserts that "human activities linked to farming had taken control of the trends of two major greenhouse gases [namely methane and carbon dioxide] thousands of years ago" and that these activities led to "a long slow rise in greenhouse-gas concentrations prior to the industrialized era, and then much more rapid increases during the last 200 years of industrialization". *Id.* at 95.

Special *legal and regulatory attention should be given to livestock production in order to reduce its contribution to global climate change*. In some countries, including the USA, livestock grazing has had a devastating effect on grassland areas, and this ecological degradation has placed little cost on the owners (or the consumers) of the livestock. Although I have highlighted and criticized this aspect of livestock production in another book,[23] my emphasis here is not on grasslands degradation but on the role of livestock in climate change. Domesticated livestock should almost surely play a significant role in natural-systems agriculture – in terms of nutrient cycling, for instance, as from manure. However, as I noted in Chapter 3, around 50 percent or more of all methane (CH_4) emissions in the world are attributable to livestock – and methane is said to be more than 30 times more potent than carbon dioxide (CO_2) in terms of its impact on climate change.[24] If, as I have already proposed, greenhouse gas emission penalties and credits should apply to agricultural operations, livestock should not be exempted. Fortunately from the perspective of small farmers incorporating livestock into a natural-systems farming operation, livestock that are raised "on the small holdings of . . . pastoralists produce much less gas than the well-fed cattle in large-scale commercial enterprises".[25]

IC. Legal changes to counter human-health risks of modern extractive agriculture

In section II of Chapter 3, I examined some risks to human health that are attributable to modern extractive agriculture. The risks that I highlighted there fall into four categories, relating to (1) agricultural chemicals, (2) genetic technology, (3) food-borne illnesses, and (4) obesity. In turning now to the legal and regulatory reforms that might be made in response to those risks, I would offer the following observations.

First, as for *agricultural chemicals*, the legal reforms that I favor would aim at correcting what I referred to in Chapter 3 as "a serious mismatch between the amount of *use* of the chemicals and the amount of *caution* that surrounds that use". Instead of following the "reactive" approach taken in many countries (including the USA) – under which restriction on the production or use of agricultural chemicals typically occurs only *after* a specific and clear injury has been shown – I believe that rules should be adopted at the national and multilateral levels to follow the more public-oriented anticipatory

---

23 See John W. Head, GLOBAL LEGAL REGIMES TO PROTECT THE WORLD'S GRASSLANDS 41–43 (2012) [hereinafter GRASSLANDS].

24 See text accompanying note 79 in Chapter 3, *supra*.

25 Kirstin Dow and Thomas E. Downing, THE ATLAS OF CLIMATE CHANGE: MAPPING THE WORLD'S GREATEST CHALLENGE 50 (2006). For further discussion of this point, see also note 79 in Chapter 3, *supra*.

approach of the Precautionary Principle as practiced in Europe and as reflected in international legal instruments.

A brief reference to how the Precautionary Principle is practiced in Europe and is reflected in international legal instruments appears in Chapter 7.[26] For present purposes, the gist of the Principle can be drawn from the 1982 World Charter for Nature and the 1992 Rio Declaration. These two instruments describe the Precautionary Principle as follows:

> Activities which are likely to pose a significant risk to nature shall be preceded by an exhaustive examination; their proponents shall demonstrate that expected benefits outweigh potential damage to nature, and where potential adverse effects are not fully understood, the activities should not proceed . . .[27]
>
> In order to protect the environment, the precautionary approach shall be widely applied by States according to their capabilities. Where there are threats of serious or irreversible damage, lack of full scientific certainty shall not be used as a reason for postponing cost-effective measures to prevent environmental degradation.[28]

If, as I have asserted in Chapter 3, the key problem relating to agricultural chemicals is that they are still used too much and with too little caution, then a legal response to that problem is to require more caution. More broad-based and formal adoption of the Precautionary Principle would be a first step in that direction; a second step would be the issuance of detailed regulations at all levels restricting the manufacture, distribution, and use of agricultural chemicals. Under those restrictions, (i) heavy testing would precede any public availability or use, (ii) the burden of proving the safety of the chemicals would rest with the entities favoring manufacture, distribution, and use, and (iii) adequate safety requirements and mitigation guarantees would serve as backstops to guard against danger that might arise despite those precautions. Naturally, the intensity of such regulations – and the stiffness of penalties imposed for their violation – would vary depending on the significance of the risk.

In discussing earlier several legal reforms to respond to the ecological unsustainability of modern extractive agriculture, I suggested in particular (i) that fossil-carbon subsidies be removed and (ii) that pesticide use be restricted. To the extent that fossil carbon is involved in the production of agricultural chemicals, the reduction in fossil-carbon subsidies[29] would also contribute to a suppression of the use of such agricultural chemicals. Likewise,

---

26 See notes 15–16 in Chapter 7, *infra*, and accompanying text.

27 World Charter for Nature, art. 11(b), G.A. Res. 37/7, U.N. GAOR, 37th Sess., Supp. No. 51, at 17, UN Doc. A/37/51 (1982); 22 ILM 455 (1983).

28 Rio Declaration on Environment and Development, principle 15, UN Doc. A/CONF.151/26 (vol. I) / 31 ILM 874 (June 14, 1992).

29 For a survey of current fossil-carbon subsidies, see Box 6.2, *supra*.

the pesticide-restriction proposal I made earlier (because of its effects on non-human species) would dovetail with the agricultural-chemicals-restriction proposal that I have made here. In aggregate, all these proposals for legal reform would help facilitate a shift to agroecological husbandry, which minimizes the use of agricultural chemicals of all sorts.

The second category of human-health risks that I identified earlier relates to *genetic technology*. I offered a "nutshell" summary of this topic in subsection IIC of Chapter 3. In that summary, I explained that although research into genetic modification ("GM") and genetic engineering ("GE") do hold promise and should not be condemned out of hand, the current trajectory of GM/GE development should cause concern for several reasons. In my view, legal reforms to address those concerns could include the following:

- *Place the overall responsibility for and control over GM/GE research in the public domain instead of in private hands,* partly in order to ensure that such research is driven not by a profit motive but by a public-benefit motive, as well as to give proper respect for the complexity of nature (and hence the supreme importance of avoiding unintended bad consequences of such research). After all, while it is true that the profit motive can inspire important social progress, its use in the context of GM/GE research can hinder research from other institutions and groups whose efforts to discover other technological advances are barred by patent law.[30]
- *Impose special regulations on transgenic manipulation,* which poses peculiar risks. As I explained in Chapter 3, recent increases in consumption of meat have added pressure to create super-crops through transgenic manipulation, so as to produce more grain for use as livestock feed. Such pressure should be counteracted both (i) by legal restrictions on transgenic manipulation for these purposes and (ii) by imposing incentives through various fiscal and regulatory means to reduce high meat consumption (and also to favor locally-sourced meat). Besides, reducing high meat consumption could have human-health benefits in its own right.[31]

A third category of human-health risks that I identified earlier relates to food-borne illnesses. I offered a "nutshell" summary of this topic in subsection IID of Chapter 3. I asserted there that the key problems relating to modern conventional agriculture's contribution to food-borne illnesses are (i) that the handling of agricultural products occurs with too much haste and waste

---

30 Such efforts can include the practice of artificial selection of the sort that farmers have employed since the founding of agriculture. See David Morrow and Colin Ingram, *Case Comment – Of Transgenic Mice and Roundup Ready Canola: The Decisions of the Supreme Court of Canada in Harvard College v. Canada and Monsanto v. Schmeiser,* 38 UNIVERSITY OF BRITISH COLUMBIA LAW REVIEW 189, 207 (2005) (discussing the case of a farmer who was barred from storing and cultivating crops that he saved from a previous year because he failed to pay a licensing fee).

31 For further factors bearing on this issue, see note 104 and accompanying text in Chapter 3, *supra*.

and (ii) that improper reliance has been placed on antibiotics, especially in livestock, without adequate caution to guard against the emergence of pathogens that are resistant to those antibiotics.

Both of these problems can be addressed by legislative and regulatory reforms. The character and intensity of the reforms will depend, naturally, on specific circumstances. These will include (i) the degree to which rules and practices are already in place in a particular location and (ii) the degree to which consumers in a particular location know how to avoid food-borne illnesses. Especially in those circumstances where rules are lax on such things as inspection of agricultural products or limits on the use of antibiotics in livestock production, and where consumers lack the ability or the understanding to protect themselves against food-borne illnesses, strong legal protections should be adopted and implemented.

As I explained in Chapter 3, the topic of food-borne illnesses is not as central to my critique of modern extractive agriculture as most other topics discussed previously, such as the economic and ecological damage that modern extractive agriculture causes, or the risk that agricultural chemicals pose to human health. The same comment applies to the last of the four categories of human-health risks that I identified in Chapter 3: obesity. In the case of obesity, however, there is an important "re-connection" to the topic of subsidies, which figures importantly in both my economic and ecological critiques of modern extractive agriculture.

As I expressed it in subsection IIE of Chapter 3, the key problem relating to agriculture's role in obesity is that the structure of agricultural subsidies – as well as that of fossil-carbon-related subsidies – is deeply flawed, in a way that encourages production and sale of unhealthful products and discourages the production and sale of healthful food. Although one approach to this issue would be to restrict and/or criminalize the sales of unhealthful products (as Mayor Michael Bloomberg attempted to do in New York City), perhaps a better approach would be to engage in what I referred to earlier as a "wholesale reorientation of agricultural subsidies". Reducing subsidies on corn, for instance, could change the price structure of sugary soft drinks by increasing the cost of producing, and therefore the cost of purchasing, high fructose corn syrup that Americans in particular consume in breathtaking volumes. Likewise, shifting public subsidy payments away from the handful of foodcrops at the core of modern extractive agriculture and toward other crops, including fruits and vegetables, could help the epidemic of obesity that now afflicts some economically developed countries and poses a near-term threat to many others.

ID. Legal changes to counter the fact that modern extractive agriculture goes "against the grain" of human development

In the last section of Chapter 3, I explored some key ways in which the most recent manifestation of modern extractive agriculture – the form that has emerged in the last several decades to become "industrial agriculture" (a term I explained further in Figure 1.1)[32] – represents a radical departure from the course that human development took through earlier centuries. I asserted that this departure has been so great, at least in the West, that we should regard today's version of agriculture as running "against the grain" of human development. I offered some observations about indigenous (especially Native American) agricultural experience, which had several features. One was the "law of return".[33] Another was a high degree of sophisticated ecological knowledge.[34] Another was a broad diversity of crops. Yet another was a low incidence of erosion, and another was an impressively high energy efficiency – what some observers call the "energy return on investment" – in terms of the number of calories of (food) energy produced from a calorie of energy expended.

Then I discussed the Green Revolution and the advent of industrial agriculture. I emphasized that today's industrial agriculture has disregarded the "law of return", has abandoned the close knowledge of nature that "bioregionality"[35] implies, has abandoned an interest in crop diversity, has abetted the development of a "consumption ethic", has changed our relationship to land by "commodifying" it, has had a similar effect on our relationship to community and shared resources and destinies, and has

---

32 See Figure 1.1, "Composite Timeline of Selected Key Technological Developments in the History of Extractive Agriculture", in Chapter 1, *supra*.

33 Recall that the "law of return", as explained by Frederick Kirschenmann, involves nutrient cycling. Nutrient cycling, Kirschenmann says, "was common until the mid-nineteenth century" in the sense that most human and animal wastes were recycled as food for the plant kingdom. However, as society became more urbanized in the nineteenth and twentieth centuries, "human waste recycling became difficult, so we began depositing it in landfills and sewers." Most recently, "[a]s agriculture became industrialized in the late twentieth century and crop and livestock production systems became isolated from each other, recycling animal wastes became difficult and costly, creating unintended consequences such as soil loss, nutrient pollution, and imperiled water systems." As a consequence, we have "shifted entirely from a nutrient cycling to an input/ output system of production" in which we "rely exclusively on exogenous inputs to supply basic nutrients". Frederick L. Kirschenmann, CULTIVATING AN ECOLOGICAL CONSCIENCE: ESSAYS FROM A FARMER PHILOSOPHER 180 (2010). See text accompanying note 117 in Chapter 3, *supra*. For a discussion of the role that one of these exogenous inputs – synthetic nitrogen fertilizers made from ammonia, which requires natural gas – see text accompanying notes 19–20 in Chapter 1, *supra*. In his discussion of the "law of return", Kirschenmann draws from Paul Hawken's book *The Ecology of Commerce* to describe the law of return: "the return of wastes to the soil creates humus, which encourages healthy crops whose remains, properly composted, return to enrich the soil's humus content". Kirschenmann, *supra*, at 179.

34 An acknowledgement of the sophisticated ecological knowledge that has developed in indigenous groups can be seen in the work done by the World Intellectual Property Organization's Intergovernmental Committee on Intellectual Property and Genetic Resources, Traditional Knowledge and Folklore. For details, see the website of that committee at www.wipo.int/tk/en/igc/.

35 For other references to bioregionalism, see note 115 in Chapter 3 and note 27 in Chapter 4, *supra*, and accompanying text.

become almost breathtakingly energy-*in*efficient through its addiction to fossil carbon – all of this while delivering economic unsustainability, environmental degradation, and worrisome dangers to human health. In short, I posited that the profound changes that have occurred in agriculture – changes that the Green Revolution accelerated and intensified so dramatically – threaten to blind us to the fact that the benefits of those changes come at the *expense* of some values and ethics and efficiencies that for thousands of years have been central to our role and identity within the ecosphere and among ourselves.

What legal reforms am I suggesting to address the fact that modern extractive agriculture runs "against the grain" of human development in these ways? I would like to think that the legal reforms I have enumerated in the preceding paragraphs would have the effect of curing some of the ills that industrial agriculture has caused. Taking legal action to encourage the rise of small farms, to encourage a dramatic diversification of crops, to reorient agricultural subsidies, to remove fossil-carbon subsidies, to stiffen agriculture-related pollution restrictions, to internalize other negative externalities . . . these and other legal reforms that I have suggested hold some promise of restoring some of the values, ethics, and efficiencies that should be central to an understanding of the role humans are to play in the ecosphere.

Beyond those legal steps, however, are several essential and fundamental legal and institutional reforms. For the most part, these other reforms must involve multilateral collective action – by which I mean action outside the realm of the nation-state. In Chapter 7, I offer a detailed description of legal reforms I propose at the global level. Before turning to that, however, I wish to highlight some proposals made by other observers for legal reform in one particular country: the USA.

II. Making the necessary legal reforms: the US context

IIA. Envisioning and enacting a 50-year farm bill

The January 4, 2009, edition of *The New York Times* contains a column written by Wes Jackson and Wendell Berry titled *A 50-Year Farm Bill*. Drawing attention first to the catastrophic soil erosion that large rains caused in Iowa in the summer of 2008, Jackson and Berry explain in that column that it is *agriculture itself*, not the rains or other natural causes, that must be blamed for the long-term degradation of the world's soil. Jackson and Berry point particularly to "industrial procedures and technologies [that are] alien to . . . nature", and then they offer this elaboration:

> Agriculture has too often involved an insupportable abuse and waste of soil, ever since the first farmers took away the soil-saving cover and roots of perennial plants. Civilizations have destroyed themselves by destroying their farmland. This irremediable loss, never enough noticed, has been

made worse by the huge monocultures and continuous soil-exposure of the agriculture we now practice.

To the problem of soil loss, the industrialization of agriculture has added pollution by toxic chemicals, now universally present in our farmlands and streams. Some of this toxicity is associated with the widely acclaimed method of minimum tillage. We should not poison our soils to save them.

Industrial agricultural has made our food supply entirely dependent on fossil fuels and, by substituting technological "solutions" for human work and care, has virtually destroyed the cultures of husbandry (imperfect as they may have been) once indigenous to family farms and farming neighborhoods.

Clearly, our present ways of agriculture are not sustainable, and so our food supply is not sustainable. We must restore ecological health to our agricultural landscapes, as well as economic and cultural stability to our rural communities.[36]

Having identified the key problems of agriculture – soil loss through the use of monocultures and soil exposure, the toxicity of agricultural chemicals, a dependency on fossil fuels, and over-reliance on technological "solutions" – Jackson and Berry then assert that a principal way of addressing those problems is through concentrating on perennials:

Any restorations will require, above all else, a substantial increase in the acreages of perennial plants. The most immediately practicable way of doing this is to go back to crop rotations that include hay, pasture and grazing animals.

But a more radical response is necessary if we are to keep eating and preserve our land at the same time. In fact, research in Canada, Australia, China and the United States over the last 30 years suggests that *perennialization of the major grain crops like wheat, rice, sorghum and sunflowers can be developed in the foreseeable future.* By increasing the use of mixtures of grain-bearing perennials, we can better protect the soil and substantially reduce greenhouse gases, fossil-fuel use and toxic pollution.

Carbon sequestration would increase, and the husbandry of water and soil nutrients would become much more efficient. And with an increase in the use of perennial plants and grazing animals would come more employment opportunities in agriculture – provided, of course, that farmers would be paid justly for their work and their goods.[37]

---

36 Wes Jackson and Wendell Berry, *A 50-Year Farm Bill*, THE NEW YORK TIMES, Jan. 4, 2009.
37 *Id.* (emphasis added).

Jackson and Berry conclude their essay by urging legislative action that reflects a national agricultural policy to bring radical change to food production and rural life:

> Thoughtful farmers and consumers everywhere are already making many necessary changes in the production and marketing of food. But we also need a national agricultural policy that is based upon ecological principles. *We need a 50-year farm bill* that addresses forthrightly the problems of soil loss and degradation, toxic pollution, fossil-fuel dependency and the destruction of rural communities.[38]

A few months following the publication of the *New York Times* column, The Land Institute prepared a brochure offering various types of elaboration, including several detailed proposals for using the system of five-year farm bills that the US Congress has tried to follow in recent decades as providing a set of "mileposts" to measure progress toward a much more ambitious 50-year farm bill that would set the USA on a course toward making a "gradual systemic change in agriculture".[39]

In the following paragraphs, I will offer an updated and enlarged description of the proposals appearing in the 2009 "50-Year Farm Bill" brochure as prepared by The Land Institute, which I refer to in the following text as the "Land Institute Brochure". My aim is to highlight the key assertions made and positions taken in the brochure but to elaborate substantially on several of them and place them in the context of points I have made in this chapter.

IIA1. Aims

The overall aim of a 50-year farm bill for the USA would be to reorient US policy on a cluster of issues. Grain production would be at the center of those issues, for the simple fact that (as already emphasized elsewhere in this book), roughly three-quarters of US acreage currently devoted to crops is devoted to grain production, and roughly 70 percent of human caloric intake in this country comes from grains. The global figures are similar, and in fact the adoption of a 50-year farm bill for the USA could help trigger similar legislative initiatives in other countries – and, as I discuss later in this book, at the global level as well.

In addition to the issue of grain production, the cluster of policy issues that a 50-year farm bill would address also includes these:

- *Biodiversity and ecosystem health.* The Land Institute Brochure points out that the Millennium Ecosystem Assessment conducted a few years ago

---

38 *Id.* (emphasis added).
39 The Land Institute, *A 50-Year Farm Bill* (June 2009), on file with author and available from The Land Institute [hereinafter "Land Institute Brochure"].

under UN auspices[40] identifies agriculture as the "largest threat to biodiversity and ecosystem function of any single human activity".[41] (Several of the following bullet-points focus on particular aspects of ecosystem health as affected by US agriculture.)

- *Soil degradation and erosion.* As I have emphasized elsewhere in this book, soil degradation is an inevitable consequence of the annual-monocultures form of agriculture that has dominated grain production for thousands of years. A new farm policy as set forth in a 50-year farm bill would aim to break that domination and transform agriculture to a perennial-polycultures model of grain production. Doing so would reduce erosion, protect soil nutrients, reduce soil toxins, and manage soil nitrogen efficiently.

- *Water pollution from agricultural run-off.* The Land Institute Brochure explains that agriculture is responsible for 70 percent of US water contamination, and that 40 percent of US waters are unfit for swimming and fishing. It also notes that the leaching of nitrogen from the agricultural lands of the Mississippi Basin "is responsible for one of the largest dead zones in the world"[42] – that is, the area just off the Mississippi delta in the Gulf of Mexico.[43] A 50-year farm bill could begin a reversal of that trend by obviating the agricultural run-off pollution.

- *Agricultural-pesticide dangers.* The Land Institute Brochure explains that pesticides "are present in nearly every water and fish sample in agricultural areas" in the USA.[44] A natural-systems agriculture policy adopted through a 50-year farm bill could drastically reduce pesticide use.

- *Fossil-carbon dependence.* The Land Institute Brochure urges that it be a goal of a 50-year farm bill to "cut fossil fuel dependence to zero".[45] As explained elsewhere in this book, most of the elimination of agriculture's current fossil-carbon dependence could be accomplished by phasing out fossil-carbon-based fertilizers and other agricultural chemicals – as would be possible with the nutrient cycling that is central to a natural-systems

---

40 Information on the Millennium Ecosystem Assessment ("MA") can be found at its website, www.millenniumassessment.org. That website offers this general explanation of the project:

> Initiated in 2001, the objective of the MA was to assess the consequences of ecosystem change for human well-being and the scientific basis for action needed to enhance the conservation and sustainable use of those systems and their contribution to human well-being. The MA has involved the work of more than 1,360 experts worldwide. Their findings . . . provide a state-of-the-art scientific appraisal of the condition and trends in the world's ecosystems and the services they provide (such as clean water, food, forest products, flood control, and natural resources) and the options to restore, conserve or enhance the sustainable use of ecosystems.

41 Land Institute Brochure, *supra* note 39, at 10.
42 *Id.*
43 *Id.* at 1.
44 *Id.* at 10.
45 *Id.* at 4.

form of food production built around perennial polycultures. If this approach is taken, the only significant remaining fossil-fuel inputs, if any, that agriculture would require 50 years from now would be those used in powering the relatively few farm implements (such as tractors) that (i) are still needed for harvesting (perennial) crops and that (ii) have not yet been converted to solar or other alternative energy sources.

- *Greenhouse gas emissions and global climate change.* The Land Institute Brochure also identifies carbon sequestration as a goal of a 50-year farm bill.[46] I urge a broader goal: to drastically *transform US agriculture's role in the trajectory of global climate change.* A 50-year farm bill could realistically set and achieve this goal by adopting a natural-systems agriculture policy that would reduce GHG emissions not only in the two main ways described above – (i) phasing out fossil-carbon-based fertilizers and other agricultural chemicals and (ii) reducing fossil-fuel inputs for mechanized farm operations – but also by (iii) reducing those forms of livestock production that produce the most damaging volumes of methane emissions[47] and (iv) increasing the carbon-sequestration capacity of farmland through the development of deep and complex below-ground root-mass that is typical of perennials.

- *Farm and rural community restoration.* A different category of goals for a 50-year farm bill would be economic and social in character. As Jackson and Berry pointed out in the last line of their *New York Times* column on a 50-year farm bill, "[w]e need a 50-year farm bill that addresses forthrightly the . . . destruction of rural communities" that modern extractive agriculture has brought to the USA in the past several decades

---

46 *Id.*
47 As noted earlier, livestock-generated methane is a major contributor to global climate change, partly because methane itself is more than 30 times more potent as a greenhouse gas than carbon dioxide. Therefore, a reversal of the globally increasing demand for meat would bring not only health benefits but also a reduction in greenhouse gas emissions of a potently dangerous kind. Livestock production has an important role to play in natural-systems agriculture – a point emphasized, in fact, by Wes Jackson and Wendell Berry in their *New York Times* column calling for a 50-year farm bill – but the *form and extent* of such livestock operations would differ substantially from those that dominate the US livestock "industry" of today. The extent (that is, the volume of meat production) would be greatly reduced, reflecting a reduced demand for meat in human diets, and the CAFOs (concentrated animal feedlot operations) would largely disappear because livestock would be integrated into farm operations more generally – as they were for thousands of years until quite recently. For references elsewhere in this book to the relationship between methane emissions and livestock, see text accompanying note 79 in Chapter 3, *supra*, as well as the discussion of livestock production in subsection IIIC of Chapter 3, *supra* (noting that in an agroecological system of farming, the grazing of cattle and other large herbivores for human consumption of their meat would occur at drastically lower levels than those that prevail today, partly to control emission of methane and partly to reflect the need for more healthful, less meat-intensive diets than those currently followed in some countries). Notably, some observers have raised the prospect of genetic modification of livestock in order to reduce the emission of methane through their excrement.

– a destruction that I have seen first-hand where I grew up in northeast Missouri. For reasons I have examined in Chapter 2 (economic aspects) and the latter part of Chapter 3 (social aspects), the most extreme form of industrial agriculture that now dominates the USA has brought numerous negative consequences for farm communities across rural America. A 50-year farm bill would have as an important goal the reversal of those consequences by adopting policies favoring natural-systems agriculture with perennial polycultures at its core.

In sum, a 50-year farm bill would aim to reorient US policy not only on grain production but also on biodiversity, soil health and conservation, water quality, human health, independence from fossil-carbon dependence, climate health, and rural restoration.

IIA2. Legal and financial initiatives

What provisions would a 50-year farm bill include? I have enumerated in section I of this chapter numerous legal actions that could be taken to address the economic, ecological, and social unsustainability of modern extractive agriculture. In a bare-bones bullet-point list, those actions include the following:

- Take action to *reduce the high entry costs and other hurdles to small farmers and beginning farmers*, through subsidies and other incentives.
- More broadly, strengthen measures to *increase the size and diversity of farm and rural populations* by improving economic and social conditions.
- Provide support also for the diversification of crops, partly through a *wholesale reorientation of agricultural subsidies*. Such a reorientation would sharply reduce financial support for the small cluster of currently-favored crops and sharply increase financial support for other crops – particularly the grains and legumes currently emerging (or to emerge) from research into perennial polycultures that lie at the heart of natural-systems agriculture.
- As one part of this subsidization, provide funding to *expand dramatically the ongoing scientific research into perennial species* of food grains and legumes that can gradually supplant the annual crops that dominate today's agriculture.[48]

---

48 While I will not attempt to enumerate specifically what the contours of that research should be, or the financial and human resources that should be devoted to it, I will offer two examples of proposals that have been made in this regard. The first example comes from Wes Jackson and some of his colleagues at The Land Institute. It includes these details:

- hiring a set of core senior researchers (perhaps 50 PhD-level scientists, plus support staff) to lead teams in various research centers around the world, concentrating exclusively on the development of perennial polycultures; and

- Likewise, provide adequate funding to *expand dramatically the ongoing scientific research into foodcrop polycultures.* Perennial grains have many advantages over annuals, but ultimately a "mimicking" of the prairie architecture requires the development of mixtures of several species in a single field – different mixtures, of course, in different climatic and soil conditions.
- *Remove fossil-carbon subsidies.*[49]
- *Stiffen agriculture-specific anti-pollution protections* to reduce the ecological damage caused by agricultural run-off and pesticide use and as part of the overall effort to internalize the negative externalities of modern extractive farming and thereby help facilitate a shift to agroecological husbandry.
- *Impose a system of penalties for greenhouse gas emissions* from agricultural operations and credits for carbon sequestration.
- *Give special legal and regulatory attention to livestock production in order to reduce its contribution to global climate change.*
- *Adopt as national policy the Precautionary Principle* as practiced in Europe and as reflected in some international legal instruments,[50] and have this policy reflected in all agriculture-related decisions – including those bearing on the manufacture, testing, and use of agricultural chemicals.
- *Place the overall responsibility for and control over GM/GE research in the public domain instead of in private hands,* partly in order to ensure that such

---

- training, through the efforts of those senior researchers, another 100 or more PhD-level scientists to work in those research centers and then establish others in various regions.

Personal correspondence with staff of The Land Institute (2012–2014). A second example comes from the Missouri Botanical Garden ("MBG"). In a recent presentation, Allison Miller (of that institution and St. Louis University) explained the importance of expanding on work already underway at the MBG. A database called "Tropicos", universally accessible, provides details about many plants from Mesoamerica. The "Tropicos" database is just a small part of "a massive, global effort to document plant biodiversity on this planet". For instance, "the herbarium at the Missouri Botanical Garden includes more than 6.4 million individual plant records". Great momentum has already been built, therefore, toward creating what Miller calls "a global inventory project" resulting in "a botanical foundation for perennial polycultures". Address by Allison Miller to Prairie Festival, Salina, Kansas, (September 2015) (on file with author), at 3. It is that "global inventory project" that could be the subject of substantial research funding, expanding on a grant recently made by the Malone Family Land Preservation Foundation to St. Louis University, The Land Institute, and the MBG. This global inventory project will also draw on the resources of the US National Center for Genetic Resources Conservation in Fort Collins, Colorado and the Millennium Seed Bank in England. As Miller explains, the project has a "long-term goal of . . . identify[ing] wild, perennial, herbaceous species as promising candidates for pre-breeding and domestication" so as to develop perennial foodcrops." *Id.* at 4.

49 For an "explanatory essay" on fossil-carbon subsidies, see Box 6.2, *supra*.
50 For a brief description of the adoption of the Precautionary Principle in Europe, see note 15 in Chapter 7, *infra*. For some details about the adoption of the Precautionary Principle in international law, see text accompanying note note 26, *supra*, and note 16 in Chapter 7, *infra*.

research is driven not by a profit motive but by a public-benefit motive, as well as to give proper respect for the complexity of nature.

- *Impose special regulations on transgenic manipulation*, which poses peculiar risks.

A 50-year farm bill should include provisions to put in place the specific types of requirements, restrictions, and initiatives listed above, in order to bring fundamental change to US agriculture. Given the urgency of the problems that such change must address, the question of timing is paramount: how soon, and along what timeline, should the changes be made?

IIA3. Implementation schedule

In their 2009 *New York Times* column, Wes Jackson and Wendell Berry point out that some move toward perennialization is possible right away. After explaining the need to "restore ecological health to our agricultural landscapes, as well as economic and cultural stability to our rural communities", Jackson and Berry urge that such restoration "will require, above all else, a substantial increase in the acreages of perennial plants." They explain that "[t]he most immediately practicable way of doing this is to go back to crop rotations that include hay, pasture and grazing animals",[51] since hay and pasture – unlike commodity crops such as corn, wheat, and soybeans – are perennials, not annuals.

The Land Institute Brochure referred to on page 267, dating from 2009, picks up on this point about hay and pasture: "Pastures and perennial forage crops are already available either in permanent stands or in rotations. We propose incentives which would maintain the present perennial acreage and increase perennials in rotations."[52]

This, then, would be a first step in implementing the changes set in motion by a 50-year farm bill: encourage through subsidy and other programs the inclusion and expansion of perennials that are already available and have been used for countless generations of farmers. The more challenging part of the implementation program for a 50-year farm bill, however, relates to the *new* perennials that are currently under development. I suggest the timeline shown below. Although I have not attempted to propose specific funding details, I have included several parenthetical notations as to some of the financial implications of certain elements of this implementation schedule. As those parenthetical observations suggest, the 50-year farm bill would require a steady stream of funding. Two points are noteworthy, though, in this regard: (i) the funding requirements would be partly met or offset by reductions in certain existing subsidy payments that the new system would

---

51 See *supra* note 37.
52 Land Institute Brochure, *supra* note 39, at 6.

phase out; (ii) the funds expended largely take the form of "investment in infrastructure", with modest up-front payments (in training new farmers, for instance, or supporting research) paying long-term dividends.

2018 After broad public discussion and a build-up of political support and leadership, a 50-year farm bill is enacted with aims and specific provisions of the sort outlined above. Pertinent regulations are drafted and adopted.

2022 Four years into implementation, the farm bill starts prompting such changes as these:

- increases in acreages devoted to food crops other than annual monoculture commodity grains, including vegetables and tree-and-vine crops;
- increases in acreages devoted to hay or forage crops, as substitutes for annual feed grains in meat, egg, and milk production;
- dramatic expansion of research into perennialization of grains;
- increase in research into polycultures based on native prairie architecture;
- expanded training and financing of new farmers through incentive programs; and
- intensifying regionalization and localization of food markets.

(Most of the funding involved at this stage relates to the expense of research efforts concentrating on perennials and polycultures, along with the costs of training and incentives for new farmers.)[53]

(The increases in acreages devoted to certain types of crops, as referred to in the first two bullet points of this list, would involve reallocation of subsidy payments away from annual-monoculture crops to change incentives.)[54]

2026 The first perennial grain, Kernza® (wheatgrass), is "farmer-ready" for use on limited acreages.[55]

2028 Ten years after its enactment, the farm bill already results in slight changes in allocation among annuals and perennials – from 80 percent

---

53 According to The Land Institute, funding in an amount less than $50 million per year "would sponsor 80 plant breeders and geneticists who will develop perennial grain, legume, and oilseed crops, and 30 agricultural and ecological scientists who will develop the necessary agronomic systems", working on "six or eight major crop species at diverse locations". *Id.* at 2. See also note 48, *supra.*

54 Current funding and other details relating to commodity programs and crop insurance available under the US Agricultural Act of 2014 are found largely in Titles I (Commodities) and XI (Crop Insurance). For summary information about those programs, see Box 6.1, *supra.*

55 For a reference to Kernza® in its current form of development, see subsection IIB in Chapter 5, *supra,* and especially the text accompanying notes 55–65 in that chapter.

annuals to about 75 percent annuals – with increases in perennials occurring mainly in tree and vine crops and in hay or forage crops. In addition, several designs of polycultures for use in various climate conditions are in an advanced stage of testing and development.

(The education and training programs referred to here would span many years and could be conducted with existing resources in the Land Grant institutions and state agricultural agencies.)

2033   Extensive education-and-training programs have succeeded by this time in gaining broad understanding among farmers and consumers of perennial grain crops, leading to expansion of production of Kernza® and selected other perennial grains.

2038   Small but significant acreages are devoted by this time to perennial grains production, contributing to an overall increase to about 30 percent in the allocation of US farmland devoted to perennials. In addition, proven systems of simple polycultures are in production in selected areas.

2048   New perennial grain varieties – beyond Kernza® – are ready by this time to be used over an expanded geographical range. Some of these perennial grains are now being produced in expanded systems of polycultures. The annuals-to-perennials ratio of acreage allocation in the USA has shifted by this time to about 60%–40%, down from 80%–20% in 2018.

2058   The 60%–40% ratio of annuals to perennials has now flipped over: the allocation of cropland acreages in the USA now stands at about 40 percent annuals to 60 percent perennials. Because research efforts into polycultures have by now yielded good results, much of the perennial-grains production takes place in polycultures, thereby providing the broad array of benefits that polycultures offer (as in native grassland areas) – including protection against pests, protections against erosion, nutrient cycling, and other forms of resilience. Contributing to that resilience will be the results of other plant-breeding research yielding varieties of corn and soybeans with higher capacities to save water and nitrogen. Moreover, another form of resilience – the resilience of rural farm-based communities – has been reintroduced in several parts of the country.

(Public support for research of the sort referred to here will take the form of continuing funding allocations.)[56]

---

56 As noted above, funding and other details regarding agricultural research are found mainly in Title VI of the Agricultural Act of 2014, selected provisions of which are summarized in Box 6.1, *supra*.

2068   Many of the goals of the 50-year farm bill have been accomplished. For instance:

- perennials grains – some of them grown in polycultures – have by now gained dominance over annual grains, so that annual crops are grown mainly on the least erodible fields as short rotations between perennial crops;
- overall, perennials (including not only grain crops but also tree-and-vine crops as well as hay and forage crops) account for 80 percent of cropland acreage in the USA;
- ecology-smart agricultural subsidies have reduced various forms of farm-based pollution and largely arrested soil degradation and erosion;[57]
- the net contribution by US agriculture to greenhouse gas emission has diminished, and carbon sequestration has increased;[58]
- rural depopulation has been reversed in many agriculture-rich areas, with corresponding strengthening of rural institutions; and
- a high degree of regionalization and localization of food markets has been achieved.

The bar-graph in Figure 6.1 offers a visual representation of one aspect of the implementation of such a 50-year farm bill. It shows the gradual expansion of perennial crops and the gradual contraction of annual crops in the overall allocation of US cropland, assuming no significant change in the overall amount of land devoted to agricultural production in this country. For each of the ten-year stacked bars, the top three components are perennials – specifically, tree-and-vine crops, hay or forage crops, and perennial grains. The bottom two components in each stacked bar are annuals. Of those, the larger by far consists of annual grains (cereals, beans, oilseeds), until about 50 years from now, by which time they have declined in importance to a volume roughly equal to that of other annuals (including vegetables).

The bar-graph in Figure 6.1 illustrates several important changes that a 50-year farm bill of the sort described here is expected to trigger in terms of the overall allocation of US farmland to various forms of food production. Those changes include:

- an overall decline – slow at first, but increasing in later years – in the proportion of US cropland dedicated to producing annual grain crops, and of course a corresponding rise in perennials allocations. From a total

---

57 According to projections offered by The Land Institute, fisheries could be restored by around 2050 in those "dead zones" caused off the Mississippi delta by eutrophication due to agricultural run-off. Land Institute Brochure, *supra* note 39, at 7.

58 According to projections offered by The Land Institute, US agriculture could, under the scenario outlined here, be "CO_2 negative" by the 2060s. *Id.*

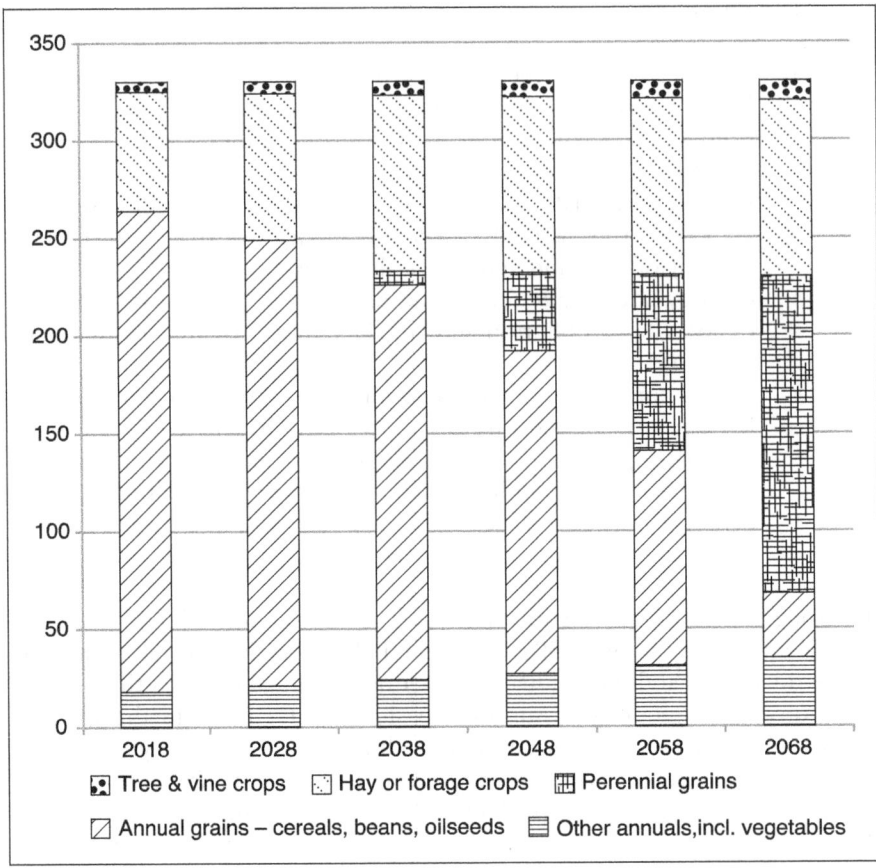

Figure 6.1 Transforming the allocations of annuals and perennials in the USA over 50 years, with constant cropland acreages in production[59]

of roughly 80 percent annuals and 20 percent perennials in 2018 (with the perennials comprising mainly hay or forage crops), the role of annuals drops to 20 percent by 2068 – and roughly half of the acreage then devoted to annuals will be in the form of vegetables, not grains.

- In fact, a reorientation of the agricultural subsidies program under a 50-year farm bill would roughly double the acreages devoted to vegetables and also roughly double the acreages devoted to tree-and-vine crops.
- By far the most important development appearing roughly halfway through the bar-graph – starting around 2038 and then dominating the

---

59 This bar graph draws in part from Land Institute Brochure, *supra* note 39, at 6.

graph by 2058 – is the emergence of perennial grains as the principal form of agriculture (in terms of acreage) in the USA. This dominance is reflected in the large medium-shaded portion of the stacked bars for 2058 and 2068, and it would be the result mainly of (i) research and development of the new grains and the polycultures in which they would be grown and (ii) education and training of farmers in the *use* of the newly-developed perennials grown in polycultures. These would also represent the two principal uses of funding required under the 50-year farm bill – in addition to whatever additional subsidy payments are directed toward perennial polycultures while they are removed from annual-monoculture crops.

Recall that in my description of a future based on agroecological husbandry, near the end of Chapter 4 of this book, I suggested that some areas of the world *should not* be "run by human beings for human purposes" in the way that Charles Mann urges. Instead, large areas should be returned carefully to wilderness and kept that way with as little human interference as possible. As I noted there, roughly 40 percent of our planet's land is currently used for food production (as compared with about 7 percent in 1700).[60] If we were to reduce that proportion by half, so that only 20 percent of the planet's land were used for food production, then the bar-graph in Figure 6.1 would change accordingly (since the USA is a major producer of world food supply). Instead of showing stacked bars of constant height (showing no significant increase or decrease in the amount of land devoted to food production in the USA over coming decades), the bars would get shorter for later years.

This speculation about the *global* allocation of land to food production, rather than focusing only on *US* land use, makes sense. After all, US agriculture is obviously part of a much larger global system. In that respect, there is of course some artificiality to the idea of a 50-year farm bill just for the USA. Many of the issues at play – global climate change and the results of scientific research are just two such issues – are global, not national, in their significance and coverage. It does make sense that the USA would take a lead in the sort of long-range project that a 50-year-farm bill represents. After all, it was US policy and pressure that created, for the most part, the agriculture of today. However, US efforts to reorient agricultural policy have global implications and require global cooperation. I will turn to these issues in Chapter 7.

First, though, I offer in the following paragraphs a survey of views expressed by other observers regarding what changes should be made in US

---

60 See text accompanying note 64 in Chapter 4, *supra*. As noted there, an area roughly the size of South America is used currently for crop production, while even more land than that is used to raise livestock.

law and policy relating to agriculture in order to achieve some of the same goals I have emphasized in the preceding pages.

IIB.   Embracing other proposals and critiques

The proposed 50-year farm bill discussed above is only one of many recommendations that have been made in recent years. The 50-year farm bill proposal happens to conform most closely to the contours of the agro-ecological husbandry proposals that I am making in this book, so I have highlighted it. Other proposals also have great value, however, and they warrant attention.

In an October 2008 "open letter" published in *The New York Times*, Michael Pollan presents a harsh critique of US agriculture and highlights in particular how monoculture farming is causing many problems. He offers several insights into what can be done – and what steps should be taken by the US President in his or her role as what Pollan calls "Farmer-in-Chief" – to promote sustainability and polycultures in US food production.[61] In the following passages, Pollan focuses especially on perennial polycultures:

> [Farmers should] be encouraged to grow as many different crops – including animals – as possible. Why? Because the greater the diversity of crops on a farm, the less the need for both fertilizers and pesticides.
>
> The power of cleverly designed polycultures to produce large amounts of food from little more than soil, water and sunlight has been proved, not only by small-scale "alternative" farmers in the United States but also by large rice-and-fish farmers in China and giant-scale operations (up to 15,000 acres) in places like Argentina. There, in a geography roughly comparable to that of the American farm belt, farmers have traditionally employed an ingenious eight-year rotation of perennial pasture and annual crops: after five years grazing cattle on pasture (and producing the world's best beef), farmers can then grow three years of grain without applying any fossil-fuel fertilizer. Or, for that matter, many pesticides: the weeds that afflict pasture can't survive the years of tillage, and the weeds of row crops don't survive the years of grazing, making herbicides all but unnecessary. . .
>
> Federal policies could do much to encourage this sort of diversified sun farming. *Begin with the subsidies: payment levels should reflect the number of different crops farmers grow or the number of days of the year their fields are green* – that is, taking advantage of photosynthesis, whether to grow

---

61 Michael Pollan, *Farmer in Chief*, THE NEW YORK TIMES: THE FOOD ISSUE (Oct. 12, 2008), http://michaelpollan.com/articles-archive/farmer-in-chief/. Michael Pollan is a professor of journalism at Berkeley and a prolific writer, perhaps best-known for his books OMNIVORE'S DILEMMA and THE BOTANY OF DESIRE.

food, replenish the soil or control erosion. If Midwestern farmers simply planted a cover crop after the fall harvest, they would significantly reduce their need for fertilizer, while cutting down on soil erosion. Why don't farmers do this routinely? Because in recent years fossil-fuel-based fertility has been so much cheaper and easier to use than sun-based fertility.

. . . Longer term, *the government should back ambitious research now under way (at the Land Institute in Kansas and a handful of other places) to "perennialize" commodity agriculture*: to breed varieties of wheat, rice and other staple grains that can be grown like prairie grasses – without having to till the soil every year. These perennial grains hold the promise of slashing the fossil fuel now needed to fertilize and till the soil, while protecting farmland from erosion and sequestering significant amounts of carbon . . . . But that is a 50-year project.[62]

After stressing the importance of perennials, and drawing attention to the "50-year project" of perennializing high-yield grain crops, Pollan then emphasizes the need for new and more farmers in order to operate a new and better form of food production:

. . . [W]ell-designed polyculture systems, incorporating not just grains but vegetables and animals, can produce more food per acre than conventional monocultures, and food of a much higher nutritional value. But this kind of farming is complicated and needs many more hands on the land to make it work. Farming without fossil fuels – performing complex rotations of plants and animals and managing pests without petrochemicals – is labor intensive and . . . will require more people growing food . . . . [This is especially true] here in America, where we have only about two million farmers left to feed a population of 300 million . . .

[Accordingly, the] sun-food agenda must include *programs to train a new generation of farmers and then help put them on the land*. The average American farmer today is 55 years old; we shouldn't expect these farmers to embrace the sort of complex ecological approach to agriculture that is called for. *Our focus should be on teaching ecological farming systems to students entering land-grant colleges today*. For decades now, it has been federal policy to shrink the number of farmers in America by promoting capital-intensive monoculture and consolidation. As a society, we devalued farming as an occupation and encouraged the best students to leave the farm for "better" jobs in the city. We emptied America's rural counties in order to supply workers to urban factories. To put it bluntly, we now need to reverse course. We need more highly skilled small

---

62 *Id.* (emphasis added).

farmers in more places all across America – not as a matter of nostalgia for the agrarian past but as a matter of national security.[63]

In short, Pollan calls for several of the legal and policy reforms I have touched on previously, including most notably a reorientation of agricultural subsidies, a strong support for polycultures and a perennialization of grain crops, and an aggressive program to restore farming in America.

Some similar points have been made by Mary Jane Angelo of the University of Florida. In a law journal article published in 2010, Angelo "explores a range of issues related to both the regulatory and incentive-based [US] federal programs that affect the crops we grow, the manner in which they are grown, and the human and environmental impacts of such programs." The ecological impacts that she examines include several I have examined in Chapter 3 of this book, such as modern extractive agriculture's impacts on water quality, its effects on biodiversity, and its implications for climate change. Angelo "evaluates the federal regulatory and incentive-based programs that encourage unsustainable, fossil-fuel-intensive, and environmentally destructive agricultural practices" and that form part of what she calls "a complex, outdated, and flawed agricultural policy that substantially interferes with the conservation of energy, water resources, and other natural resources, and substantially contributes to climate change." She then "proposes alternative approaches to agricultural policy that could dramatically reduce environmental impacts, be more protective of public health, result in a more nutritious food supply, and be more environmentally and economically sustainable, while helping to address the challenges of climate change and dependence on foreign fossil fuels."[64]

One set of specific proposals Angelo makes relates to the Clean Water Act ("CWA") of 1972. Under that statute's National Pollutant Discharge Elimination System ("NPDES") program, a permit must be obtained for any discharge of a pollutant from a point source into waters of the USA. However, the NPDES program explicitly exempts most agricultural discharges, in particular agricultural stormwater and irrigation return flow,

---

63 *Id.* (emphasis added). In other passages of his long "Farmer-in-Chief" column, Pollan also proposes (i) decentralizing the US food market, (ii) subsidizing four-season farmers' markets, (iii) establishing a Strategic Grain Reserve, (iv) supporting localized Agricultural Enterprise Zones (in which "[f]ood-safety regulations [would be] sensitive to scale and marketplace, so that a small producer selling direct off the farm or at a farmers' market is not regulated as onerously as a multinational food manufacturer"), and (v) establishing a definition of "food" for legal (and food-stamp) purposes, so that "in order to be regarded as a food by the government, an edible substance must contain a certain minimum ratio of micronutrients per calorie of energy. At a stroke, such a definition would improve the quality of school lunch and discourage sales of unhealthful products, since typically only 'food' is exempt from local sales tax." *Id.* Some of those same points are made by other observers whose views are summarized below.

64 See generally Mary Jane Angelo, *Corn, Carbon, and Conservation: Rethinking U.S. Agricultural Policy in a Changing Global Environment*, 17 GEORGE MASON LAW REVIEW 593, 594–595 (2010).

from the definition of "point sources". Later amendments to the Clean Water Act also exempted agricultural stormwater runoff. Consequently, Angelo explains, "most of the current significant water quality problems with the nation's waters are caused by these unregulated nonpoint sources discharges", and one of the two "greatest contributors to nonpoint sources water pollution [is] runoff from agriculture". Indeed, she says, "[i]n many areas of the country, agricultural runoff is considered to be the greatest challenge of water pollution control efforts."[65]

Angelo also turns her attention to agricultural subsidies. She cites numerous sources for the proposition that "[n]owhere else does US government policy create as perverse incentives as with our current system of agricultural subsidies".[66] In part, this perversity arises because the subsidies system largely disregards the fundamental precepts of sustainable agriculture. In this regard, Angelo points out that "[t]he Union of Concerned Scientists identifies five key techniques of sustainable agriculture: crop rotation, cover crops, soil enrichment, natural pest predators, and biointensive integrated pest management."[67] She explains further that the Union of Concerned Scientists uses an ecology-based definition of sustainable agriculture:

> [S]ustainable agriculture views a farm as a kind of *ecosystem* – an "agro-ecosystem" – made up of elements like soil, plants, insects, and animals. These elements can be enriched and adjusted to solve problems and maximize yields. This integrated approach is both practical and scientific: it relies on modern knowledge about the interactions within natural systems, as well as cutting-edge technologies, to achieve its results. It is a powerful approach that can produce high yields and profits for farmers while protecting human health, animal health and the environment.[68]

Having identified various legal and regulatory shortcomings of US agricultural policy, Angelo offers several proposals, including these:

- Impose stormwater treatment requirements on agricultural discharges that are currently not subject to Clean Water Act regulation because they are not defined as point sources.[69]
- Also under the Clean Water Act, eliminate the exemption from Section 404 wetlands permit requirements for normal agricultural activities, at

---

65 *Id.* at 614–616.

66 *Id.* at 638.

67 *Id.* at 641, citing Union of Concerned Scientists, Sustainable Agriculture Techniques, www.ucsusa.org/foodandagriculture/science_and_ Impacts/science/sustainableagriculture.

68 Angelo, *supra* note 64, at 641. Citing Union of Concerned Scientists, Sustainable Agriculture-A New Vision, www.ucsusa.org/food_and_agriculture/solutions/bigpicture-solutions/sustainable-agriculture-a.html.

69 Angelo, *supra* note 64, at 642.

least for large-scale farming operations. This, Angelo says, could protect many jurisdictional wetlands that are currently allowed to be plowed with impunity.[70]

- In order to address some of the environmental impacts stemming from agricultural pesticide use, amend the Federal Insecticide, Fungicide and Rodenticide Act ("FIFRA") by revising the standard for registering pesticides under FIFRA in a way that makes it clear that "high-risk pesticides may only be registered if there are overriding public health, social, or economic benefits that justify registration".[71]

- Further amend FIFRA so as to "create a mechanism for localized decision making" that "can take into account geographic factors to protect wildlife, natural resources, and ecosystem services." This could, Angelo says, be implemented through "a permitting system for largescale releases of pesticides into the environment wherein permit conditions could be imposed to maximize protection of natural resources and ecosystem services." Such conditions might include, for instance, the observance of buffer areas around certain habitats, or waterbodies, or nests, or the imposition of "restrictions on spraying certain pesticides during certain times of years to avoid [interference with] migration, breeding, or nesting".[72]

- Institute a proposal offered earlier by J. B. Ruhl to adopt a "Farm Release Inventory". Angelo explains that this would be "an approach similar to the Toxics Release Inventory . . . , which would require farms to publicly report releases of agro-chemicals" – a form of publicity that has been found effective in some circumstances to reduce careless releases.[73]

- Make big changes to the current US system of commodity subsidy programs. Angelo says in order to "make a dramatic shift to a more sustainable system of agriculture", the USA must engage in more than "mere tinkering with existing regulatory regimes" but must instead undertake a "complete overhaul of existing agricultural policy", including "a complete rethinking of commodity subsidy programs."[74] Doing so must, she says, fundamentally change an agricultural system that she cites Devon Peña as having called "one composed primarily of 'industrialized monocultures controlled by a handful of transnational corporations.'"[75]

- Specifically, decouple agricultural subsidies from production levels and tie them instead to ecological measures. Angelo points out that the European Union took a good step in 2003 when it decoupled its subsidies from production. Indeed, she says, a part of the EU's 2003 overhaul of

---

70 *Id.*
71 *Id.* at 643.
72 *Id.* at 644–645.
73 *Id.* at 645.
74 *Id.* at 646.
75 *Id.*

agricultural policy was the imposition of requirements that growers must comply with certain specified environmental practices in order to receive subsidies".[76]

- More specifically, impose a "tiered" system of subsidies under which "subsidy levels are tied to the adoption of varying levels of sustainable practices". Angelo suggests, for instance, that one level of subsidy would be available to "growers who continue to grow in a monoculturistic industrial fashion, but reduce their use of fertilizers, pesticides, and water and employ certain best management practices to limit erosion, depletion of organic matter in soils, contamination of ground and surface water, and harm to surrounding biodiversity", whereas a higher level of subsidy could be "provided to growers who meet existing USDA organic certification growing standards".[77]

In addition to Michael Pollan and Mary Jane Angelo, countless other observers have offered criticisms and proposals designed to reorient US agricultural policy in a way that will be more ecologically and environmentally sustainable and bring other benefits relating to human health, climate change, and improvement of rural and agrarian life. A small sampling of them yields the following viewpoints and recommendations:

- Matthew Walker urges a "regionalization" of food economies geared toward promoting sustainable practices in part through Agricultural Enterprise Zones.[78]
- Julie Foster surveys various proposals for a twenty-first-century farm bill for the USA, including putting an end to price supports, promoting polycultures, amending existing conservation programs, and other initiatives.[79]
- In two separate articles, William Eubanks focuses also on the failures of the farm bill conservation programs[80] and offers proposals for subsidizing sustainable agriculture.[81]

---

76 *Id.*
77 *Id.* at 654–655. Angelo explains that yet another level of subsidy might go to those "growers who, while not going as far as to meet organic standards, engage in a set of identified sustainable practices, which would require a very different approach to farming than large-scale monoculture industrialized production." *Id.* at 655.
78 Matthew J. Walker, *Exploring Regionalization of United States Agriculture: A Glance at Vermont Initiatives*, appearing in *Small, Slow, and Local: Essays on Building a More Sustainable and Local Food System*, 12 Vermont Journal of Environmental Law 353, 381–385 (2011).
79 Julie Foster, *Subsidizing Fat: How the 2012 Farm Bill can Address America's Obesity Epidemic*, 160 University of Pennsylvania Law Review 235, 255–277 (2011).
80 William S. Eubanks II, *A Rotten System: Subsidizing Environmental Degradation and Poor Public Health with Our Nation's Tax Dollars*, 28 Stanford Environmental Law Journal 213, 240–251 (2009).
81 William S. Eubanks II, *The Sustainable Farm Bill: A Proposal for Permanent Environmental Change*, 39 Environmental Law Reporter News & Analysis 10493, 10505–10507 (2009).

- Renee Johnson and Jim Monke discuss policy changes that the US federal government could adopt in order to deal with a planting restriction on fruits and vegetables (and the corresponding promotion of modern extractive agriculture).[82]
- Matthew Bradshaw discusses the "urban agriculture" movement in some cities and suggests ways in which a farm bill should address this issue – including a shift in subsidies away from monoculture-producing farms and toward small efforts to "agriculturize" urban areas.[83]
- Similarly, Sarah Schindler discusses "urban agriculture", with special emphasis on the need to lift bans on certain types of agriculture in rural areas.[84]
- William Even also concentrates on subsidies and proposes how they might be applied to ecologically-sound agricultural practices.[85]
- Melanie Wender likewise focuses on subsidies, urging that they be aimed at supporting ecologically sustainable farming practices.[86]

Proposals have come also not just from individual observers but from institutions. The National Sustainable Agriculture Coalition ("NSAC"), founded by my long-time friend Ferd Hoefner, issued an extensive policy framework for the proposed 2012 farm bill that included numerous recommendations on agricultural subsidies, conservation programs, and rural redevelopment.[87] The only specific references in the NSAC document to

---

82 Renee Johnson and Jim Monke, *Eliminating the Planting Restrictions on Fruits and Vegetables in the Farm Commodity Programs*, CRS Report for Congress (May 25, 2007).
83 Matthew Bradshaw, *The Rise of Urban Agriculture: A Cautionary Tale – No Rules, Big Problems*, 4 William & Mary Business Law Review 241, 257–258 (2013).
84 Sarah B. Schindler, *Of Backyard Chickens and Front Yard Gardens: The Conflict Between Local Governments and Locavores*, 87 Tulane Law Review 231, 276 (2012).
85 William J. Even, *Green Payments: The Next Generation of U.S. Farm Programs?*, 10 Drake Journal of Agricultural Law 173 (2005).
86 Melanie J. Wender, *Goodbye Family Farms and Hello Agribusiness: The Story of how Agricultural Policy is Destroying the Family Farm and the Environment*, 22 Villanova Environmental Law Journal 141 (2011). Wender offers this summary:

> Instead of eliminating the Farm Bill subsidies, Congress could shift a fair portion of those payments to farmers who are implementing sustainable agricultural methods. This policy would offer subsidies to all farmers based on their farming practices rather than the crops they cultivate, allowing smaller farmers to receive these payments.

Id. at 163. Wender also explains the proposed Food From Family Farms Act that the National Family Farm Coalition has urged Congress to enact. That legislation would replace agricultural subsidies with measures to ensure that farmers would receive fair prices for the sale of commodity crops; these would include a Strategic Reserve and a Food Security Reserve, similar to the "reserve" proposals referred to in Michael Pollan's "open letter" of 2008, discussed above. *Id.* at 164–165.
87 See generally *Farming for the Future: A Sustainable Agriculture Agenda for the 2012 Food & Farm Bill*, National Sustainable Agriculture Coalition (2012), http://sustainableagriculture.net/wp-content/uploads/2008/08/2012_3_21NSACFarmBillPlatform.pdf.

perennial polycultures appear in the context of "biofuel" – that is, the use of perennials in creating alternative (non-fossil-carbon) energy. However, the policy document does emphasize the disadvantages of current policies favoring monoculture concentrations.

Key points from the executive summary of the NSAC report, excerpted below, reflect the coalition's overall recommendations that US agricultural policy undergo some important changes:

> The overarching theme coming from farm fields and communities was that the investments from the 2008 Farm Bill in conservation, business development, and research took a step in the right direction. That step enabled farmers to meet the burgeoning consumer demand for sustainably produced foods. Simultaneously, farmers made it abundantly clear that gross inequalities persist through skewed farm subsidies, inadequate risk management tools, and continued consolidation in key agricultural markets. The NSAC farm bill platform both builds on the last farm bill by prioritizing programs that foster economic development and environmental stewardship, and proposes to reform our outdated farm subsidy system . . .
>
> *Farm Programs and Policies*
> Agriculture is a growing sector of our nation's economy, yet barriers to entry make farming one of the hardest careers to pursue. Current producers are then faced with policies that favor mega-farms over family farms, provide risk management tools for monocultures and not diversified systems, and facilitate market consolidation instead of independent production. Together with a strategy to remove barriers for beginning and socially disadvantaged farmers, reforms that restore commonsense rules to farm programs and crop insurance will help build a next generation of family farmers and will eliminate wasteful spending in agriculture. . .
>
> *Conservation Programs and Policies*
> As stewards of forty percent of the landmass in the United States, American farmers and ranchers are important managers of our natural resources. A suite of distinct but interrelated conservation programs work together to help producers protect and rebuild soil, improve water and air quality, and reverse habitat loss while maintaining productive farms and ranches. Paired with strong conservation requirements for the receipt of crop insurance, these programs reward good land stewardship, boost agricultural productivity, and contribute to a robust farm safety net . . .
>
> *Marketing, Food Systems, and Rural Development Programs and Policies*
> Local and regional agriculture is a major driver in the farm economy,

creating a unique set of opportunities and challenges for farmers and rural communities. Producers are responding to skyrocketing demand for local and regional food by increasing production and creating new markets. Rural entrepreneurs are launching new businesses. Despite the opportunities, significant infrastructure, marketing, and training barriers are limiting growth . . .

Research, Education, and Extension Programs and Policies
Investment in agricultural research is vital to continued productivity and innovation in diverse and expanding sectors of American agriculture. Consumer demand for sustainably and organically produced foods has fueled strong, continued growth in these sectors, yet federal investments in sustainable and organic farming research have not kept pace. The continued growth of these systems depends on a strong investment in sustainable and organic research, education, and extension programs . . .

The NSAC's themes of conservation, sustainability, rural development, research investment, and agricultural subsidies reform are elaborated in the remainder of its policy document, which details numerous specific proposals for legislative and regulatory changes that should be brought to US farm policy through the "farm bill" approach. Those themes and proposals are hardly new, of course. Two decades ago, in 1997, a task force established by President Bill Clinton reached consensus on nine key policy recommendations, many of which are echoed in the NSAC's policy document and in the various proposals offered recently by individual observers, as summarized on pages 278–284: (i) integrate pollution prevention and natural resource conservation into agricultural production; (ii) increase the flexibility for participants in commodity programs to respond to market signals and adopt environmentally sound production practices and systems, thereby increasing profitability and enhancing environmental quality; (iii) expand agricultural markets; (iv) revise the pricing of public natural resources; (v) keep prime farmlands in agricultural production; (vi) invest in rural communities' infrastructure; (vii) continue improvements in food safety and quality; (viii) promote the research needed to support a sustainable US agriculture; and (ix) pursue international harmonization of intellectual property rights.[88]

In sum, there is no shortage of proposals for changing US laws and regulations in ways that would transform agriculture in this country. Several that I have summarized immediately above, including those highlighted by the Presidential Task Force from two decades ago, strike me as lacking the necessary boldness – even though they were apparently far too bold to

---

88 The President's Council on Sustainable Development, *Sustainable Agriculture: Task Force Report* 1 (1997), http://clinton2.nara.gov/PCSD/Publications/TF_Reports/ag-top.html, cited in *Sustainable Farm Bill, supra* note 81, at 10509.

be incorporated to any significant degree in the series of "farm bill" legislation enacted in the last two decades. The problems of modern extractive agriculture are fundamental enough to warrant fundamental change. No such change has yet been embraced by US law and policy, including in the most recent farm bill signed into law in February 2014.[89]

In contemplating the prospects for fundamental change, especially in light of the rootedness of the interests favoring the status quo, it might be worth reflecting on just how the current system of modern extractive agriculture, and particularly the "industrial" form of agriculture that has characterized the last several decades, came into being. In a presentation he made a few years ago, Wes Jackson recounted some of that narrative,[90] drawing substantially from the account provided by Angus Wright in his 1990 book *The Death of Ramón González*.[91]

---

89 The various proposals summarized in the preceding pages for changes in US agriculture-related law and regulation were, for the most part, made before the enactment of the 2014 farm bill. However, this most recent US legislation made relatively few substantial changes in US farm policy and programs (although it did reduce food-stamp funding considerably). For instance, in its approach to financial support through agricultural subsidies, the 2014 farm bill changed the form of some of those subsidies (replacing some direct payments with crop insurance), but little else. In a *New Republic* article posted in early February, just before the 2014 farm bill was adopted and signed into law, David Dayen acknowledged the fact that the new legislation eliminated a two-decade-old program of so-called Direct Payments, but he emphasized that the new law did very little in fact to reduce subsidies or change their beneficiaries:

> Democrats and Republicans alike have pointed to the repeal of $4.5 billion in annual direct cash payments, a long disfavored policy where farmers receive a fixed amount of money for every acre they owned, regardless of whether it was planted. [This was a big change.] But don't be fooled: The politicians patting themselves on the back for [this change] have found a surreptitious way to deposit these savings right back in the pocket of agribusiness. That's because the farm bill will expand subsidies for crop insurance, which . . . actually hands over virtually the same amount of taxpayer money to farmers, mostly wealthy ones, as the old direct payment program. What's more, the shift from direct payments to crop insurance ensures that those handouts can be distributed in a hidden, more politically palatable way, making it more difficult to ever dislodge them.

David Dayen, *The Farm Bill Still Gives Wads of Cash to Agribusiness (It's Just Sneakier About It)*, THE NEW REPUBLIC, Feb. 4, 2014. In another analysis, a *Washington Post* reporter likewise emphasized that the 2014 farm bill would produce little change in the extent or recipients of agricultural subsidies; indeed, he surmised that under the 2014 farm bill, "subsidies to farmers may go up, not down . . . [so] there's a lot of debate about whether the new system is actually an improvement." Explaining that "direct payments" to farmers ("whether they planted at all") "had become untenable politically" because of their cost. In fact, though, the insurance payouts under the new system might "soar unexpectedly". Brad Plumer, *The Farm Bill Is Up for a Final Vote Soon (Here's Why So Many People Hate It)*, THE WASHINGTON POST, Feb. 3, 2014. For a synopsis of selected provisions of the 2014 farm bill, see Box 6.1, *supra*.

90 Wes Jackson, *Thirty Five Years: The Past and Beyond, the Future and Beyond* (address to The Land Institute Prairie Festival, Sept. 30, 2012) (on file with author).

91 See Angus Wright, THE DEATH OF RAMÓN GONZÁLEZ: THE MODERN AGRICULTURAL DILEMMA 171–178 (rev. ed. 2010).

In his summary of Wright's account, Jackson focused particularly on the confluence of four factors. The first factor was a powerful individual: Henry A. Wallace, in his role as US Secretary of Agriculture and then as Vice President under Franklin Roosevelt, had a passion for agriculture and technological innovation. The second factor was a tangle of political problems: Mexico was facing economic distress and social unrest and the West was (by some accounts) facing a Communist "Red Menace" that it wanted to repulse. In order to examine the Mexico situation first-hand, Henry Wallace traveled there from Washington, DC, in late 1940. The third factor was the build-up over several decades of the land-grant college system in the USA, with its close attention on agricultural development.[92] The fourth factor was the involvement of a private foundation: the Rockefeller Foundation was prepared to finance an ambitious project of scientific research to create new strains of high-yielding crops.[93]

From the confluence of those four factors came the bulk of modern US agricultural policy. Specifically, those factors led to both (i) the Green Revolution (perceived originally as an analogous and countervailing force to the "Red Revolution", and labeled with that analogy in mind), and (ii) a revolution also in US farm operations, rural life, and food production. The second of those revolutions – the one that transformed the economics and the ecology of agriculture in the USA – rested on a high-production, high-inputs, "get-big-or-get-out" philosophy. The results were dramatic. As I have noted earlier, the number of US farms and farmers plummeted over the course of just a few decades. Farming based on natural nutrient cycling was largely abandoned. Agriculture took on the attributes of industry. Farming

---

92 Regarding that land-grant college system, Jackson highlighted four legislative actions – the Homestead Act in May 1862, the Morrill Act in July 1862 (providing for the establishment of a land grant college in every state), the Hatch Act in 1887 (making possible numerous agricultural experiment stations), and the Smith-Lever Act in 1914 (creating the extension service to give farmers access to the most recent knowledge and research results – thus laying the foundations for "the land grant college system [in which the] USDA and the state colleges interacted, almost as one." *Id.* at 5.

93 Jackson explains:

Henry [Wallace] arrived home [from his trip to Mexico] in time for Christmas [1940]. Early in the new year of 1941, now vice president, he did not waste time to act on what he thought Mexico needed. He met with Raymond Fosdick, president of the Rockefeller Foundation in New York. He told Fosdick "If the Rockefeller Foundation would undertake to help the Mexican people increase the yield per acre of corn and beans it would mean more to the future of Mexico than anything else that government or philanthropy would devise." The answer was more bushels per acre . . . [and it was an answer the Rockefeller Foundation was ready to support].

Id. at 3–4.

was transformed to "agribusiness". The production of food became thoroughly dependent on fossil carbon.[94]

Given the fact that agriculture was transformed in such profound ways over the course of just a few decades, and given the fact that this transformation came about as a consequence of *conscious policy* adopted and implemented by a relatively small cluster of leaders in politics and technology with a clear vision of achieving certain goals, it strikes me as absurd – I would say even juvenile and small-minded – to think that radical change cannot be brought to US agriculture over the course of the coming half century. What I have tried to do in this section II of Chapter 6 is to sketch out the main features that such a radically changed approach to agriculture in this country might have.

As I have explained, I believe one of the most effective ways to do this would be to create a 50-year farm bill of the sort proposed a few years ago by Wes Jackson and Wendell Berry. Doing so would provide a legislative and policy template for addressing the ecological, economic, social, and other problems that plague modern extractive agriculture; doing so would put agriculture on a track toward agroecological husbandry that would feature perennial grains grown in polycultures and would restore farming life and farming communities to a state of health. The same approach could and should be taken in other countries as well, so that the same sort of national-level legislative and policy template would be put in place in capitals around the world.

However, action at the *national* level is not enough. I have concentrated in this chapter on changes required at the national level, and I have emphasized that the USA is a good place to start making such changes because of the central role that this country played in putting in place the most extreme form of extractive, "industrial" agriculture. However, in view of the fact that the catalogue of problems that I surveyed in Chapters 2 and 3 are global in character, not just regional or national, an aggressive *multilateral* approach will also be needed. What I turn to next, in Chapter 7, is an examination of some legal aspects of such a multilateral approach.

---

94 As Jackson explains, it was President Dwight Eisenhower's "Secretary of Agriculture, Ezra Taft Benson, [who] told farmers to, 'Get big or get out.'". As a consequence of this and other policies, Jackson says, "[f]armers added commercial fertilizer and more industrial farm equipment. Farms got bigger. Dairies declined, chickens and cows and hogs moved to feedlots, fences came down. As farm families left, rural communities declined. Longtime food security fell." *Id.* at 4.

7 A new collaborative global legal framework for agroecological husbandry

Having concentrated in Chapter 6 on a range of substantive legal changes that would be required at the *national* level – using the USA as an illustration – in order to make a transition from modern extractive agriculture to agroecological husbandry, I now turn to the *global* arena. This is a setting in which I feel most comfortable, having spent over 30 years practicing, exploring, and explaining international law. In this chapter I draw on that experience to address this question: What initiatives can and should be taken in *international law* in order to facilitate this sort of transformation of agriculture?

In order to set the stage for an analysis of that question, I would highlight three themes that I have tried to establish thus far in this book. All three of these themes can provide useful context and perspective on the topics to be explored in close detail in the following pages.

I. Collective ambition, natural reinvigoration, and the necessity of restraint

One theme is about *collective ambition,* and in particular whether it is realistic to strive for fundamental change in the existing agricultural system. At the end of Chapter 6, I posited that it is entirely realistic to seek and expect fundamental change in US agricultural law and policy. After having summarized several specific factors that combined to create modern US agricultural policy,[1] I made this assertion:

> Given the fact that agriculture was transformed in such profound ways over the course of just a few decades, and given the fact that this transformation came about as a consequence of *conscious policy* adopted and implemented by a relatively small cluster of leaders in politics and technology with a clear vision of achieving certain goals, it strikes me as absurd – I would say even juvenile and small-minded – to think that radical change cannot be brought to US agriculture over the course of the coming half century.

The same holds true at the global level as well. Although more parties and more points of view need to be involved and accommodated in the context of international negotiations, it would be naïve and short-sighted to assume that a collaborative system of rules and policies cannot be built at the global level to support a shift from modern extractive agriculture to agroecological husbandry. After all, such systems have been built before on other subjects. As I will describe in Chapter 8, modern history is replete with examples in which collective action on important issues was taken by nation-states (which I will refer to from this point on as just "states", in keeping with usual usage of that term in international law).[2]

---

1 These included the ambitions of Henry Wallace, a cluster of foreign-relations issues involving Mexico and Communism, the evolution of the land-grant college system, and the availability of private-foundation funding for agricultural research. See text accompanying notes 90–93 in Chapter 6, *supra.*

2 In international law, a state is an entity that has four features: a permanent population, a functioning government, a defined territory, and recognized capacity to conduct foreign relations. See Convention on the Rights and Duties of States, art. 1 (1933, entered into force 1934).

This first theme is pertinent to the discussion that follows in this Chapter 7, which includes numerous proposals for creating a new legal framework at the global level, particularly a new treaty system.[3] Some observers, especially in the USA, might hold the view that creating such a legal framework is impossible and unprecedented because states regularly disregard international law. People who hold this view are evidently unaware of the fact that most international legal rules, especially treaty commitments, are regularly honored[4] – more so, I would assert, than is true with national and local legal rules in many parts of the world. Many news programs emphasize, of course, those instances in which law-breaking occurs at the international level, as with military adventures by Russia in the Ukraine or by the USA in Iraq. Those instances are noteworthy, no doubt, but they tend to conceal the fact that states have collaborated for decades in establishing and administering among themselves a wide range of rules and procedures touching on important topics in the areas of trade regulation, commercial transactions, foreign investment, cross-border finance, legal process, telecommunications, navigation, shipping, diplomatic relations, and much more.[5]

Such rules and procedures are usually formalized by states negotiating, signing, and ratifying treaties, which is why the Statute of the International

---

3 For the most part, I try to be consistent in the terminology I use in describing legal and institutional initiatives at the global level that are under consideration in the discussions that follow. With some exceptions, I use the term "legal framework" (or "legal frameworks") to encompass both (i) a treaty system and (ii) an institutional structure regarding agroecological husbandry or agroecological integrity. I regard "framework" as a term having a larger compass than "system" or "structure". While I often use the term "treaty" (or "convention") on its own in the pages that follow, I also frequently use the broader term "treaty system" to signify matters lying outside the specific terms of a particular treaty; these matters include issues of participation, underlying philosophy, dispute resolution, and relations with other treaties and the states that participate in them. I have typically steered clear of the term "treaty regime" in order to avoid the connotation that "regime" might involve some sort of global government; but see my further comments on the term "global governance" in note 28, *infra*. When I use the term "institutional structure", as I will do in the two books that will follow this one, I aim to encompass not just a particular organization and its features but also the larger setting in which that organization fits and the political principles or other values it is meant to reflect.

4 One of the most highly respected international-law scholars of the twentieth century made this observation about the degree of compliance with international legal rules: "It is probably the case that almost all nations observe almost all principles of international law and almost all of their obligations almost all of the time. Louis Henkin, How Nations Behave – Law and Foreign Policy 47 (2nd ed., 1979).

5 Some of the main contours of these rules and procedures can be found by conducting an electronic search for the General Agreement on Tariffs and Trade, the Subsidies and Countervailing Measures Agreement, the Convention on Contracts for the International Sale of Goods, the New York Convention (governing international commercial arbitration), the Agreement on Trade-Related Investment Measures, the Convention on Stand-By Credits, the Hague–Visby Rules, the International Convention on the Settlement of Investment Disputes, the Hague Convention on the Taking of Evidence Abroad, the OECD Anti-Bribery Convention, the Vienna Convention on Diplomatic Relations, and scores of other treaties to which most states in the world are parties.

Court of Justice (like that of its predecessor, the Permanent Court of International Justice) puts treaties at the top of the list of those sources of law that the Court will look to in addressing disputes brought before it.[6] The significance of treaties in international relations is also reflected in the fact that most countries in the world have agreed to a detailed set of rules and principles – themselves found, not surprisingly, in a treaty – about how treaties are to be negotiated, observed, amended, interpreted, and enforced.[7]

In sum, it is possible and indeed rather commonplace in the modern age for collective action to be taken at the global level – that is, by states as the key "players" on that stage – for the purpose of establishing and implementing rules considered binding on, and mutually beneficial to, those states.

The first theme that I wish to emphasize, then, concerns the *possibility* of significant legal change through formal multilateral action. Such change is possible and precedented. To that first theme I would add a second one that relates to the *character and foundation* of the changes that I discuss in the following pages: those changes reflect the very same natural realities that underlie the shift I have described so far in this book from modern extractive agriculture to agroecological husbandry. The need to make that shift rests largely on the fact that modern extractive agriculture constitutes an attempt to either ignore or subdue nature, based on a profoundly flawed view that *Homo sapiens* somehow exist outside or above nature rather than comprising just one component of the ecosphere. A shift from modern extractive agriculture to agroecological husbandry would represent an acknowledgement that this is fiction. Moreover, an *urgent* shift of that sort – as I have suggested earlier in this book – would reflect the urgency of such issues as soil degradation, diversity collapse, and rapid climate change.

At the beginning of Chapter 4, where I started exploring the details of agroecological husbandry and how it differs from modern extractive agriculture, I emphasized this "natural-systems" foundation to agroecological husbandry. I also explained that certain aspects of it are in fact not really new:

> What I wish to do . . . is to explain how a different form of food production, one based on agroecological husbandry, is both possible and preferable. I begin . . . with a general description of agroecological husbandry . . . and then I turn . . . to a more detailed account of the advantages that [it] has over the form of agriculture that prevails in the world today . . .
>
> In describing below the main contours of agroecological husbandry, I emphasize the extent to which this alternative approach differs from

---

6 STATUTE OF THE INTERNATIONAL COURT OF JUSTICE, June 26, 1945, art. 38, 59 Stat. 1031, 1055, 1976 U.N.Y.B. 1052. For a discussion of the pertinent provision of the Statute of the International Court of Justice, see note 55 and accompanying text in Chapter 8, *infra*.
7 See the 1969 Vienna Convention on the Law of Treaties, *infra* note 30.

the approach it would replace . . . However, . . . some features of agroecological husbandry itself are not really new. In important ways, [it] reflects perspectives and values – particularly those that relate to how humans fit in, and depend upon, the natural world – that have merely been temporarily suppressed by the industrialization of agriculture and now need reinvigoration. Fortunately, some key scientific developments of just the past few decades make it possible now to forge a way *out* of the grip of modern extractive agriculture and to claim, or reclaim, the benefits that can come from a natural-systems approach to food production.

The essential point expressed in that explanatory paragraph is that agroecological husbandry is a new, different, and better approach to food production than modern extractive agriculture *but* that it does draw on certain "natural-systems" values that are part of our past and can be revived in revised form. The same essential point applies in respect of the collaborative system of international law that I describe in this chapter relating to global agriculture. In order to show how that point applies in this latter respect, I repeat here the paragraph just quoted, but this time with a few words substituted:

> What I wish to do . . . is to explain how a different form of ~~food production~~ <u>international law regarding agricultural and environmental matters</u>, one based on ~~agroecological husbandry~~ <u>a new collaborative framework of international law featuring a revised version of sovereignty</u>, is both possible and preferable. I begin . . . with a general description of ~~agroecological husbandry~~ <u>this new collaborative framework</u> . . . and then I turn . . . to a more detailed account of the advantages that [it] has over the form of ~~agriculture~~ <u>international law (regarding agricultural and environmental matters)</u> that prevails in the world today . . .
>
> In describing below the main contours of ~~agroecological husbandry~~ <u>this new collaborative framework of international law, and the principles and concepts it reflects</u>, I emphasize the extent to which this alternative approach differs from the approach it would replace . . . . However, . . . some features of ~~agroecological husbandry~~ <u>the new collaborative approach</u> itself are not really new. In important ways, [the approach] reflects perspectives and values – particularly those that relate to how humans fit in, and depend upon, the natural world – that have merely been temporarily suppressed by the ~~industrialization of agriculture~~ <u>persistence of a "monolithic" concept of sovereignty</u> and now need reinvigoration. Fortunately, some key ~~scientific~~ <u>legal</u> developments of just the past few decades make it possible now to forge a way *out* of the grip of ~~modern extractive agriculture~~ <u>the old framework</u> and to claim, or reclaim, the benefits that can come from a ~~natural-systems approach to food production~~ <u>collaborative framework of international law featuring a revised version of sovereignty</u>.

My *second* theme, then, is that there is a structural and logical similarity between (i) the changes I have discussed earlier in this book regarding agriculture and ecology and (ii) the changes I will discuss in this chapter and in Chapter 8 regarding international law. Importantly, the collaborative approach that my proposed international legal framework will require is not entirely new; instead, it has impressive precedents, particularly in the post-war years of the twentieth century.

A third theme that I wish to highlight here is also one that I have developed earlier in this book. It concerns *limits and restraint*. In Chapter 4, for instance, I explored some ideas about limits to growth. In several segments I also have referred to an "agriculture of restraint", citing Frederick Kirschenmann and others for the proposition that until recently, farming typically exhibited (especially in indigenous systems) some restraint in the extent of intensity and ambition, so as to remain prudently within operating limits of the ecosystem in which the farming occurred.[8] Examples abound, of course, of *un*restrained agriculture, and many of them have ended quite badly.[9]

The relevance of the notion of *limits* or *restraint* in the context of international law is twofold. First, the same natural ecological limits that agroecological husbandry aims to respect as a scientific matter will also need to be reflected in the substance of a collaborative framework of international law regulating and supporting agriculture. In other words, the survival of the human species will require some sort of legal system that implements and enforces an "agriculture of restraint". Moreover, and perhaps more importantly from a legal perspective, the natural jealousies and self-interest that states exhibit in their international relations will need to be tempered significantly in order to ensure that the human species stays within the natural limits that the ecosphere imposes.

Expressed differently, there must in my view be a "sovereignty of restraint" that parallels an "agriculture of restraint". Such a sovereignty of restraint would rest on the same ground, literally and figuratively, as an agriculture of restraint. In both cases, a recognition of systemic limits would prompt the players to refrain from overplaying their hands: states would serve long-term national interests by cooperating more and competing less; farmers would serve their own long-term interests (and those of other components of the ecosystem they share) by nurturing soil through the sort of husbandry I have discussed earlier.

How is such restraint – whether in the form of an agriculture of restraint or in the form of a sovereignty of restraint – to be achieved? From a scientific perspective, the answer can be found through continued research, such as the efforts to develop herbaceous perennial seed-producing polycultures. From a legal perspective, the answer lies in persistent efforts to negotiate new

---

8 For discussions of an "agriculture of restraint", see text accompanying notes 117 and 118 in Chapter 3, *supra*. See also note 118 in Chapter 5, *supra*. Some element of such an "agriculture of restraint" are nutrient cycling and an observance of the "law of return".

9 This is a theme developed in David Montgomery, DIRT: THE EROSION OF CIVILIZATIONS (2007). See also note 52 in Chapter 4, *supra*, and accompanying text.

rules that serve the mutual long-term benefit of the negotiating parties and the full array of constituents and interests that they are supposed to represent. At the international level, this effort takes the form of treaty-making. It is that subject which I address in this Chapter. Then, in Chapter 8, I offer some observations about a sovereignty of restraint – or, as I describe it there, a "pluralistic sovereignty" concept involving blended authorities.

II. A Global Convention on Agroecology: designing a new treaty system

Treaties – that is, negotiated contract-like agreements among states – constitute the principal means of establishing formal mechanisms for cooperation between those states. In the following paragraphs I enumerate many details about a proposed Global Convention on Agroecology. I explain, among other things, its overall outline and aims, the elements that might appear in its Preamble, certain specific principles that it would announce, the obligations that participating states would commit themselves to meet, and numerous other aspects. I do so on grounds that the *idea* of a treaty is of little use without some elaboration of the specific *content* of the treaty.

IIA. Preamble and Principles

Box 7.1 shows the outline of a proposed treaty intended to help facilitate a transition away from modern extractive agriculture and toward agroecological integrity. The treaty, styled "Global Convention on Agroecology", would consist mainly of a statement of principles and an enumeration of the responsibilities of Contracting States. One of those responsibilities would be to participate in the creation of a Global Corporate Trust for Agroecological Integrity, an organization that would have legal personality in international law and that would serve as the coordinating institution for various entities responsible for ecological protection and agricultural production.

IIA1. Preamble

The Preamble of the Global Convention on Agroecology (the "GCA", or the "Convention") would look both backward and forward. It would look *backward* by drawing attention to prior steps taken by the international community in the direction of a sustainable form of human life on Earth. For instance, the preamble could include clauses on these topics:

* *Key mileposts in the evolution of international environmental law.* The Preamble could cite the *Trail Smelter* arbitral decision from the 1940s,[10] the

---

10 The arbitral panel's decision in this case involving transborder pollution from Canada to the USA was an important early statement of liability for environmental harm by one state to another. For the text of the decision, see http://legal.un.org/riaa/cases/vol_III/1905-1982.pdf.

> ## Box 7.1 Outline of a Global Convention on Agroecology
>
> Preamble
> >   Context
> >   Aims
>
> Statement of Principles
> >   Integration and cooperation
> >   Natural-systems agriculture
> >   Meeting specific structural challenges
>
> Responsibilities of Contracting States
> >   Facilitating the advance of the principles
> >   Reporting and cooperation
> >   Participating in the creation of a Global Corporate Trust
> >   for Agroecological Integrity
>
> Definitions and interpretation
> >   Definitions
> >   Interpretation of provisions
> >   Dispute settlement and external review
>
> Closing provisions
> >   Reservations and denunciation
> >   Entry into force and provisional application

Stockholm Declaration emerging from the 1972 Stockholm Conference on the Human Environment, the various "MarPol" treaties on marine pollution, the 1982 UN Convention on the Law of the Sea (many of the provisions of which concentrate on protection of ocean resources, both living and nonliving), the Common Heritage of Mankind principle announced in that 1982 treaty and elsewhere, the Rio Declaration emerging from the 1992 Rio Conference on Environment and Development, both of the treaties that emerged from that Conference – namely the Convention on Biological Diversity ("CBD") and the UN Framework Convention on Climate Change ("UNFCCC") – the Vienna and Montreal Conventions on protection of the Earth's ozone layer, and other developments contributing to international environmental law.[11] Several such international legal instruments would

---

11 All of the treaties and declarations mentioned here are easily found online or in printed sources relating to international environmental law. For several other such international instruments, relating especially to trustee-like duties over resource management, see *infra* note 60.

provide both inspiration and guidance, of course, in the formulation of the Global Convention on Agroecology.

- *Recent developments in agriculture*, and a growing understanding today of the significance of those developments. These recitals would concentrate on certain unfortunate human actions in the past – agriculture-related environmental destruction generally, with the Great Plow-up as a specific illustration[12] – and on some particular ills generated by the Green Revolution and the industrialization of agriculture around the world. This portion of the Preamble could also identify important global institutions created in the past several decades to address these and other problems, including for instance the FAO and IFAD, as well as numerous research institutions and the Consultative Group on International Agricultural Research ("CGIAR") – but acknowledge the inadequacy of these efforts to date.
- *The special horrors presented by global climate change, and its relation to agriculture.* This portion of the preamble could refer to the two-way path of causation between agriculture and climate change: agriculture is one of the largest contributors to greenhouse gas emission, and hence to anthropogenic climate change, and climate change poses one of the greatest threats to agriculture in coming years.
- *The almost equally-immediate danger posed by species extinction* – a trend traceable both to climate change and to agriculture.
- *The inadequacy of attention given at the global level to issues of human population* – issues that cry out for attention now because of the strains that a dramatically-increasing global human population (careening toward 9 billion by mid-century) has already placed on the Earth's natural systems, especially in terms of greenhouse gas emissions, air and water pollution, soil degradation, biodiversity decline, and food insecurity.

The recitals in the preamble to the Global Convention on Agroecology would not only look backward. They would also look *forward* by announcing these general aims that the Convention would be designed to achieve:

- to announce a robust set of Principles of behavior pertaining to ecological protection and agricultural life and productivity – in essence, a new ecological conscience and a new form of agriculture;
- to reflect the commitment of the Convention's Contracting States to honor and implement those Principles;
- to give specific attention to the preeminent danger of global climate change by the contribution that a new form of agriculture – one that sequestered dramatically larger amounts of carbon – could make to the efforts at mitigating some of the worst consequences of climate change;

---

12 For an explanatory reference to the Great Plow-up, see text accompanying notes 14–15 in Chapter 1, *supra*.

- to anticipate the opening of a new chapter in international law that will create a multilayered system of protection and management that (i) will be founded on a more sophisticated, flexible, and beneficiary-oriented form of sovereignty (what I explain in Chapter 8 as "pluralistic sovereignty") and (ii) will have as its legal and structural manifestation a new Global Corporate Trust for Agroecological Integrity; and
- to undertake these reforms in a way that will attract the greatest possible participation by states and other entities, in order to assure broad global acceptance of changes that are essential to ecological protection, agricultural sustainability, and food security while honoring key principles of human rights and freedoms that must never be sacrificed.

IIA2. Statement of Principles

The heart of the Global Convention on Agroecology will be its Statement of Principles. As suggested in the outline in Box 7.1, the principles may be enumerated in three broad categories: those that focus on integration and cooperation, those that focus on a transition to a new form of agriculture, and those that focus on meeting special structural challenges.

As for the first of these – principles focusing on integration and cooperation – a rough draft of this portion of the Convention might include the following elements:

- *Principle 1. Reintegration of humans with the rest of the natural world.* The provision announcing this principle would emphasize the need for humans to recognize their shared ecological destiny with other species, giving special attention to soil, water, air, and biodiversity. The provision could draw from Principle 1(a) of the Earth Charter, which recognizes "that all beings are interdependent and every form of life has value regardless of its worth to human beings".[13]

---

13 See "The Earth Charter 2000", as reprinted in Klaus Bosselmann and J. Ronald Engel, eds., THE EARTH CHARTER: A FRAMEWORK FOR GLOBAL GOVERNANCE 257–261 (2010). In a similar vein, the Preamble of the Earth Charter also asserts that "we must recognize that in the midst of a magnificent diversity of cultures and life forms we are one human family and one Earth community with a common destiny." *Id.* at 257. The Earth Charter was launched in 2000 at the Peace Palace in The Hague by the Earth Council in cooperation with Green Cross International and UNESCO. Although it does not take the form of a treaty and is not generally regarded as binding on its own in international law, some observers assert that it "is now widely accepted as a foundational document for future international law and global governance." *Id.* at 15. For more details on the drafting and adoption of the Earth Charter, see *id.* at 7–19. The Earth Charter is, of course, only one of several instruments designed to prescribe principles and muster support for global environmental protection. Others include the Stockholm Declaration and the Rio Declaration referred to in text accompanying note 11, *supra*, the Aarhus Convention (Convention on Access to Information, Public Participation in Decision-making and Access to Justice in Environmental Matters), the Berne Convention on the Conservation of European

- *Principle 2. Responsibility of humans to restore the Earth's ecological integrity.* This second principle would build on the first. It would recognize that because of the special status and capacities of human beings, our reintegration with the rest of the natural world necessarily involves a duty of restoration.[14] That duty would be expressly identified in this Principle 2 as having the character of a trustee's duty. (I address issues of trusts and trustees near the end of this chapter.) More specific details regarding the discharge of this duty would appear in Principles 8 and 9 of the Convention (see below at pages 304–305).

- *Principle 3. Precaution in human activities affecting the natural world.* This third principle would unambiguously adopt the Precautionary Principle as it has developed in Europe[15] and as it has been reflected thus far in some international legal instruments,[16] including the Earth

---

Wildlife and Natural Habitats, the Convention on International Trade in Endangered Species of Wild Fauna and Flora (CITES), the Convention on Biodiversity (CBD), the Convention on the Conservation of Migratory Species of Wild Animals (CMS), the Espoo Convention (Convention on Environmental Impact Assessment in a Transboundary Context), the Convention Concerning the Protection of the World Cultural and Natural Heritage, the Ramsar Convention (Convention on Wetlands of International Importance, especially as Waterfowl Habitat) – and even the Convention for the Preservation of Wild Animals, Birds and Fish in Africa, which is recognized as one of the very earliest agreements on natural conservation even though it never entered into force after being signed by European colonial powers in 1900.

14 As expressed in the Earth Charter, humans have a duty to "[p]rotect and restore the integrity of Earth's ecological systems, with special concern for biological diversity and the natural processes that sustain life". EARTH CHARTER, *supra* note 13, at 258 (Principle 5).

15 The European version of the Precautionary Principle is described on various EU websites, including (i) the one for Eur-Lex (giving access to EU Law) at http://eur-lex.europa.eu/legal-ontent/EN/TXT/?uri=URISERV%3Al32042 (noting the importance of Article 191 of the Treaty on the Functioning of the European Union as amended in 2009 by the so-called Treaty of Lisbon, and summarizing the principle as requiring that "if there is the possibility that a given policy or action might cause harm to the public or the environment and if there is still no scientific consensus on the issue, the policy or action in question should not be pursued"); and (ii) the website for the February 2000 European Commission press release announcing the adoption of a "Communication on Precautionary Principle" at http://europa.eu/rapid/press-release_IP-00-96_en.htm (explaining the need for caution "where scientific evidence is insufficient, inconclusive or uncertain and preliminary scientific evaluation indicates that that there are reasonable grounds for concern that the potentially dangerous effects on the environment, human, animal or plant health may be inconsistent with the high level of protection chosen by the EU"). For commentary on the European version of the Precautionary Principle, see Kenneth Foster, Paolo Vecchla, and Michael Repacholi, *Science and the Precautionary Principle,* 288 SCIENCE, issue 5468, pp. 979–981 (May 2000); Robert v. Percival, *Who's Afraid of the Precautionary Principle?,* 23 PACE ENVIRONMENTAL LAW REVIEW 21 (2006); and Noga Morag-Levine, *Is Precautionary Regulation a Civil Law Instrument? Lessons from the History of the Alkali Act,* 23 JOURNAL OF ENVIRONMENTAL LAW, issue 1, pp. 1–43 (Jan. 2011).

16 As noted briefly earlier in this book, there are several instances in which the Precautionary Principle has been incorporated into international law. An early significant international legal reference to the Precautionary Principle appeared in the World Charter for Nature in 1982, adopted by the UN General Assembly. Article II, part 11(b), of that instrument asserts that "activities which are likely to pose a significant risk to nature shall be preceded by an exhaustive

Charter.[17] It would emphasize the importance of a minimalist approach ("light touch") that humans should take in their relations with the natural world – reminiscent of both (i) the approach described by Willa Cather in *Death Comes to the Archbishop*, from which I quoted in Chapter 4[18] and (ii) the approach shown in some forms of indigenous agriculture of the sort I discussed in subsection IIIB of Chapter 3.[19] Later in this chapter I will examine more closely this adoption of the Precautionary Principle

---

examination; their proponents shall demonstrate that expected benefits outweigh potential damage to nature, and where potential adverse effects are not fully understood, the activities should not proceed . . ." G.A. Res. 37/7, U.N. GAOR, 37th Sess., Supp. No. 51, at 17, U.N. Doc. A/37/51 (1982); 22 ILM 455 (1983). The Precautionary Principle was expressly endorsed five years later in the Second International Conference on the Protection of the North Sea. The Ministerial Declaration, within the meeting's resulting document, explains that "a precautionary approach is necessary" to protect the North Sea "from the damaging effects of the most dangerous substances." Second International Conference on the Protection of the North Sea, Nov. 24–24, 1987, www.seas-at-risk.org/1mages/1987%20London%20Declaration.pdf. That declaration goes on to say that prudence is to be applied "especially when there is reason to assume that certain damage or harmful effects on the living resources of the sea are likely to be caused by such substances, even when there is no scientific evidence to prove a causal link between emissions and effects ('the principle of precautionary action')." *Id.* at art. 1. In that same year (1987), the Montreal Protocol's Preamble contained a passage asserting that the Protocol's signatories are "*Determined* to protect the ozone layer by taking precautionary measures to control equitably total global emissions of substances that deplete it . . ." Montreal Protocol, 1987, 1522 U.N.T.S. 3. The year of 1992 then had four official recognitions of the precautionary principle in the international sphere. First, Principle 15 of the Rio Declaration on Environment and Development reads as follow:

In order to protect the environment, the precautionary approach shall be widely applied by States according to their capabilities. Where there are threats of serious or irreversible damage, lack of full scientific certainty shall not be used as a reason for postponing cost-effective measures to prevent environmental degradation.

UN Doc. A/CONF.151/26 (vol. I) / 31 ILM 874 (June 14, 1992). In addition, the UN Convention on the Protection and Use of Transboundary Watercourses and Lakes provides (in Article II) that parties shall be guided by the precautionary principle in minimizing transboundary impacts and protecting water ways. Moreover, the precautionary principle also appeared in the Convention on the Protection of the Marine Environment of the Baltic Sea. Likewise, it appeared again that same year in the Convention for the Protection of the Marine Environment of the North-East Atlantic. A more recent appearance of the Precautionary Principle in international law was in Article II of the Fifth International Conference on the Protection of the North Sea, the so-called Bergen Declaration, in 2002. Focusing on genetically modified organisms and sustainable fishing, that Declaration underscores "the need to apply the precautionary principle." *Id.*

17 See, e.g., EARTH CHARTER, *supra* note 13, at 259 (announcing the requirement in Principle 7 to "apply a precautionary approach" that would "[p]lace the burden of proof on those who argue that a proposed activity will not cause significant harm").

18 See text accompanying note 55 in Chapter 4, *supra*.

19 That subsection focused on key components of Native American food production before the European invasion began in the late fifteenth century.

in Principle 3 of the Convention, and especially the question of whether it is realistic to think that this can be accomplished.[20]

- *Principle 4. Cooperation of humans with other humans.* Whereas Principles 1, 2, and 3 would focus on the relationship of humans with the other (non-human) elements of the physical and living environment of the Earth, this Principle 4 would concentrate on humans' relations among themselves. It would, in recognition of our shared humanity, announce and endorse a duty of cooperation that would go far beyond the rather shallow and indirect duty appearing in Article 56 of the UN Charter.[21]

As for the second of the three categories of principles to be announced in the GCA – that is, principles focusing on a transition to a new form of agriculture – a rough draft of this portion of the Convention might include the following elements:

- *Principle 5. Creation of a new form of agricultural life and production.* Unlike Principles 1, 2, and 3, which concentrate on foundational propositions about the role of humans in nature (and Principle 4, about the relations among humans), Principles 5, 6, and 7 concentrate directly on agriculture. After all, the Convention would have as its *raison d'etre* the transformation away from modern extractive agriculture and toward agroecological husbandry of the sort described at some length in Chapters 4 and 5.

- *Principle 6. Recognition of the principal objectives of such a new form of agricultural life and production.* The four principal objectives – reflecting the specific problems I surveyed in Chapters 2 and 3 of this book – are:

  i   to overcome the economic unsustainability of modern extractive agriculture,
  ii to overcome the ecological unsustainability of modern extractive agriculture,
  iii to mitigate and reverse certain human-health dangers of modern extractive agriculture, and
  iv to ensure that agriculture develops in a way that is not "against the grain" of human development.[22]

Achieving those objectives would require successful effort on several fronts. It would require a reorientation of the economics of farming and

---

20 See subsection IIIA, *infra*.
21 Article 56 of the UN Charter asserts that member states "pledge themselves to take joint and separate action in co-operation with the [UN] Organization for the achievement of the purposes set forth in Article 55". Article 55, in turn, provides that the UN itself "shall promote" such things as higher standards of living, social progress, solutions to international economic and health problems, and universal respect for human rights and fundamental freedoms.
22 I addressed those four specific aspects of modern extractive agriculture in Chapter 2 and sections I, II, and III of Chapter 3, respectively.

almost surely also a reorientation of economic theory more generally, in order to reflect the mismatch between (i) a planet of limited resources and (ii) an economic model premised on the need for perpetual growth. (I will give further attention to these matters of economic theory later in this chapter).[23] It would require also the development of new grain crops based on a natural-systems approach of the sort I explored in Chapters 4 and 5. It would also require adoption of an approach of caution and restraint in agriculture (see Principle 3) and an appreciation for what Frederick Kirschenmann has called "the community component of agricultural sustainability".[24]

- *Principle 7. Protection of "safeguard rights" of all people during the transition to a new form of agriculture.* This Principle would assert that certain values are to be considered inviolable. These include the value of basic civil and political rights – to life, to fair treatment before the law, to gender equality, to participation in governance, and the like – so that during the dramatic changes that are called for in Principles 5 and 6 to create a new form of agriculture, powerful interests will not be permitted to act with inhumanity toward anyone, whether those powerful interests take the form of autocratic political systems such as that in China or the form of intolerant majorities.[25]

---

23 See subsection IIIB, *infra*.

24 After examining the economic and the ecological components of agricultural sustainability, Kirschenmann discusses its "community component", saying that "it is impossible to achieve either the economic or environmental goals of agriculture without the 'proper functioning of those social institutions which are essential to satisfactory farm life'". Frederick L. Kirschenmann, Cultivating an Ecological Conscience: Essays from a Farmer Philosopher 183 (2010) (quoting from Herbert C. Hanson, *Ecology in Agriculture*, 20 Ecology (1939). Kirschenmann continues:

> Thus, the economic, ecological, and community components of agricultural sustainability are interdependent. If we focus only on economic viability while ignoring farm ecology, soil quality will degrade, requiring more fertilizer, which affects the economic viability of the farm. Likewise, if we ignore community welfare, the public services that support the farm economy, such as public roads and quality local education and local research, will deteriorate and affect farm viability.
>
> Similarly, if we only attend to the ecology and the community and ignore the farm's economic viability, then the deteriorating economy of the farm makes it impossible for the farmer to actively support the community or properly care for the ecology. These three components are inextricably linked, and one cannot imagine a sustainable agriculture without acknowledging the interdependent whole.

Kirschenmann, *supra*, at 183–184.

25 In like fashion, the Earth Charter includes principles calling on its adherents to "[e]nsure that economic activities and institutions at all levels promote human development in an equitable and sustainable manner", to "[a]ffirm gender equality and equity as prerequisites to sustainable development", to "[u]phold the right of all [persons], without discrimination, to a natural and

I suggested above on page 299 that the third of the three categories of principles to be announced in the Global Convention on Agroecology would focus on meeting "special structural challenges". The structural challenges I have in mind include those that relate (i) to climate change, which involves not only agriculture but also the structure of global energy production and use more generally, (ii) to the structure of human presence on Earth (and humans' use of the Earth's resources), with specific attention to the size, distribution, and impact of human population on the planet, and (iii) to the structure of international relations, particularly under the Westphalian model of nation-state sovereignty. A rough draft of this portion of the Convention might include the following elements (in Principles 8, 9, and 10):

- *Principle 8. Urgent necessity to address climate change issues*, with special reference to agriculture. As I noted above in discussing the elements of the Convention's Preamble, global climate change – and the existential dangers that it presents – can be regarded as a prominent and immediate triggering cause for undertaking a reform of global agriculture, especially if that new form of agriculture promises to sequester dramatically larger amounts of carbon. Hence it seems appropriate that one of the Principles announced in the GCA would give special attention to global climate change. Specifically, it could establish a set of targets regarding climate change, such as a target to avoid at all costs a "four-degree world" of the sort that the World Bank has warned against in dire terms,[26] or some specific targets emerging from the climate-change-treaty negotiations in Paris in late 2015.[27] I will give further attention to the issue of climate change in subsection IIIC.
- *Principle 9. Urgent necessity also to address problems of population, preservation, and food security.* Reflecting points I made above regarding the Preamble to the GCA, the Convention's aim to reform agriculture would not be premised on a presumed need to feed an ever growing global human population. Rather, it would be to help address serious economic, ecological, and other problems posed by the form of agriculture that has prevailed in the world for the past several thousand years. Ecological

---

social environment supportive of human dignity, bodily health, and spiritual well-being", and to "[s]trengthen democratic institutions at all levels, and provide transparency and accountability in governance, inclusive participation in decision making, and access to justice". EARTH CHARTER, *supra* note 13, at 259–260 (Principles 10, 11, 12, and 13).

26 See World Bank, *Turn Down the Heat: Why a 4°C Warmer World Must Be Avoided* (2012). First released in 2012, this World Bank publication has now become a series, available at www.worldbank.org/en/topic/climatechange/publication/turn-down-the-heat.

27 For information about the late-2015 global climate change conference in Paris, hosted by the Conference of the Parties to the Framework Convention on Climate Change, see www.cop21.gouv.fr/en.

problems play a central role, and those ecological problems result in part from the breathtaking increases in human populations. Accordingly, the Convention should reflect a consensus that population growth must be seen as a problem to be solved, not as a "given" to be accommodated. The same logic applies to the related problems of species extinction, habitat degradation, and other ecological ills associated with the combination of modern extractive agriculture plus runaway human population growth: the natural world can be preserved, and food security can be attained, only if human population issues are addressed. This Principle 9 would acknowledge that reality.

- *Principle 10. Restructuring global governance,*[28] *particularly as it relates to ecological and agricultural affairs.* The Convention would offer in this Principle an express acknowledgement that effective reform necessary to implement the provisions of the Convention requires fundamental restructuring of the "international community" and how it is governed. This principle would recognize the legitimacy of some complaints and criticisms raised over many years about global governance – for instance, complaints made in the 1970s as part of the effort to create a New International Economic Order, complaints made against the global economic institutions (the IMF, the World Bank, and others) regarding a "democracy deficit", criticisms of the UN for lacking effectiveness, and indeed criticisms of international law in general for its apparent inability to prevent blatant disregard of its rules. Echoing one of the Preamble recitals I suggested above, this Principle would anticipate the opening of a new chapter in international law involving a multilayered, transparent, and widely-participatory system of governance, particularly as it relates to ecology and agriculture.

The ten Principles I have just outlined for inclusion in a Global Convention on Agroecology would naturally require careful formulation and clarification; these descriptions are only preliminary in character. As noted earlier, further elaboration would be inspired by the broad variety of international legal

---

28 For an explanation of my use of certain terms in this book – "international legal framework", "treaty system", and "institutional structure" – see *supra* note 3. As indicated there, I have shied away from the term "regime" in this context because it can have some negative connotations that I wish to avoid, including the specter of a repressive global government. However, I do not shy away from the term "governance", which I construe as a label for the entire body of mechanisms – messy and inefficient and unjust as they may be – by which international relations lurch forward. In my view, "global governance" does not connote the existence or desirability of any singular "world government" of some form. Whether there should be such a thing ("world government") is a different matter and not at issue here. In asserting that there should be a "restructuring of global governance" relating to ecological and agricultural affairs, this Principle 10 is urging changes that will make the mechanics for protecting the Earth against modern extractive agriculture less messy, less inefficient, and less unjust – for the benefit of all living and nonliving members of the ecospheric community, not just for humans.

instruments – some in the form of treaties, mostly in the form of resolutions and other non-binding instruments – that have emerged in recent years regarding international cooperation concerning environmental protection, agricultural development, biodiversity, and related issues.[29]

IIB. Responsibilities of Contracting States

The principal obligations of those states that would become parties to the Global Convention on Agroecology would reflect the Statement of Principles that I have described above on pages 299–305. Specifically, states would, by signing and ratifying (or acceding to)[30] the Convention, commit themselves to facilitate the implementation of those Principles in several explicit ways. There would be four ways in particular: legislation, regulation, education, and research. Beyond this responsibility to implement the Principles, states that become parties to the Convention would be committed to honor a duty of cooperation and to participate in institutional innovation. These various treaty obligations are summarized in the following paragraphs.

IIB1. Implementation of Principles – legislation and regulation

As is the case in many other types of treaties (for instance, human rights treaties), a key obligation of a Contracting State to the Global Convention on Agroecology would be to enact *legislation* that would breathe life into the Principles appearing in the Convention. Such legislation could in some cases involve constitutional amendments. These might be required, for instance, to clarify the state's recognition of certain "safeguard rights" mentioned under Principle 7. In most cases, though, they would involve promulgation of statutory enactments through regular parliamentary processes. Examples could include the following, some of which echo measures that I discussed in subsections IA and IB of Chapter 6:[31]

---

29 See text accompanying note 11, *supra*. See also notes 13 and 60, *supra*.

30 In international treaty law, a state that signs a multilateral treaty during a prescribed period while the treaty is "open" for signature will not be bound fully and immediately to the treaty's provisions merely by that act. It will complete its "consent to be bound" by ratification of the treaty (or, more precisely, by a ratification of its earlier signature). By contrast, a state that has not signed a treaty (as when the period during which the treaty was "open" for signature expired before that state even existed) may typically express its consent to be bound by the treaty through the process of "accession" ("acceding" to the treaty). States expressing their consent to be bound in either fashion – that is, either through ratification or through accession – are equally committed to honor the provisions of the treaty under the doctrine of *pacta sunt servanda*. For details, see VIENNA CONVENTION ON THE LAW OF TREATIES, arts. 11, 14, 26 (1969, entered into force Jan. 27, 1980) [hereinafter "1969 VCLT"].

31 Those measures addressed such things as support for small farmers and beginning farmers, increasing the size and diversity of farm and rural populations, crop diversification, subsidies reform, and more.

- legislation defining and adopting the Precautionary Principle,
- legislation setting aside "nature reserves" for the preservation of endangered or threatened species and to restore biodiversity more generally,
- legislation prohibiting agricultural practices (such as nitrate discharge) causing serious and direct ecological damage,
- legislation requiring that entities responsible for environmental injury pay compensation – the "Polluter Pays Principle",[32]
- legislation reorienting agricultural subsidies – perhaps in ways that I explained previously for correcting certain economic and ecological problems stemming from such subsidies,
- legislation removing fossil-carbon subsidies, and
- legislation facilitating land reform, particularly by using the "constructive trust" (or "public trust") doctrine to protect the interests of various participants in the ecosystems falling within the jurisdiction of the Contracting States – a matter I shall expand on in subsection IIB3.

To these examples of legislation that the Convention might require Contracting States to enact could be added many others types of laws. In most countries, the implementation of such laws is conducted by administrative and regulatory agencies. Accordingly, *regulation* of various forms would also be required by the terms of the Convention. Prominent among these forms of regulation would be the administration of (reoriented) agricultural subsidies, the handling of environmental impact assessment procedures (in accordance with the Precautionary Principle), the management of "nature reserve" areas set aside for biodiversity protection, and so forth.

IIB2. Implementation of Principles – education

In addition to legislation and regulation, Contracting States would commit under the Convention to enhance broad *public education* aimed at implementing the Principles that it announces. One purpose of such education – particularly applicable in those economically developed countries in which agriculture is no longer visible in the lives of most people – should be to provide everyone with at least a basic grounding in food production. As noted earlier, one of the reasons it is difficult for many humans to recognize their dependence on local ecosystems is that they are so disconnected from

---

32 The "Polluter Pays Principle" was adopted by Principle 17 of the Stockholm Declaration, and it provides that all environmental costs of a product or activity be internalized, so that, as one resource expresses it, "the actor who might cause pollution or environmental harm bears the costs of avoiding the harm, cleaning up the mess and compensating for any injuries". EARTH CHARTER, *supra* note 13, at 10. See also Peter P. Rogers, Kazi F. Jalal, and John A. Boyd, AN INTRODUCTION TO SUSTAINABLE DEVELOPMENT 190 (2008) (speculating on whether the Polluter Pays Principle is already part of international law).

the natural world; in some societies, most people do not even know the most basic details of their own food production.[33]

In *all* countries, though, a basic grounding should also be provided in Earth sciences – a part of what Wes Jackson of The Land Institute is fond of calling "ecospheric studies"[34] – so that people would understand the importance, both in their own ecoregion and for the world as a whole, of climate cycles, soil conservation, the "law of return",[35] biodiversity, and the like. Widespread popular knowledge of such matters is essential, and the Convention should record the commitment of Contracting States to educate their people accordingly. The Earth Charter reflects the importance of such a commitment by positing in general terms that we should "[i]ntegrate into formal education and life-long learning the knowledge, values, and skills needed for a sustainable way of life".[36]

IIB3. Husbandry, land, and constructive trusts

It bears emphasis that such an education would go beyond science. It would also impart to the populace at large what Frederick Kirschenmann calls an "ecological conscience".[37] One element of this is the concept of "husbandry" that serves as the centerpiece of agroecological husbandry as I have described it previously, especially in Chapter 4. This element warrants special emphasis here, partly because the need for husbandry – aimed at placing agricultural production on a sustainable foundation for the future – forms the rationale for the type of land reform that I referred to above in enumerating the forms of legislation that the Global Convention on Agroecology would call on its Contracting States to undertake.

In introducing the concept of "husbandry" in Chapter 4, I drew a connection between that concept and the concept of the trust. I explained there that a trust (as a legal matter) involves an equitable obligation that legally binds a person (the trustee) who has legal title and control over certain

---

33 See text accompanying note 56 in Chapter 4, *supra*, drawing from the writings of Frederick Kirschenmann.

34 The Land Institute has sponsored two three-day conferences in 2015 and 2016 in order to explore the meaning and significance of "ecospheric studies", and to organize educational initiatives that could encourage a worldview in which the ecosphere – as distinct from merely the biosphere (life on Earth) or the ecosystem (confined to a particular territory) – would be the primary frame of reference for policy-making. For views that resemble those discussed at the conferences mentioned above, see the writings of J. Stan Rowe, cited and summarized in notes 100–102 in Chapter 8, *infra*.

35 For explanatory references to the "law of return", see text accompanying note 117 in Chapter 3, *supra*, and text accompanying note 33 in Chapter 6, *supra*.

36 EARTH CHARTER, *supra* note 13, at 260 (Principle 14).

37 Kirschenmann urges the USA "to launch a nationwide educational program to foster a national ecological conscience", and he quotes Aldo Leopold for this observation: "Obligations have no meaning without conscience from people to land." Kirschenmann, *supra* note 24, at 187.

property (the trust property) to manage that property *not* for his or her own direct benefit but rather for the benefit of a specified group of persons named as the beneficiaries of the trust.[38]

Viewed from an ecological perspective, the concept of the trust could be used to characterize the humans alive today as trustees who bear a responsibility to protect the Earth (the "trust property") for the benefit of two main classes of beneficiaries: (i) other (non-human) species that are alive today and (ii) all species who are to be our successors and inheritors here in the future. These views and values are, I believe, entirely consistent with what Kirschenmann calls an "ecological conscience" and what Aldo Leopold referred to as a "land ethic", but with an additional legal perspective. Indeed, applying the concept of the trust to the land and other elements of the natural world would require the enactment of legislation and regulation and would need to be a part of the broad public education effort that the Global Convention on Agroecology would require Contracting States to make if the Convention is to meet its goals.

The prospect of using the concept of the trust in the fashion I have suggested above is an important enough topic to warrant some further attention, especially to address concerns that might arise for legal professionals who are familiar with the concept of the trust as it developed in English law. In his recent book *Stewarding the Earth*, Sanjay Kabir Bavikatte explores the notion of "biocultural rights", which he describe as a cluster of "third generation group rights" under which indigenous peoples, tribal, and other traditional communities play a "stewardship role . . . over their lands and waters".[39] His explanation of biocultural rights is intriguing; it relates at least indirectly to natural-systems agriculture as I am exploring it in this book. What is particularly important about Bavikatte's book for present purposes, though, is that he examines (i) the notions of property, private property, and communal property, and (ii) the concept of the trust, and especially the "constructive trust".

Bavikatte offers the following explanation about property – in particular, property claims over land – as a legal matter that the passage of time has substantially altered:

> The understanding of property in law has never been constant, and discussions around the concept are among the most contested areas of jurisprudence. In fact, if we approach law as politics, then the discourse of property is perhaps the epicenter. The very notion of personhood,

---

38 See note 12 in Chapter 4, *supra*, and accompanying text.

39 Sanjay Kabir Bavikatte, Stewarding the Earth: Rethinking Property and the Emergence of Biocultural Rights 1–2 (2014). Bavikatte is an environmental lawyer who founded the international organization Natural Justice, has served as legal advisor to several African countries, and has worked through the United Nations and other entities in coordinating grassroots initiatives with indigenous peoples on environmental matters.

and hence the juridical subject in liberal democracies, is based on an assumption that a right to property is integral to what we understand as 'person' . . . [since such liberal democracies typically characterize] personhood as incorporating a bundle of individual rights including the right to property.

[Some modern observers, however, are attempting] to counter the two main fallacies regarding property in capitalist societies.

The first fallacy is one where property is understood as a thing rather than a right. The influential Canadian political scientist C. B. Macpherson in his analysis of property notes: 'In current common usage, property is *things*; [but] in law and in the [works of legal] writers, property is not things, but *rights*, rights *in* or *to* things'. The second fallacy is one where property is understood as predominantly private property. It is, therefore, treated as an individual exclusive right; a right to exclude others from use or benefit . . . The reason why these two dominant perceptions of property are fallacies is because both of them are a product of historical circumstances, specifically of the dominance of market economies, which . . . [despite their recent development] are passed off as ahistorical and natural.[40]

Bavikatte then quotes Macpherson further in pinpointing when the first of these developments – that is, the equating of property with things rather than rights – arose:

> The change in common usage, to treating property as things themselves, came with the spread of the full capitalist market economy from the seventeenth century on, and the replacement of the old limited rights in land and other valuable things by virtually unlimited rights. As rights in land became more absolute, and parcels of land became more freely marketable commodities, it became natural to think of the land itself as property . . .[41]

According to Bavikatte, the second development – the "fallacy" of conflating property with private property – also goes back only to the seventeenth century, and is attributable in important part to the writings of John Locke:

> In fact, the predominant conception of property until the seventeenth century was that of common property, with private property being a subset . . . Locke had to work . . . [very hard] to justify unlimited

---

40 *Id.* at 116–117, citing C. B. Macpherson, *The Meaning of Property*, in C. B. Macpherson, ed., PROPERTY: MAINSTREAM AND CRITICAL POSITIONS (1978), at 2.

41 Macpherson, *supra* note 40, at 7–8, as quoted in Bavikatte, *supra* note 39, at 118.

private property at a time when the moral climate leaned in favour of common property.[42]

Having offered these comments about the concept of property as it has developed in relatively recent times, Bavikatte then explains how the concept of the equitable trust, and especially that of the "constructive trust", provides a technique for ameliorating some disadvantages that he sees in these emphases on (i) the treatment of property as things rather than rights and (ii) the replacement of public property with private property:

> One of the problems with the conflation of property with private property is the misconception that legal title, control, and benefit would all have to vest in the titleholder. Common law has produced legal instruments such as 'trusts'[,] however, that allow for the separation between title, control, and benefits[,] thereby allowing the coexistence of multiple claims over things. The history of trusts in common law extends back to the time of the crusades in the thirteenth century. They were developed to allow feudal landowners who went off to fight in foreign lands to pass the legal rights over their estates to someone they trusted (a trustee), who would then manage the lands on the understanding that the original landowner and his family would still have ultimate rights over the land as the beneficiaries. The trustee was expected to act in the best interests of the beneficiaries and in accordance with the agreement reached with the original landowner (the deed of trust).
>
> The concept of [the] trust has obviously evolved since then and the modern trust can be created by anyone who is the absolute owner of a property (the settlor) by appointing a trustee to manage the property in accordance with the trust deed for [the benefit of] a beneficiary (or beneficiaries, some of whom may be yet to be born). Once the trust is created, the legal title vests in the trustee, but the actions of the trustee are clearly limited by the trust deed and have to serve the beneficiaries who are said to have an 'equitable or beneficial interest' in the trust property. Thus, while the trustee has legal title over the property, the beneficiary has both the 'equitable property right' against the trust property and personal claims against the trustee to ensure that the trustee acts in accordance with the terms of the trust.
>
> The development of trusts owes its origin to common law's efforts to create the necessary flexibility in property jurisprudence to ensure equitable management of property. [Such a need for equitable management of property can arise, of course, in circumstances *other* than those in which an owner of land creates a trust – as where the owner himself is treating or threatens to treat the land in an offensive or

---

42 Bavikatte, *supra* note 39, at 119, 126.

inequitable manner.] Courts have, therefore, imposed trusts in situations where the direct application of property law was likely to create an injustice. Such trusts, *created irrespective of the intention of the parties*[,] are referred to as 'constructive trusts'.[43]

Bavikatte, having introduced the concept of "constructive trusts" and linked it to notions of property, then explains how the concept of the constructive trust has been expressed and applied in English law. He cites several court judgments defining constructive trusts, including this one:

> A constructive trust arises by operation of law whenever the circumstances are such that it would be unconscionable for the owner of the property . . . to assert his own beneficial interest in the property and deny the beneficial interest of another.[44]

Bavikatte explains that "perhaps the boldest exposition of a constructive trust" in English law appears in a 1972 case, in which Lord Denning offered this description:

> [A] constructive trust . . . is a trust imposed by law whenever justice and good conscience require it. It is a liberal process, founded upon large principles of equity, to be applied in cases where the defendant cannot conscientiously keep the property for himself alone, but ought to allow another to have the property or share in it. The trust may arise at the outset when the property is acquired, or later on, as the circumstances may require. It is an equitable remedy by which the Court can enable an aggrieved party to obtain restitution.[45]

In asserting that a constructive trust is one that is "imposed by law", Lord Denning is referring to circumstances in which *a court* takes action, under the sort of equitable powers that English courts developed through a complicated historical process.[46] A constructive trust – or some similar form of mechanism designed to protect the interests that various persons and other entities are considered to have in land – could likewise be "imposed by law"

---

43 *Id.* at 216 (emphasis added), citing Alastair Hudson, Understanding Equity and Trusts 13–14 (2008).

44 *Paragon Financial PLC v. D. B. Thakerar & Co.*, [1999] 1 All England Law Reports 400, as cited in Hudson, *supra* note 43, at 98.

45 *Mrs. Emily Hussey v. P. Palmer*, [1972] 3 All England Law Reports 747, as cited in Bavakatte, *supra* note 39, relying also on Hudson, *supra* note 43.

46 For a summary description of that historical process leading to the development of the rules, the remedies, and the courts of equity in the English legal system, see John W. Head, Great Legal Traditions: Civil Law, Common Law, and Chinese Law in Historical and Operational Perspective 354–369, 442–444 (2011).

through action by *a legislature*. It is that type of action that Contracting States would promise to take by participating in the Global Convention on Agroecology. They would commit to enacting legislation establishing constructive trusts or similar structures for the purpose of arresting further degradation of ecosystems by private owners acting contrary to the public interest, broadly defined.

Such proposals and legal doctrines are not new. In US law, the so-called "public trust doctrine" asserts that "courts should imply restrictions when private development threatens to destroy public use".[47] In my earlier book examining global legal regimes to protect the world's grasslands, I explained pertinent provisions of the Federal Land Policy and Management Act of 1976 ("FLPMA") enacted in the USA. That legislation imposes on the Bureau of Land Management ("BLM") a "multiple use" mandate – that is, the mandate that the BLM is to manage the lands under its control in ways that will permit multiple uses of it. These include the uses listed in the Multi-Use, Sustained Yield Act of 1960 ("MUSY"), which refers to "outdoor recreation, range, timber, watershed, and wildlife and fish purposes".[48] Although the FLPMA and the MUSY *might* together be seen as providing a legal foundation for the development of a public trust doctrine – under which the lands under the BLM's jurisdiction are held in trust for the benefit of the public – this has not occurred.

Recently, the public trust doctrine gained support in the form of court judgments both in Pennsylvania and in Oregon – the first interpreting an environmental-rights provision in Pennsylvania's constitution and the second finding that children have standing to raise a climate-change lawsuit against the US government.[49] More generally, though, efforts at adopting

---

47 George Cameron Coggins, Charles F. Wilkinson, and John D. Leshy, Federal Public Land and Resources Law 338 (4th ed., 2001).

48 For further details, see George Cameron Coggins, Charles F. Wilkinson, John D. Leshy, and Robert L. Fischman, Federal Public Land and Resources Law 687 (6th ed., 2007) [hereinafter Coggins-2007]. For a critical assessment of the MUSY approach, see George Cameron Coggins, Restoration 17 (2011) (on file with author) [hereinafter Coggins-2011], noting that "[t]he MUSY legislation tries to be all things to all people. In essence, however, it is a standardless delegation of decision-making authority to unelected bureaucrats."

49 For an assessment of the Pennsylvania ruling, see Ellen M. Gilmer, *Enviros Push 'Public Trust' as Trump Card over Oil and Gas Influence*, EnergyWire, Aug. 15, 2014, www.eenews.net/stories/1060004530. The Pennsylvania constitutional provision asserts that "[t]he people have a right to clean air, pure water, and to the preservation of the natural, scenic, historic and aesthetic values of the environment." It goes on to posit that Pennsylvania's public natural resources "are the common property of all the people, including generations yet to come" and that in its capacity "[a]s trustee of these resources, the Commonwealth shall conserve and maintain them for the benefit of all the people." See also Matthew Thor Kirsch, *Upholding the Public Trust in State Constitutions*, 46 Duke Law Journal 1169 (1997). For a description of the Oregon ruling, see the website of Our Children's Trust, ourchildrenstrust.org, whose so-called "atmospheric litigation" urges courts to use the public trust doctrine. See also James Conca, *Federal Court Rules on Climate Change in Favor of Today's Children*, Forbes, Apr. 10, 2016, www.forbes.com/sites/

such a public trust doctrine in the USA have met with fierce resistance,[50] perhaps in part because it does not fit the pattern of equitable trusts in US law more generally,[51] but also because it has been regarded by some as an assault on private property rights and doctrines of due process, separation of powers, and rule of law.[52] As a consequence of this resistance, NGOs and private actors have taken separate initiatives (independent of the public trust doctrine) to establish trusts and similar structures in order to protect various areas in the USA[53] and elsewhere.[54]

In her 2014 book *Nature's Trust*, Mary Christina Wood urges a reliance on the "public trust" doctrine to create a new framework for ecological protection, not just in the USA but more generally in the world. Near the end of her book she offers this summary:

---

jamesconca/2016/04/10/federal-court-rules-on-climate-change-in-favor-of-todays-children/ #6ff719f06219, and also *Judge Denies Motions by Fossil Fuel Industry and Federal Government in Landmark Climate Change Case*, Forbes, Apr. 10, 2016, http://readersupportednews.org/news-section2/318-66/36234-judge-denies-motions-by-fossil-fuel-industry-and-federal-government-in-landmark-climate-change-case.

50 See John W. Head, GLOBAL LEGAL REGIMES TO PROTECT THE WORLD'S GRASSLANDS 104 n.77 (2012) [hereinafter GRASSLANDS]. For a discussion of the public trust doctrine in US public land law, see Coggins-2007, *supra* note 48, at 22. That text by Coggins and three co-authors is the preeminent work on the law of US public lands. For an abbreviated treatment, see generally Robert L. Glicksman and George Cameron Coggins, MODERN PUBLIC LAND LAW (2006). As noted in the 2007 treatise, efforts at adopting a public trust doctrine in US public lands have met with strong objections, so that the doctrine has only very limited applicability in this setting.

51 For instance, in an equitable trust as envisioned under US law, a grantor executes a trust agreement that is to be administered pursuant to the statutory and common laws of a particular state and that names and describes the role of a trustee and a successor trustee who have legal title to and specified control over trust assets, which the trustee is then obligated to administer for the benefit of named beneficiaries, some of whom may be unborn and who might in fact not be born for over 100 years in the future. A trustee so named can be sued under state law by specified beneficiaries or their representatives and is subject to oversight and enforcement by the state's courts – and can typically obtain fiduciary liability insurance to offset some of the liability related to these responsibilities. Even in this abbreviated summary, it is clear that several elements of this form of equitable trust as elaborately developed in US law would be difficult or impossible to replicate in the context of a "public trust doctrine" of the sort that some observers have urged.

52 See, e.g., James L. Huffman, *Why Liberating the Public Trust Doctrine is Bad for the Public*, 45 ENVIRONMENTAL LAW 337 (2015).

53 An example of such initiatives can be seen in the work of the Ranchland Trust of Kansas. See, e.g., Ranchland Trust of Kansas, *Agreement Reached for Smoky Hills Conservation Project*, Ranchland Trust of Kansas News 2, issue 4, pp. 1–2 (December 2008), www.ranchlandtrustofkansas.org/Articles/RTK%20News-Dec08.pdf. For other examples, see the website of the Kansas Land Trust, http://www.klt.org/ and the website of The Nature Conservancy, www.nature.org/.

54 See, e.g., Colin Barraclough, *Back to Nature*, FINANCIAL TIMES, Sep. 11, 2010, p. 9 (describing Conservación Patagonica, a land trust established to combat ecological damage). For a survey of other efforts by NGOs and individuals to use trusts and similar structures to protect grasslands, see GRASSLANDS, *supra* note 50, at 171–177.

Arising primarily from statutes passed in the 1970s, the field of environmental law stands as a failed legal experiment. The administrative state vests agencies with breathtaking power that came justified by one simple assumption: officials will deploy public resources and invoke their technical expertise *on behalf of the public interest*. Instead, too many environmental agencies today use their power to carry out profit agendas set by corporations and singular interests . . . . Nearly across the board, environmental statutory processes do not prohibit harm: they permit it.

The problem lies not in the statutes themselves but in [a political reality that] continually bends agencies into serving those parties holding the most political power, which in today's world often means corporations rather than the general public. [For instance,] . . . political operatives use their discretion to favor the industries they continue to serve from inside the agency. Staff scientists and permit writers operating within highly charged political cultures suffer varying degrees of pressure to fall in line with the agency's political agenda. These pressures remain obscured from the public, taking place in procedural fortresses made nearly impenetrable by their sheer complexity . . .

Humanity cannot hope for a livable planet if government agencies continue to license industries to pollute and destroy the remaining natural resources. Environmental law becomes profoundly relevant to the daily life and future wellbeing of every citizen alive today. In words that now reverberate truth in nearly every nation, Ansel Adams once said: "It is horrifying that we have to fight our own government to save the environment." Citizens living in all parts of the world stand on common ground as never before, in both their challenge and mission.[55]

With that as background, Wood explains the "paradigm called Nature's Trust" that she proposes as a legal grounds on which "to reconstitute environmental law in countries throughout the world":

[This paradigm] calls forth an ancient duty embodied in the public trust doctrine, a legal principle that has flowed through countless forms of government through the ages. *At its core, the doctrine declares public property rights originally and inherently reserved through the peoples' social contract with their sovereign governments. The trust remains an attribute of sovereignty that cannot be alienated by any legislature. This principle designates government as trustee of crucial natural resources and obligates it to act in a fiduciary capacity to protect such assets for the beneficiaries of the trust, which include both present and future generations of citizens.* Unlike the permissive bent of administrative discretion under statutory law, the public trust imposes a strict duty to

---

55 Mary Christina Wood, Nature's Trust: Environmental Law for a New Ecological Age 335–336 (2014).

protect the people's commonwealth. At a time when government actions worldwide threaten to rob today's youth of an ecologically secure future, the trust breathes legal rights into the aspiration of inter-generational equity.

While the public trust has long offered a theoretical ideal for environmental law, until now it has lacked the precision necessary to apply it to a broad realm of practical conflicts arising before modern legislatures and agencies. This book [*Nature's Trust* has] sought to illuminate a fiduciary path by setting forth the substantive and procedural obligations incumbent on government entities as trustees of public resources. They must protect the trust, not allow waste, maximize the societal value of trust assets, restore assets where they have been damaged and recoup monetary damages from third parties that have injured the assets. Further, a legislature must not alienate public trust assets where doing so would not serve public trust purposes, or would substantially impair the ecological wealth remaining in the trust. These are active duties, leaving no room for idle management. Government's failure to protect the planet's climate system on which all life depends amounts to the most dangerous perversion of this fiduciary responsibility.[56]

Even though the "public trust" doctrine that Mary Christina Wood says should be revived has its roots (as does the "constructive trust" doctrine) in English common law, it need not be confined to common-law countries. The introductory remarks to a 2012 conference focusing on "the trust in Europe" highlighted growing European interest in the flexibility of trusts or other trust-like instruments. Those remarks gave special attention to the fact that "the law of trusts in Anglo-American jurisdictions [extends] to trusts imposed by law such as resulting and constructive trusts. The constructive trust, in particular, has the potential to serve as a broad and flexible remedy."[57]

Examples of equitable trusts – whether styled as "constructive trusts" or "public trusts" – also appear in international law. Although I will examine these more in my next book,[58] it is worth emphasizing here that the entire International Trusteeship System established in 1945 as part of the United Nations reflected the main contours of an equitable trust. That system was established by public action in the form of the negotiation and adoption of a treaty (the UN Charter) by official representatives of most of the countries

---

56 *Id.* at 336–337. For a commentary on Wood's book, see a posting on the website of Bill Moyers & Company from September 19, 2014, at http://billmoyers.com/2014/09/19/natures-trust-new-approach-environmental-law/.

57 Thomas P. Gallanis, *The Trust in Continental Europe: A Brief Comment from a U.S. Observer*, in 18 Columbia Journal of European Law Online iii (2012), www.cjel.net/wp-content/uploads/2012/08/CJEL-Trust-Law-Final1.pdf.

58 See John W. Head, A Global Corporate Trust for Agroecological Integrity: Managing a New Agriculture in a World of Eco-States (forthcoming).

in the world at that time. It placed large portions of the planet's territory under the management of "trustee" states – that is, certain long-established, highly-developed states responsible for handling the affairs of various non-self-governing territories, most of which emerged from decolonization but had not yet gained political independence.[59]

Also at the international level, structures and arrangements bearing the marks of a trust – whether carrying that label or not – have been developed for various other purposes as well, including for environmental protection. Peter Sand has recently provided a comprehensive survey of the concept of public trusteeship as it has developed in international law – dating back, in his account, to the late 1800s with the Pacific Fur Seal Arbitration.[60] Similarly, in a recent book examining sustainable development and international law, Elisabeth Bürgi Bonanomi has emphasized the continuing relevance of the legal framework for inter-generational equity that Edith Brown Weiss proposed in the late 1980s.[61] According to that framework, "[e]ach generation receives the natural and cultural legacy [of the world] in trust from previous generations and holds it in trust for future generations."[62]

My reason for devoting the preceding several paragraphs to the concept of the constructive trust – with special attention to its historical roots and its emergence thus far in both national law and international law – is to highlight some crucial responsibilities to be shouldered by Contracting States to the Global Convention I am describing here. The Global Convention on Agroecology will require such Contracting States to take legal measures aimed at protecting the public interest in land, along with the interests of other species and components of the ecoregions under those states' jurisdiction. Mustering the broad public support required for the enactment

---

59 For details about the International Trusteeship System, see the UN website at www.un.org/en/decolonization/its.shtml. As noted there, all of the "trust territories" originally under the system – Togoland, Somaliland, Cameroon, Tanganyika, Ruanda-Urundi, Western Samoa, Nauru, New Guinea, and the Trust Territory of the Pacific Islands – ultimately gained formal political and legal independence. A precedent for the International Trusteeship System was the "mandate" system established under the League of Nations. For UN Charter provisions establishing and governing the International Trusteeship System (including the operation of the Trusteeship Council, one of the six organs of the UN), see UN CHARTER, arts. 75–91.

60 Peter H. Sand, *The Concept of Public Trusteeship in the Transboundary Governance of Biodiversity*, appearing as chapter 3 in Louis J. Kotzé, Thilo Marauhn, eds., TRANSBOUNDARY GOVERNANCE OF BIODIVERSITY (2014). Sand touches on such structures as those found in the Global Environment Trust Fund dating from the 1990s, the 1982 Convention on the Law of the Sea, the 1946 Whaling Convention, the proposed 1972 World Heritage Trust Convention, the 2001 International Treaty on Plant Genetic Resources for Food and Agriculture, and those at issue in the Children's Atmospheric Trust litigation – and even in the 1997 Danube Dam case adjudicated before the International Court of Justice. *Id.* at 60–62.

61 Elisabeth Bürgi Bonanomi, SUSTAINABLE DEVELOPMENT IN INTERNATIONAL LAW MAKING AND TRADE: INTERNATIONAL FOOD GOVERNANCE AND TRADE IN AGRICULTURE 125–127 (2015).

62 *Id.* at 126, quoting Edith Brown Weiss, IN FAIRNESS TO FUTURE GENERATIONS: INTERNATIONAL LAW, COMMON PATRIMONY, AND INTERGENERATIONAL EQUITY (1989).

of these legal measures must involve a comprehensive campaign to provide all people with a new basic grounding in Earth sciences and a "land ethic" or "land conscience" of the sort I referred to above. Widespread attention to "ecospheric studies" is essential, and the Convention should record the commitment of Contracting States to educate their people accordingly.

IIB4. Specialized training and research

In addition to providing their people with this basic grounding, however, Contracting States to the Convention described above will commit to give *specialized* education and training to the persons most closely involved in farming. Frederick Kirschenmann, in advocating for natural-systems farming (what he refers to as "ecosystem-sensitive farming") calls for "a new generation of farmers who are highly skilled in ecology, husbandry, and evolutionary biology, and who seek opportunities to work closely with nature". In order to train this new generation of farmers, Kirschenmann insists, "[w]e need to introduce more college courses in agroecology and provide internship opportunities for experience-based learning in ecosystems management on real farms."[63]

In addition to legislation, regulation, and education – both general and specialized – a Contracting State to the Global Convention on Agroecology would need to support *research* in pertinent subjects. Numerous observers have emphasized the need for expanded research, and particularly the need for *public sponsorship and funding* of such research. Kirschenmann, for instance, has asserted that making agriculture more sustainable requires that we "redesign our food and agriculture system so that its functions are more consistent with our best understanding of how the biotic community works". This requires, he says, that we "*refocus* our public-research agenda to investigate the synergies and synchronies of the diverse species in each agricultural watershed" and that we "evaluate how they can be employed to increase our agricultural productivity, while simultaneously enhancing the capacity of local ecologies to renew themselves".[64]

Contracting States to the proposed Global Convention on Agroecology could commit themselves to robust public support for the kind of research that Kirschenmann urges. For those Contracting States that lack the financial resources to do so, the Convention would offer the prospect of financial support from other Contracting States under a "special and differential treatment" provision described below under "other provisions of the Convention".

Beyond receiving this sort of financial support, Contracting States that lack other kinds of resources necessary for robust agricultural research would also benefit from a provision in the Convention requiring advanced countries to

---

63 Kirschenmann, *supra* note 24, at 222–223.
64 *Id.* at 186.

give assistance in that area. That provision is also described under "other provisions of the Convention". Such a provision appears in the Earth Charter, which calls for "the open exchange and wide application of the knowledge acquired" through studies of ecological sustainability. The Earth Charter provision goes on to require "special attention to the needs of developing nations".[65]

In sum, principal commitments undertaken by those states that would become parties to the Global Convention on Agroecology would be to facilitate the implementation of the Convention's Principles in four explicit ways: through legislation, regulation, education, and research. However, Contracting States would also have two other important obligations of an "outward-facing" character – that is, in their relations with other states.

IIB5. Duty of cooperation

First, Contracting States would be required to report to, and cooperate with, other Contracting States. An approach taken in many modern multilateral treaties with reporting requirements (such as those involving human rights obligations) is an "every-country-every-year" approach. That is, each Contracting State is to issue a report each year evaluating and documenting its performance during the preceding year in meeting its commitments under the treaty in question. The Convention would provide in detail the formal and substantive requirements for the preparation and submission of such reports.

In many cases, such a reporting requirement will be accompanied by a consultation requirement – with the consultation to occur between the Contracting State and a body of experts whose task is to study the Contracting State's report and offer its own independent assessment. For example, in the case of the international treaty commitments that states make as members of the International Monetary Fund ("IMF"), each state engages (every year) in consultations with a group of IMF staff members whose role is (i) to examine in depth that state's performance of its IMF Charter obligations through the preceding year, and (ii) to issue a public report with candid assessments and recommendations.[66]

---

65 EARTH CHARTER, *supra* note 13, at 259 (Principle 8).
66 Such consultations take place under the provisions of Article IV of the IMF Charter. For further details about these "Article IV consultations", see Head, LOSING THE GLOBAL DEVELOPMENT WAR, *infra* note 68, at 112, 115, 229, and 268. Incidentally, the IMF Article IV consultation report regarding the USA for 2007 criticized that country for the risks that US authorities were creating by a lowering of lending standards (as occurred in the case of "subprime" mortgages), and it suggested that "the same could be happening in other market segments, such as leveraged loans". *Id.* at 268 n.44.

IIB6. Participation in institutional innovation

What would be the "body of experts" that might review reports issued each year by Contracting States under the GCA I am describing here? A new international organization – the Global Corporate Trust for Agroecological Integrity. While saving a discussion of that organization until later, I will simply explain here that another important obligation of Contracting States to the GCA would be to participate in good faith in the establishment of that new organization.

Doing so will require fresh thinking about the character of international institutions, because the Global Corporate Trust for Agroecological Integrity would depart from the structural pattern used in the mid-1940s to create such post-war institutions as the UN, the IMF, and the World Bank. Unlike those institutions, the Global Corporate Trust for Agroecological Integrity would have a diverse membership that would include non-state actors. For this and other reasons, its establishment would require states (in keeping with Principle 10 of the Convention, noted above) not only (i) to accept and endorse, as they did in the mid-1940s, some collective security and management arrangements designed for mutual benefit but also (ii) to learn from the experience of other institutions both at the international level and at the regional level (such as the EU),[67] and (iii) to fashion and implement a more sophisticated and beneficiary-oriented concept of sovereignty than the one that has developed over the past four centuries. Because sovereignty sits at the very center of modern public international law, I will give it separate consideration in Chapter 8.

As I will also explain in Chapter 8, I have over the past two decades or so directed many criticisms at the international institutions that emerged in the 1940s. Still, I believe their creation reflects the capacity of modern humans to design global solutions in response to global crisis. In the face of today's crises, we can build on the mix of strengths and weaknesses we see in earlier generations of institutions in order to create better ones. Under the Global Convention and Agroecology, Contracting States would commit themselves to participating in that effort.

---

67 For instance, the details regarding the structure and functions of the Global Corporate Trust for Agroecological Integrity would need to reflect the experience of the EU as a particularly durable "supranational" institution, having grappled over time with issues of externalities, tragedy of the commons, race to the bottom, and NIMBY ("not in my back yard") attitudes – all as explored in an article by my colleague Rick Levy. See Richard E. Levy, *The Law and Economics of Supranationalism*, 40 EUROPEAN JOURNAL OF LAW AND ECONOMICS 1 (2015), http://link.springer.com/article/10.1007/s20657-015-9508-x.

IIC. Other provisions of the Convention

IIC1. Special and differential treatment

As reflected in the outline appearing above in Box 7.1, the Global Convention on Agroecology would also include several other provisions. One of them would prescribe the details of a "special and differential treatment" system of the sort referred to briefly above in connection with the obligation of Contracting States to expand publicly-supported research. Such a "special and differential treatment" ("SDT") approach is commonly seen in modern treaties involving international trade and finance.[68] The central idea of the SDT approach is to provide a formal means by which countries that are *more* developed in particular aspects – for instance, those that are more economically advanced, or that have stronger financial systems or government institutions, or that are better endowed with certain natural resources (fisheries, minerals, forests, etc.) – will assist countries that are *less* developed in these aspects.

One way to provide a formal means of such assistance is through a division of Contracting States into two or more categories. The GCA could provide for a distinction between Category A Contracting States (economically and technologically more advanced) and Category B Contracting States (less so), with a requirement that Category A Contracting States provide assistance to Category B Contracting States in particular areas. A more sophisticated system would differentiate between different *types* of assistance. For instance, Country #1 might be much more highly advanced than Country #2 in developing perennial strains of sorghum, but have much less than Country #2 does in the way of financial resources to fund public education. In that case, Country #1 might be classified as a Category A Contracting State in respect of agricultural research – and therefore have an obligation to provide assistance to Country #2 in that respect – but be classified as a Category B Contracting State for purposes of funding expanded public education into "ecospheric studies" or for purposes of funding specialized training for persons most closely involved in farming.

---

68 For a description of "special and differential treatment" in the context of the General Agreement on Tariffs and Trade and the WTO, see John W. Head, Losing the Global Development War: A Contemporary Critique of the IMF, the World Bank, and the WTO 198–199 (2008) [hereinafter Head-2008]. An earlier version of the "special and differential treatment" approach formalized during the Uruguay Round of trade negotiations was the Generalized System of Preferences ("GSP"), under which economically developed countries allowed products from economically less developed countries to be imported at lower (sometimes zero) tariff rates. For details on this GSP approach, see *id.* at 140, 220, and 304. Other global economic institutions have followed similar approaches. For a description of "concessional lending" in the context of the International Monetary Fund and the multilateral development banks – aimed at providing especially favorable treatment for economically less-developed countries borrowing from those institutions – see *id.* at 120–125.

In short, a provision of the Convention might establish either a simple or a sophisticated system of "special and differentiated treatment" among the Contracting States, with an eye to facilitating the effective implementation of the Convention. As noted on page 323, one provision of the Convention might expressly allow a state to opt out (by way of a "reservation") from any requirement that it provide financial support under this SDT system.

IIC2. Formalities of treaty application

In addition to this "SDT" set of provisions, the Convention would also include provisions addressing a number of other issues commonly handled in multilateral treaties. I list them here with brief descriptions:

- *Definitions.* As in most contractual instruments, certain terms carrying special significance will be defined. In the context of the GCA, these might include such terms as modern extractive agriculture, agroecological husbandry, global climate change, safeguard rights, the Precautionary Principle, and the Polluter Pays Principle, as well as the abbreviations or names of certain organizations and international legal instruments (such as FAO, the UN Framework Convention on Climate Change, etc.)

- *Interpretation of provisions.* The Convention would prescribe a procedure for establishing the meaning of other terms, and the appropriate interpretation of particular provisions of the Convention, in the case of ambiguity or disagreement. In some treaty systems, a committee established under the treaty is entrusted with such an interpretation function. If, as anticipated in this case, there will be an international organization responsible for facilitating the implementation of the treaty, a governing body of that organization would carry out such an interpretation function.

- *Dispute settlement and external review.* For matters of conflict among Contracting States that rise to a level more significant than a disagreement over the Convention's provisions, a mechanism for dispute settlement can be prescribed. In keeping with the practice in some other treaty systems, the Convention might include provisions calling for a two-step procedure for dispute settlement. In the first step, the disputing parties would be required to engage in good-faith negotiations and mediation in order to avoid a litigation-like process. If that approach were to fail, the second step would be to refer the dispute to the International Court of Justice under the second part of Article 36(1) of the ICJ Statute.[69]

---

69 Article 36(1) provides that the jurisdiction of the ICJ "comprises all cases which the parties refer to it and all matters specially provided for in the Charter of the United Nations or in treaties and conventions in force". See ICJ Statute, *supra* note 6, art. 36(1). The first part of that sentence serves as the foundation for what is commonly called "*compromis*" jurisdiction, in which the two disputing states decide *after the dispute has arisen* to submit to the ICJ a formal legal instrument

- *Reservations and denunciation.* As is the case with many multilateral treaties, the GCA could provide explicitly for a Contracting State to enter a reservation to its participation in the treaty. The 1969 Vienna Convention on the Law of Treaties defines a "reservation" as "a unilateral statement . . . made by a State, when signing, ratifying, accepting, approving or acceding to a treaty, whereby it purports to exclude or to modify the legal effect of certain provisions of the treaty in their application to that State."[70] However, the GCA should provide that *no* reservation may be made with respect to any portion of the treaty except for one: the provision imposing a requirement of "special and differential treatment" financial assistance under the SDT (special and differential treatment) system described above. In other words, the Convention might allow an economically advanced country to become a Contracting State without committing itself to provide funding to economically less-developed Contracting States. The principal reason for permitting such an exception (which obviously could undermine the effectiveness of the SDT system) would be to encourage as much participation as possible from economically advanced countries.
- *Entry into force and provisional application.* Closing provisions of the Convention would address such details as (i) the manner in which a state would express its consent to be bound to the Convention (probably through signature plus ratification for original participants, or accession for later participants), (ii) designation of a depository institution (probably the UN Secretariat) and requirements for depositing of instruments of ratification or accession with that institution, (iii) specification of the number of ratifications or accessions to be gained before the Convention may be entered into force, (iv) specification also of a "waiting period" between achievement of the requisite number of ratifications or accessions and the Convention's entry into force, and (v) the effect, if any, of the Convention on a provisional basis pending its entry into force.[71]

(a *compromis*) stipulating the facts of the dispute and identifying the specific legal questions they are asking the ICJ to address. The last part of the Article 36(1) sentence – and particularly "all matters specially provided for . . . . in treaties and conventions in force" – serves as the foundation for a different type of ICJ jurisdiction, in which the parties to a treaty decide *at the time of preparing the treaty* (and therefore *before* any dispute has arisen) to grant jurisdiction to the ICJ over any dispute that might arise in connection with that treaty. It is this second type of Article 36(1) jurisdiction that, in my proposal, the Global Convention on Agroecology would specify.

70 1969 VCLT, *supra* note 30, art. 2(1)(d). Other provisions of the 1969 Vienna Convention prescribe when reservations may or may not be made, and what effect they will have on other parties to a multilateral treaty. See *id.* at arts. 19–23.
71 Default provisions for some of these matters appear in the Vienna Convention on the Law of Treaties.

III. A question of practicalities: is such a treaty system impossible or essential?

The foregoing paragraphs have given what seems like a great deal of detail about a Global Convention on Agroecology. I have explained its overall outline and aims, suggested the elements that might appear in its Preamble, identified ten specific Principles that it would announce, described the obligations that Contracting States to the treaty would commit themselves to meet, given particular attention to the "special and differential treatment" approach that the Convention would follow, and sketched out a few points about dispute settlement, reservations, and entry into force.

I have addressed these and other issues with some specificity because it is these details – the building blocks and particular commitments that the treaty memorializes – that would dominate the attention of individuals authorized to engage in the hard work of negotiating such a treaty. As I noted at the outset of this discussion, the *idea* of a treaty is of little use without some elaboration of the specific *content* of the treaty.

Even with this amount of detail, I have only scratched the surface. Extensive drafting and negotiation efforts would be involved in bringing such a treaty into existence. Before anyone would agree to undertake such efforts, though, a threshold two-part question surely must be addressed: (i) Would such a treaty have any chance of attracting enough support among states to become a reality and (ii) if it were actually put in place, would such a treaty have any chance of achieving its aims – that is, of being effective in facilitating a transition to a new "natural-systems" form of global agriculture?

Some observers cite many reasons for answering "no" to each part of that question. After all, the Convention as I have described it here is, according to its own terms, in part a transitional instrument. Recall that a key responsibility that Contracting States to the Convention would have is to participate in efforts to create a new international organization – the Global Corporate Trust for Agroecological Integrity – that would serve as the coordinating institution for various entities that will exercise authority for ecological protection and agricultural production based not along territorial political nation-state lines but rather along other lines. As I will explain further in Chapter 8, these would be reflective of the world's ecoregions and biomes. The establishment and operation of such an institution would require changes and limitations to the concept of sovereignty. Given this, it might be easy to conclude that as a *practical* matter, the sort of Global Convention on Agroecology that I have described above is simply impossible.

In the following paragraphs I will explore that question. My approach in doing so will be to examine four particularly nettlesome issues that are central to the Convention: precaution, economics, climate change, and cooperation. My analysis concludes that creating and implementing such a Convention is not impossible. Faced with the reality of crisis, responsible parties will, I predict, regard it instead as *essential*.

IIIA. Precaution, technology, and research in an agroecological context

Two of the Principles I have proposed above for a Global Convention on Agroecology bear on the issues of precaution and technology. Principle 3 would unambiguously adopt the Precautionary Principle as it has developed in Europe and as it has been reflected thus far in some international legal instruments.[72] Principle 6 provides that one of the four key objectives of creating a new model of agricultural life and production would be to overcome the ecological unsustainability of modern extractive agriculture, and that doing so would require the development of new grain crops based on a natural-systems approach of the sort I explored above in Chapters 4 and 5. Moreover, Contracting States would, under the Convention, have specific obligations to engage in much more robust publicly-funded agricultural and ecological research.

Therefore, one way of evaluating the feasibility of the Convention is to consider in further detail the interplay of these three elements: precaution, technology, and research. As Frederick Kirschenmann has observed, "[e]veryone agrees that biotechnology has the ability to make dramatic changes in nature . . . . But if powerful technologies have the potential to radically change complex relationships, thereby potentially upsetting delicate interactions that have evolved over millennia, shouldn't it inspire caution?"[73] A simple answer to his question would be "yes, of course". But how much caution, and in what form? Bear in mind that the transition to agroecological husbandry requires agricultural research aimed at creating new strains of crops. How can this research be undertaken consistently with the Precautionary Principle? And in any event, is it realistic to expect in this context that some countries, including the USA, would adopt a full-throated version of the Precautionary Principle?

Kirschenmann has offered a detailed set of observations on this point. He starts by emphasizing the complexity of nature, and therefore the problems that can arise from efforts to manipulate "just one thing" in nature:

> A [key] underlying question we might ask is this: is it possible to do "just one thing" at the molecular level? We know it is not possible to do "just one thing" in the ecosystems in which we live. Even when we have made good-faith efforts to improve the resilience of our ecological homes, we have often miscalculated the extent to which, and the manner in which species within ecosystems are interdependent.
>
> Ecologist Yvonne Baskin provides a chilling example of this kind of miscalculation. In an effort [several years ago] to boost the numbers

---

72 For a reference to the reflection of the Precautionary Principle in international instruments, see *supra* note 16. For a reference to the application of the Precautionary Principle in Europe, see *supra* note 15.

73 Kirschenmann, *supra* note 24, at 177.

of salmon that swim upstream from Montana's Flathead Lake to spawn in Glacier National Park's McDonald Creek, state fisheries officials stocked the upstream portions of the watershed with exotic opossum shrimp to provide food for the salmon, which would in turn provide more food for eagles, bears, gulls, mallards, golden eyes, coyotes, minks, otters, and other species that feed on salmon and their eggs.

But, "the plan overlooked an important bit of natural history of both shrimp and fish." The salmon, it seems, feed on zooplankton near the surface during the day, while the shrimp spend the day near the bottom, pretty much out of reach of the fish. At night, the shrimp migrate upwards to feed on zooplankton themselves – the same zooplankton, unfortunately, that serve as the chief food for [the salmon]." Consequently, "rather than supply a new food resource for the [salmon], humans . . . unwittingly introduced a competitor." As a result, "zooplankton quickly declined, . . . [and in] just a few years, the [salmon] population in the lake had collapsed, too. One hundred kilometers upstream in McDonald Creek, the disappearance of the spawning [salmon] eliminated a food resource that had once fortified eagles for their winter migration and fattened bears for hibernation. It also brought to an end a wildlife spectacle that had boosted off-season tourism revenues for the park and neighboring communities."

In less than nine years, the population of 100,000 salmon was reduced to fifty. If our judgment is this bad, are we really ready to begin modifying the genome?[74]

Kirschenmann then shifts his focus away from this large-scale and rather obvious (perhaps stupid) sort of mistake and toward the issue of ecosystem dynamics at a much smaller level, and therefore toward the issue of genetic manipulation and transgenic research:

The same ecosystem dynamics that are at work on the organism level are most likely at work at the molecular level as well. In fact, diversity and interconnectedness of the world of single-celled microbes is astonishing. Robert Service has . . . said, a "pinch of soil can contain 1 billion microbes or more," and the world of microbes is a "thimble-sized rainforest." Moreover, he said it is "virtually impossible" to describe the "ecological structure" of this biodiversity.

. . . I am not suggesting transgenic research be abandoned. All species, after all, do modify their environments. Environments are constructed by living organisms out of the bits and pieces of the external world available to them; the environment wouldn't even exist if it were not for organisms modifying it. But if we continue to ignore the ecological

---

74 *Id.* at 163–164, citing Yvonne Baskin, The Work of Nature (1997).

dimensions of our modifications, as we seem to regularly do with genetic engineering, we are likely to experience many unpleasant surprises.[75]

For Kirschenmann, then, agricultural research and experimentation requires close attention to technological risks. He quotes Stephen Schneider as saying that "the bigger the technological solution, the greater the chance of extensive, unforeseen side effects", and he quotes Aldo Leopold as proclaiming that "the greater the rapidity of human-induced changes, the more likely they are to destabilize complex systems of nature".[76]

Kirschenmann then widens the focus and emphasizes the wisdom of precaution as reflected in myth and literature:

> Literary artists have, of course, urged such caution for millennia. Ancient as well as modern writings are replete with warnings about the indiscriminate introduction of new technologies unaccompanied by the exercise of appropriate wisdom. Greek mythology recounts the story of Daedalus . . . and his son, Icarus . . . who became intoxicated with this new power [to fly and] flew too near the sun, and plunged to his death. New technology can be useful – it can even save us from previous technologies – but without appropriate caution, it can destroy us.
>
> The ancient Hebrews were even more explicit in their warnings. The Garden of Eden story in the book of Genesis portrays a bit of Hebrew mythology about how people should live in creation: *on creation's terms*. If they succumb to temptation and arrogantly assume they know better (the Tree of Knowledge) than the ecology of nature (the Tree of Life) how the garden should be managed, then they will experience one curse after another and eventually be expelled from creation's fecundity.
>
> Many similar literary examples are familiar to us. The tale of the sorcerer and the servant who misuses his master's magic has been told in many forms . . . [as in] a poem for children titled "The Sorcerer's Apprentice." This tale depicts a wise magician who has the power to make dramatic changes in his environment. The servant or apprentice releases these powers indiscriminately, causing great destruction. The lesson as . . . [the poem] suggests, is that . . . unnatural changes must be undertaken with care.[77]

---

75 *Id.* at 164, citing Robert F. Service, *Microbiologists Explore Life's Rich, Hidden Kingdoms*, 275 SCIENCE 1740 (1997).

76 *Id.* at 172, citing Stephen H. Schneider, THE GENESIS STRATEGY: CLIMATE AND GLOBAL SURVIVAL 14 (1976) and Aldo Leopold, A SAND COUNTY ALMANAC 220 (1949).

77 Kirschenmann, *supra* note 24, at 206–207.

With that as background, Kirschenmann then enumerates four principles regarding the need for precaution in the introduction of technologies in agriculture (or, for that matter, more generally):

- If the *magnitude* of potential harm is limited, we might say yes to the technology. If the effect of the introduction of a material or practice could last for generations, if it doesn't disappear in one generation, we should say no.
- If the *geography* of the potential harm from the technology is limited, we might say yes to the technology. The larger the area affected by the introduction of a material or practice, the more safeguards we must use. For example, if the technology to be introduced is airborne, water borne, bioaccumulative, or in other ways ubiquitous to the environment, we should say no.
- If the *biology* of the potential harm is limited, we might say yes to the technology. Because all species in a biotic community have coevolved, we consider the welfare of all affected species. If the introduction of a material or practice potentially threatens the integrity of capacity for renewal of the biotic community, we must say no.
- If the potential *social cost* of the technology is limited, we might say yes. Often we say yes to a new material or practice because the short-term economic gain is attractive, or because of potential economic gains for one sector of society. But rarely are these gains weighed against the long-term economic costs. If the introduction of a technology compromises future economic well-being or is achieved in one sector of society at the expense of another sector, we should say no.[78]

Kirschenmann writes that such precautionary factors "are seldom considered" and that "indeed, some sectors of our society are hostile toward them". We should, he urges, "be asking 'And then what?' at least seven times before we proceed much further" in introducing new technologies into agriculture.[79]

Exercising caution of the sort that Kirschenmann urges – and, indeed, adopting a robust version of the Precautionary Principle as the GCA would require – does *not* amount to a rejection of technology in agriculture, or to a dumbing-down of research. Quite the opposite. As Kirschenmann insists, "we should emphasize [agricultural and ecological] research that [is] . . . multidisciplinary and increases our understanding of evolving complexities

---

78 *Id.* at 212–213.

79 *Id.* at 213. In another passage, Kirschenmann expands on this "seven times" point: "Asking 'and then what?' at least seven times recognizes that we can never introduce innovations into the world without setting in motion a series of related consequences rarely apparent at the time or place of introduction. The future welfare of the human species may therefore hinge on such cautionary action." *Id.* at 206.

and interdependence of our social and biological lives"[80] – in other words, a more sophisticated, more integrated form of agricultural research than has been achieved thus far.

Indeed, Kirschenmann goes into some detail in explaining how the general scientific research agenda in the USA over the past seven decades or so is too narrow, too closely aligned with high productivity rather than high quality of life, and too intent on creating partnerships among industry, government, and universities for the primary purpose of increasing economic competitiveness of American industry.[81] Instead of pursuing these goals, Kirschenmann says, agricultural research should aim at gaining a better understanding of how to farm ecologically, and this requires a better understanding of those ecosystems that are intrinsically suited to crop production – or, in Kirschenmann's words, "a new vision of the prairie".[82] Undertaking this research, with a goal of gaining what Kirschenmann calls an "agroecological literacy", is a task that "probably won't be funded by commodity groups or the agribusiness industry. If the popular media see the need for such a transformation, then public support for this task might be generated."[83]

Kirschenmann's reference to "public support" triggers the question of *where* such research would occur. The answer reflected in the provisions of a Global Convention on Agroecology of the sort I have described above is this: the research should take place in *public* institutions, and not predominantly in the research laboratories of private industry. The research should be heavily supported by government promotion and public funding.

In this regard, it is instructive to consider how agricultural research has developed in recent decades, particularly in the USA. Wes Jackson has pointed out that "the Green Revolution needed government promotion. It required a campaign to gain institutional support, and that required great coordination. The market could not make it happen."[84] As a Rockefeller

---

80 *Id.* at 193.

81 See *id.* at 191.

82 *Id.* at 130. Kirschenmann asserts that "[w]hat farmers need from an ecological perspective, rather than a purely production perspective, is [an understanding of] the ecosystem in which the farming is practiced. In the Great Plains [of North America], that ecosystem is a prairie . . . . As [Herbert] Hanson argued, the prairie must serve as the standard against which we evaluate the effects of farming." *Id.* These ideas resemble those of Wes Jackson, particularly in such works as NEW ROOTS FOR AGRICULTURE (1980) and his more recent CONSULTING THE GENIUS OF THE PLACE (2010). In the latter of those, Jackson deplores the "deficit spending of the earth's capital" that has occurred "since we started agriculture ten thousand years ago" and asserts that "[t]he only good examples of an alternative way [to engage in agriculture] are among nature's ecosystems that feature material recycling and run on contemporary sunlight." *Id.* at x. The prairie or grassland is such an ecosystem.

83 Kirschenmann, *supra* note 24, at 131.

84 Wes Jackson, *Thirty Five Years: The Past and Beyond, the Future and Beyond* (address to The Land Institute Prairie Festival, Sept. 30, 2012), at 10 (on file with author). See also text accompanying notes 90–93 in Chapter 6, *supra*.

Fellow in the 1950s, Jackson himself joined many others in US land-grant universities established under the Morrill Act of 1862.[85] Since that time, however, much agricultural research in the USA has moved away from such public universities, reflecting "a shift toward the belief that the market will find solutions".[86] Reversing that shift, and returning universities and other public institutions to a prominent role in agricultural research, would better serve the aim of promoting "agroecological literacy" of the sort that Kirschenmann has urged.

Restoring the prominence of public institutions in agricultural research could also have another benefit that Kirschenmann has highlighted: it could "help reduce some of the secrecy associated with the nondisclosure, trade secrets, and publication delays characteristic of industry-supported research. This secrecy has become increasingly worrisome." Focusing on interdisciplinary ecology-oriented research could, he says, resurrect "free and open development and exchange of information" in such institutions.[87]

I have gone into some detail in examining these three related points – precaution, technology, and research – to provide adequate grounds for my conclusion. The conclusion is this: although it might seem at first impossible to achieve consensus behind the Precautionary Principle (as the Global Convention on Agroecology would require) and thereby to redirect research in a way that insists that technological input to agriculture and agricultural research be accepted only if it can be justified as running only minimal risk of creating more problems than it solves, in fact it might *not* be impossible to achieve such a consensus. Why? Because when properly understood, the research agenda that would be involved in a move to agroecological husbandry can be seen as striking an appropriate balance between two extreme positions. The first of those extreme positions is that "reductive" technology, or "industrial" technology – working at the molecular or genetic level – can be used to solve individual problems without creating more problems than it purports to solve. The second of those extreme positions is that an ecology-based form of agriculture somehow rejects technology altogether and insists that so much precaution should be exercised that *no* technological inputs or research could *ever* be allowed.

Both of those positions are unrealistic, for reasons that I have tried to explain previously. Instead, the three interconnected issues of precaution, technology, and research should be properly understood as follows:

- The Precautionary Principle must be honored in all research and in the application of all new technologies to agriculture.
- However, what this means in practice in the agricultural context is that greater attention should be given to ecology, so as to gain an

---

85 For details on the Morrill Act, see text accompanying note 15 in Chapter 6, *supra*.
86 Jackson, *Thirty Five Years*, *supra* note 84, at 11.
87 Kirschenmann, *supra* note 24, at 135–136.

"agroecological literacy" that offers a "new vision of the prairie" (grasslands) that should serve as the model for a new form of agriculture. With this approach, the negative effects of introducing new technologies can more readily be identified and quantified.

- The trajectory of agricultural research for the future, therefore, should (i) depart from the production orientation of the past several decades, particularly in the USA, (ii) adopt instead an ecological orientation, and (iii) embrace new technologies in the conduct of the research, and also in the adoption of new technologies into crops *if* they pass muster under the Precautionary Principle.

I quoted Kirschenmann earlier as having observed that the kind of agricultural research that I have described here "probably won't be funded by commodity groups or the agribusiness industry". Indeed, the greatest *obstacle* to a reorientation of agricultural research might be those very interest groups who resist such a reorientation, urging instead a continuation of the status quo. However, as Kirschenmann also points out, "[i]f the popular media see the need for such a transformation, then public support for this task might be generated."[88] That, in my view, is the principal grounds for hope: that strong public support can be generated for a reorientation of agricultural research. Proper explanation of the issues involved, and particularly the way that such research would embrace *both* the Precautionary Principle *and* technological innovation, could generate such public support and therefore overcome objections that might otherwise be raised to these aspects of a Global Convention on Agroecology.

IIIB. Reorienting economics

Principle 6 in the Global Convention on Agroecology as I have envisioned it above would assert that Contracting States recognize that one of the four principal objectives to be achieved in adopting a new form of agricultural life and production is "to overcome the economic unsustainability of modern extractive agriculture". As I explained there, achieving that objective would require not only a reorientation of the economics of farming but also a reorientation of economic theory more generally, in order to reflect the mismatch between (i) a planet of limited resources and (ii) an economic model premised on the need for perpetual growth.

At the very end of Chapter 4, I identified three main components in such a reoriented economic theory. I wrote there that it would be "one in which (i) endless growth is not an essential element, (ii) consumption is recognized as being subject to limits, not celebrated as a central aim of the economy,

---

88 Kirschenmann, *supra* note 24, at 131.

and (iii) all externalities are properly accounted for in the pricing of goods, particularly agricultural goods".[89]

These are three central components of a new economic theory that the Convention would require its Contracting States to embrace. Is this a real possibility or a pipe-dream? If it is merely a pipe-dream, then it would seem futile to try putting in place a Convention of the sort I have described above for the purpose of facilitating a transformation from modern extractive agriculture to agroecological husbandry.

I do not believe that a reorientation of economic theory is a pipe-dream. Instead, I believe that when properly understood, such a reorientation will be considered not impossible but critically essential.

The central issue is one of *limits*. In subsection IIIF of Chapter 4, I summarized views offered by Kenneth Boulding (known for his "spaceship Earth" analysis), Herman Daley (writing on the "steady-state economy"), and Richard Heinberg, whose book *The End of Growth* explains that there is now "at least a score of think tanks, institutes, and publications advocating fundamentally revising economic theory in view of ecological limits".[90] For these and other authors, an under-appreciated element of economics is the element of limits.

Tim Jackson, in his book *Prosperity Without Growth*, expands on the issue of limits, starting with a series of questions:

> What can prosperity possibly look like in a finite world, with limited resources and a population expected to exceed 9 billion people within decades? Do we have a decent vision of prosperity for such a world? Is this vision credible in the face of the available evidence about ecological limits? How do we go about turning vision into reality?
>
> The prevailing response to these questions is to cast prosperity in economic terms and to call for continuing economic growth as the means to deliver it. Higher incomes mean increased choices, richer lives, an improved quality of life for those who benefit from them. That at least is the conventional wisdom.[91]

Having posed key questions and given the "conventional wisdom" answer, Tim Jackson then explains that "the single most important [economic] policy goal across the world for most of the last century" has been to achieve an ever-increasing per capita gross domestic product (GDP).[92] This policy goal is unrealistic, however, according to Jackson: "[A]ny credible vision of prosperity has to address the question of limits. This is particularly true of a

---

89 See the closing paragraph in Chapter 4, *supra*.
90 See text accompanying notes 82, 83, and 84 in Chapter 4, *supra*.
91 Tim Jackson, PROSPERITY WITHOUT GROWTH: ECONOMICS FOR A FINITE PLANET 3 (2009).
92 *Id.*

vision based on growth. *How – and for how long – is continued growth possible without coming up against the ecological limits of a finite planet?*[93]

Jackson is not asking a trivial question. It is certainly worth examining the central role that economic growth plays in national and international policy, in an effort to supplant the "prosperity as growth" model with a new and different model that can serve as the centerpiece of national and international economic policy.

Another advocate of such a new and different model is Richard Heinberg, who opens his book *The End of Growth* with a "central assertion" that he calls "both simple and startling: *Economic growth as we have known it is over and done with.*"[94] He identifies three specific factors that "stand firmly in the way of further economic growth". These are:

- The *depletion* of important resources including fossil fuels and minerals;
- The proliferation of *negative environmental impacts* arising from both the extraction and use of resources (including the burning of fossil fuels) – leading to snowballing costs from both these impacts themselves and from efforts to avert them; and
- *Financial disruptions* due to the inability of our existing monetary, banking, and investment systems to adjust to both resource scarcity and soaring environmental costs – and their inability (in the context of a shrinking economy) to service the enormous piles of government and private debt that have been generated over the past couple of decades.[95]

In part, Heinberg echoes the points made in the 1972 book *Limits to Growth*. Calling that book "the best-selling environmental book of all time",[96] Heinberg explains that "the original *Limits to Growth* scenarios have held up quite well", and he cites recent extensive studies as authority for that assertion.[97] However, Heinberg extends his argument much further, to assert that "we have created monetary and financial systems that *require* growth"[98]

---

93 *Id.* at 5–6 (emphasis added). In addition to criticizing the "prosperity as growth" model on grounds that it disregards the resource limits of the Earth, he asserts that it (i) features unfairness in the form of inequality in benefits among the purported beneficiaries (noting that "[a] fifth of the world's population earns just 2 per cent of global income" while the "richest 20 per cent by contrast earn 74 per cent of the world's income"), and (ii) disregards the fact (recognized as far back as the time of Aristotle) that beyond a certain point, continued economic growth does not appear to advance and may even impede human happiness. *Id.* at 5, 36–37.

94 Richard Heinberg, THE END OF GROWTH: ADAPTING TO OUR NEW ECONOMIC REALITY 1 (2011).

95 *Id.* at 2–3.

96 *Id.* at 4, citing Donella H. Meadows, Dennis L. Meadows, Jørgen Randers, and William W. Behrens III, THE LIMITS TO GROWTH (1972).

97 See Heinberg, *supra* note 94, at 4–5, citing a 2008 work by Graham Turner and a 2004 "update" by Donella Meadows, Jørgen Randers, and Dennis Meadows.

98 Heinberg, *supra* note 94, at 6.

in a way that is a fairly recent phenomenon. He offers a sweeping explanation of this point, starting with two historical accounts.

The first historical account Heinberg offers focuses on the history of economies – that is, the systems humans create to create and distribute wealth:

> Throughout over 95 percent of our species' history, we humans lived by hunting and gathering in what anthropologists call *gift economies*. People had no money, and there was neither barter nor trade among members of any given group. Trade did exist, but it occurred only between members of different communities.
>
> It's not hard to see why sharing was the norm within each band of hunter-gatherers, and why trade was restricted to relations with strangers. Groups were small, usually comprising between 15 and 50 persons, and everyone knew and depended on everyone else within the group. Trust was essential to individual survival, and competition would have undermined trust. Trade is an inherently competitive activity: each trader tries to get the best deal possible, even at the expense of other traders. For hunter–gatherers, cooperation – not competition – was the route to success, and so innate competitive drives (especially among males) were moderated through ritual and custom, while a thoroughly entangled condition of mutual indebtedness helped maintain a generally cooperative attitude on everyone's part . . . . Freeloading [in such a group] was occasionally a problem, and when it became a drag on the rest of the group it was punished by subtle or not-so-subtle social signals – ultimately, ostracism. But otherwise no one kept score of who owed whom what; to do so would have been considered very bad manners.[99]
>
> Here is economic history compressed into one sentence: As societies have grown more complex, larger, more far-flung, and diverse, the tribe-based gift economy has shrunk in importance, while the trade economy has grown to dominate most aspects of people's lives, and has expanded in scope to encompass the entire planet. Is this progress or a process of moral decline? Philosophers have debated the question for centuries. Approve or disapprove, it is what we have done.
>
> [In short, although we] still enjoy some of the benefits of the old gift economy in our families and churches . . . increasingly, the market rules our lives. Our apparent destination in this relentless trajectory toward

---

99 Heinberg explains that this account of gift economies is confirmed by the accounts of twentieth-century anthropologists who visited surviving hunter-gatherer societies. In those societies, Heinberg reports, the giving of a gift to an outsider signified his or her acceptance into the group, as part of the family; if the recipient of the gift then immediately gave another gift in return, it was regarded as offensive to the original gift-giver because "the immediate offering of a gift in return smacked of trade – something only done with strangers". *Id.* at 29.

expansion of trade is a world in which everything is for sale, and all human activities are measured by and for their monetary value.[100]

Heinberg then turns from the history of *economies* to the history of *economics* – beginning with the eighteenth century. In several paragraphs he summarizes how economic theory came gradually to give such emphasis to growth, nothwithstanding natural limits of the planet's resources:

> "Classical" economic philosophers such as Adam Smith (1723–1790), Thomas Robert Malthus (1766–1834), and David Ricardo (1772–1823) introduced basic concepts such as supply and demand, division of labor, and the balance of international trade. As happens in so many disciplines, early practitioners were presented with plenty of uncharted territory and proceeded to formulate general maps of their subject that future experts would labor to refine in ever more trivial ways.
>
> Importantly, these early philosophers had some inkling of natural limits and anticipated an eventual end to economic growth. The essential ingredients of the economy were understood to consist of *land, labor,* and *capital.* There was on Earth only so much land (which in these theorists' minds stood for all natural resources), so of course at some point the expansion of the economy would cease.
>
> But, starting with Adam Smith, the idea that continuous "improvement" in the human condition was possible came to be generally accepted. At first, the meaning of "improvement" (or *progress*) was kept vague, perhaps purposefully. Gradually, however, "improvement" and "progress" came to mean "growth" in the current economic sense of the term – abstractly, an increase in Gross Domestic Product (GDP), but in practical terms, an increase in consumption.
>
> A key to this transformation was the gradual deletion by economists of *land* from the theoretical primary ingredients of the economy (increasingly, only labor and capital really mattered, land having been demoted to a sub-category of capital).[101]

Having offered these historical and conceptual observations, Heinberg then turns to the future, concentrating especially on the notion of "post-growth economics". As he explains, "conventional economics starts with certain

---

100 *Id.* at 28–30. Heinberg hastens to point out that he does not yearn for a return to the gift economy. Noting that Communism was one of many attempts throughout history to restore a sort of gift economy, "trying to institutionalize a gift economy at the scale of the nation state introduces all kinds of problems, including those of how to reward initiative and punish laziness in ways that everyone finds acceptable, and how to deter corruption among those whose job it is to collect, count, and reapportion the wealth." *Id.* at 30.

101 *Id.* at 34–36. For an elaboration on this last point – the "demotion" of land – see text accompanying note 85 in Chapter 4, *supra.*

basic premises that are clearly, unequivocally incorrect: that the environment is a subset of the economy; that resources are infinitely substitutable; and that growth in population and consumption can continue forever." Specifically, in conventional economics, "natural resources like fossil fuels are treated as expendable income, when in fact they should be treated as capital, since they are subject to depletion."[102]

Heinberg then posits "four fundamental principles" that he says must stand at the core of economic theory if, in his words "economics is to stop steering society into the ditch":

- Growth in population and consumption rates cannot be sustained.
- Renewable resources must be consumed at rates below those of natural replenishment.
- Non-renewable resources must be consumed at declining rates (with rates of decline at least equaling rates of depletion), and recycled wherever possible.
- Wastes must be minimized, rendered non-toxic to humans and the environment, and made into "food" for natural systems or human production processes. [103]

I would emphasize that the specific details of the economic-system models that some "alternative-economics" advocates such as Tim Jackson and Richard Heinberg have offered are less important than is the overarching point that I made above and will repeat here: it is time to examine more closely the central role that economic growth plays in national and international policy, in an effort to supplant the "prosperity-as-growth" model with a different model that can serve as the centerpiece of national and international economic policy. The logical grounds for undertaking such a reorientation of economic theory are surely strong enough, and have earned enough respect, to gain acceptance among many policy-makers.[104]

This is what Principle 6 of the Convention that I have described above would require: acceptance among the Contracting States of a reorientation of economic theory, away from a growth model and toward a model that

---

102 Heinberg, *supra* note 94, at 246.

103 *Id.* at 246–247. Further, Heinberg asserts, "economists must aim for a dynamic balance between efficiency (maximizing throughput) and resilience (adaptability, redundancy, diversity, and interconnectivity) – whereas today economists focus almost entirely on efficiency". *Id.* at 247.

104 A manifestation of the respect that has been earned by those who advocate a reorientation of economic theory *away* from the "prosperity-as-growth" model appears in an early-August 2015 issue of *The Economist*. A one-page written debate is offered there on this question: "If the rich world aimed for minimal growth, would it be a disaster or a blessing?". Tim Jackson, whose work I explained above, writes on the "blessing" side of the debate. Adam Posen (of the Peterson Institute for International Economics) writes on the "disaster" side of the debate. THE ECONOMIST, Aug. 1, 2015, at 12. It is highly questionable whether the question posed to those two experts would have been regarded until quite recently as even a matter of debate.

recognizes the reality of *limits*. Although it might seem at first glance to be impossible to achieve consensus for such a reorientation of economic theory, I believe that when properly understood, such a reorientation will be considered essential.[105]

IIIC. The most urgent emergency: climate change and agriculture

Principle 8 of the Convention that I described in section II of this chapter would require that Contracting States recognize the "urgent *necessity to address climate change issues*, with special reference to agriculture". As I suggested above in discussing that Principle and some related provisions that could appear in the Convention's Preamble, we can regard global climate change – and the existential dangers that it presents – as a prominent and immediate triggering cause for undertaking a reform of global agriculture, especially if that new form of agriculture promises to sequester dramatically larger amounts of carbon.

It is common knowledge, however, that the action taken thus far at the international level has been inadequate to halt climate change or take effective mitigation or adaptation actions.[106]

---

105 Lester Brown also urges that a reorientation of economic theory is essential, especially in its treatment of ecological issues. He offers these observations that draw a similarity between today's need to formulate a new economic worldview and the need in Copernicus's day to formulate a new astronomical worldview:

> [M]ainstream economics pays little attention to the sustainable yield thresholds of the earth's natural systems. Modern economic thinking and policymaking have created an economy that is so out of sync with the ecosystem on which it depends that it is approaching collapse. How can we assume that the growth of an economic system that is shrinking the earth's forests, eroding its soils, depleting its aquifers, collapsing its fisheries, elevating its temperature, and melting its ice sheets can simply be projected into the long-term future? What is the intellectual process underpinning these extrapolations?
>
> We are facing a situation in economics today similar to that in astronomy when Copernicus arrived on the scene, a time when it was believed that the sun revolved around the earth. Just as Copernicus had to formulate a new astronomical worldview after several decades of celestial observations and mathematical calculations, we too must formulate a new economic worldview based on several decades of environmental observations and analyses.

Lester R. Brown, WORLD ON THE EDGE: HOW TO PREVENT ENVIRONMENTAL AND ECONOMIC COLLAPSE 8–9 (2011).

106 See, e.g., Alex Morales, *Kyoto Veterans Say Global Warming Goal Slipping Away*, BLOOMBERG BUSINESS, Nov. 4, 2013 (quoting a former UNFCCC executive secretary as saying that "[t]here is nothing that can be agreed in 2015 that would be consistent with the 2 degrees [target]"). More recent observations confirm that. See, e.g., Fred Pearce, *Will the Paris Climate Talks Be Too Little and Too Late?*, YALE ENVIRONMENT 360 (Sep. 14, 2015) (quoting one expert's assertion that "if the Paris meeting locks in present climate commitments for 2030, holding warming below 2 degrees could essentially become infeasible"). See also Karl Ritter, *New Climate Pledges Invigorate UN Talks but Deemed Insufficient to Halt Dangerous Warming*, U.S. NEWS & WORLD REPORT, Sep 28, 2015 (citing an analysis concluding that that even the tentative commitments

Given that fact, why would there be any reason to expect that the Convention I have described – focusing on agroecological husbandry – would somehow succeed in prompting officials around the world to take action to attack climate change vigorously? Expressed differently, is it simply impossible to secure global consensus and effective action on this issue?

In my view, such consensus and action is not impossible. A persuasive presentation of (i) *why* global climate change poses an immediate existential threat requiring bold and immediate multilateral measures and (ii) *how* a dramatic transformation of agriculture will help address that threat can, I believe, succeed in mustering the public support necessary to force policy-makers at the national and international levels to take action – and that action can include the conclusion of a Global Convention on Agroecology of the sort I have described above.

A late-2012 World Bank report documents the dangers of a "Four Degree World". The report, titled *Turn Down the Heat: Why a 4°C Warmer World Must Be Avoided*, [107] opens with an explanation of why it makes sense to try anticipating the danger of such a development:

> Without further commitments and action to reduce greenhouse gas emissions, the world is likely to warm by more than 3°C above the preindustrial climate. Even with the current mitigation commitments and pledges fully implemented, there is roughly a 20 percent likelihood of exceeding 4°C by 2100. If they are not met, a warming of 4°C could occur as early as the 2060s. Such a warming level and associated sea-level rise of 0.5 to 1 meter, or more, by 2100 would not be the end point: a further warming to levels over 6°C, with several meters of sea-level rise, would likely occur over the following centuries.
>
> Thus, while the global community has committed itself to holding warming below 2°C to prevent "dangerous" climate change, and Small

---

made by leading countries in advance of the late-2015 climate negotiations would still "leave the world on a path toward 3.5 degrees C (6.3F) of warming compared with pre-industrial times"). This view was widely confirmed following the Paris climate talks. See, e.g., George Monbiot, *Grand Promises of Paris Climate Deal Undermined by Squalid Retrenchments*, GUARDIAN Dec. 12, 2015, www.theguardian.com/environment/georgemonbiot/2015/dec/12/paris-climate-deal-governments-fossil-fuels (asserting that "[b]y comparison to what [the outcome at Paris] should have been, it's a disaster" and noting that the outcomes at Paris "are likely to commit us to levels of climate breakdown that will be dangerous to all and lethal to some" because the negotiators there "agreed to burden our successors with a . . . dangerous legacy: the carbon dioxide produced by the continued burning of fossil fuels, and the long-running impacts this will exert on the global climate").

107 See World Bank, *Turn Down the Heat: Why a 4°C Warmer World Must Be Avoided* (2012). As explained in note 26, *supra*, the 2012 report has now given rise to an ongoing series. See www.worldbank.org/en/topic/climatechange/publication/turn-down-the-heat. As noted there, the yearly reports are prepared for the World Bank by the Potsdam Institute for Climate Impact Research and Climate Analytics. The third report, issued in 2014, finds that about 1.5°C of warming is already locked in.

Island Developing states (SIDS) and Least Developed Countries (LDCs) have identified global warming of 1.5 °C as warming above which there would be serious threats to their own development and, in some cases, survival, *the sum total of current policies – in place and pledged – will very likely lead to warming far in excess of these levels*. Indeed, present emission trends put the world plausibly on a path toward 4 °C warming within the century . . .

A world in which warming reaches 4 °C above preindustrial levels (hereafter referred to as 4 °C world), would be one of unprecedented heat waves, severe drought, and major floods in many regions, with serious impacts on human systems, ecosystems, and associated services . . . . [Indeed,] a global mean temperature increase of 4 °C approaches the difference between temperatures today and those of the last ice age, when much of central Europe and the northern United States were covered with kilometers of ice and global mean temperatures were about 4.5 °C to 7 °C lower. And this magnitude of climate change – human induced – is occurring over a century, not millennia . . . .[108]

The report goes on to offer observations and predictions about what specific effects such changes would bring, beginning with the sheer increase in heat in many portions of the world:

The effects of 4 °C warming will not be evenly distributed around the world, nor would the consequences be simply an extension of those felt at 2 °C warming. The largest warming will occur over land and range from 4 °C to 10 °C. Increases of 6 °C or more in average monthly summer temperatures would be expected in large regions of the world, including the Mediterranean, North Africa, the Middle East, and the contiguous United States . . .

Projections for a 4 °C world show a dramatic increase in the intensity and frequency of high-temperature extremes. Recent extreme heat waves such as in Russia in 2010 are likely to become the new normal summer in a 4 °C world. Tropical South America, central Africa, and all tropical islands in the Pacific are likely to regularly experience heat waves of unprecedented magnitude and duration. In this new high-temperature climate regime, the coolest months are likely to be substantially warmer than the warmest months at the end of the 20th century. In regions such as the Mediterranean, North Africa, the Middle East, and the Tibetan plateau, almost all summer months are likely to be warmer than the most extreme heat waves presently experienced. For example, the warmest July in the Mediterranean region could be 9 °C warmer than today's warmest July.

---

108 *Id.* at 1–3 (emphasis added).

Extreme heat waves in recent years have had severe impacts, causing heat-related deaths, forest fires, and harvest losses. The impacts of the extreme heat waves projected for a 4°C world have not been evaluated, but they could be expected to vastly exceed the consequences experienced to date and potentially exceed the adaptive capacities of many societies and natural systems.[109]

The implications of a "Four Degree World" on ocean acidification, on rising sea levels, on coastal inundation, and on various other "human support systems" then receive attention in the World Bank report. For instance, it asserts that the change in ocean acidity over the next century, with a global warming of 4°C or more, will likely "be unparalleled in Earth's history", and this will bring adverse consequences for marine organisms and ecosystems.[110] Another of the repercussions of such global temperature increases is severe water scarcity.[111]

The report gives special attention to the impact that a "Four Degree World" would have on food and agricultural output:

> Maintaining adequate food and agricultural output in the face of increasing population and rising levels of income will be a challenge irrespective of human-induced climate change. The IPCC [has] projected that global food production would increase for local average temperature rise in the range of 1°C to 3°C, but may decrease beyond these temperatures.
>
> New results published since 2007, however, are much less optimistic. These results suggest instead a rapidly rising risk of crop yield reductions as the world warms. Large negative effects have been observed at high and extreme temperatures in several regions including India, Africa, the United States, and Australia. For example, significant nonlinear effects have been observed in the United States for local daily temperatures increasing to 29°C for maize and 30°C for soybeans. These new results and observations indicate a significant risk of high-temperature thresholds being crossed that could substantially undermine food security globally in a 4°C world.[112]

The World Bank report concludes with a dire warning of the "risks of disruptions and displacements in a 4°C world", noting that "[t]he projected impacts on water availability, ecosystems, agriculture, and human health could lead to large-scale displacement of populations and have adverse

---

109 *Id. at* 4–5.

110 *Id.* at 5.

111 *Id.* at 7.

112 *Id.* at 7. The report goes on to note that these risks would be compounded by the adverse effects that projected sea-level rise would have on agriculture in important low-lying delta areas, such as in Bangladesh, Egypt, Vietnam, and parts of the African coast.

consequences for human security and economic and trade systems."[113] The closing paragraphs provide this grim summary:

> With pressures increasing as warming progresses toward 4°C and combining with nonclimate-related social, economic, and population stresses, *the risk of crossing critical social system thresholds will grow.* At such thresholds existing institutions that would have supported adaptation actions would likely become much less effective or even collapse . . . . [For instance,] stresses on human health, such as heat waves, malnutrition, and decreasing quality of drinking water due to seawater intrusion, have the potential to overburden health-care systems to a point where adaptation is no longer possible, and dislocation is forced.
>
> Thus, given that uncertainty remains about the full nature and scale of impacts, *there is also no certainty that adaptation to a 4°C world is possible.* A 4°C world is likely to be one in which communities, cities and countries would experience severe disruptions, damage, and dislocation, with many of these risks spread unequally. It is likely that the poor will suffer most and the global community could become more fractured, and unequal than today. The projected 4°C warming simply must not be allowed to occur – the heat must be turned down. Only early, cooperative, international actions can make that happen.[114]

In introducing this discussion of global climate change, I suggested that official support might be mustered for a Convention to facilitate a transition to agroecological husbandry if a persuasive presentation could be made about (i) *why* global climate change poses an immediate existential threat requiring bold and immediate multilateral measures and (ii) *how* a dramatic transformation of agriculture can help address that threat. The World Bank report referred to above is just one of many recent projections showing the

---

113 *Id.* at 8.

114 *Id.* at 8 (emphasis added). Since the time that the World Bank report was published in 2012, further emissions of greenhouse gases have prompted some observers to ask whether the buildup of such gases has already placed the world past a "point of no return". For a 2015 blogpost on the website of *Scientific American* offering views on those issues, see *Have We Passed the Point of No Return on Climate Change?*, www.scientificamerican.com/article/have-we-passed-the-point-of-no-return-on-climate-change/ (April 13, 2015) (emphasizing that "another 0.4 degree Fahrenheit rise in temperature, representing a global average atmospheric concentration of carbon dioxide . . . of 450 parts per million . . . , could set in motion unprecedented changes in global climate and a significant increase in the severity of natural disasters – and as such could represent the dreaded point of no return". According to some reports, March 2015 was the first month in which CO_2 concentrations exceeded 400 parts per million for the entire month. See Adam Vaughan, *Global Carbon Dioxide Levels Break 400 ppm Milestone*, GUARDIAN (May 6, 2015). The conversion from degrees Fahrenheit to degrees Centigrade is 1°F = 0.56°C or 1°C = 1.8°F.

consequences climate change will bring,[115] and surveys now show a surge of support around the world for the propositions that global climate change is a serious and imminent threat and that formal global action is needed to address it.

For instance, a recent Pew research survey found that majorities of persons in all of the 40 nations polled about global climate change "say it is a serious problem, and a global median of 54 percent consider it a *very* serious problem. Moreover, a median of 78 percent support the idea of their country limiting greenhouse gas emissions as part of an international agreement in Paris." The Pew report continues:

> Climate change is not viewed as a distant threat. Across the nations surveyed, a median of 51% believe people are already being harmed by climate change and another 28% think people will be harmed in the next few years. More than half in 39 of 40 countries are concerned it will cause harm to them personally during their lifetime . . . and a global median of 40% are *very* worried this will happen.
>
> . . . . [A]ccording to most respondents, confronting climate change will entail more than just policy changes; it will also require significant changes in how people live. A global median of 67% say that in order to reduce the effects of climate change, people will have to make major changes in their lives. A median of just 22% believe technology can solve this problem without requiring major changes. Even in the U.S., a country known for its technological innovations, 66% believe people will need to significantly alter their lifestyles . . . . In most countries, publics tend to believe that much of the burden for dealing with climate change should be shouldered by wealthier countries. Across the nations polled, a median of 54% agree with the statement "Rich countries,

---

115 See, e.g., *Climate Change 2014 Synthesis Report: Summary for Policymakers* 13, Intergovernmental Panel on Climate Change (Mar. 13, 2015), www.ipcc.ch/pdf/assessment-report/ar5/syr/AR5_SYR_FINAL_SPM.pdf ("Climate change will amplify existing risks and create new risks for natural and human systems. Risks are unevenly distributed and generally greater for disadvantaged people and communities in countries at all levels of development"); *Future Climate Change*, US Environmental Protection Agency (Mar. 4, 2014), www.epa.gov/climatechange/science/future.html#ref4; *Climate Change Impacts in the United States: Highlights* 5, U.S. Global Change Research Program (2014), www.globalchange.gov/sites/globalchange/files/NCA3_Highlights_LowRes-small-FINAL_posting.pdf ("The amount of warming projected beyond the next few decades is directly linked to the cumulative global emissions of heat-trapping gases and particles"). For a very recent assessment, see James Hanson *et al.*, *Ice Melt, Sea Level Rise and Superstorms*, 40 ATMOSPHERIC CHEMISTRY AND PHYSICS 3761 (2016), www.atmos-chem-phys.net/16/3761/2016/acp-16-3761-2016.pdf (explaining feedback mechanisms likely to result in higher sea level rise than earlier predicted, resulting in a "loss of all coastal cities, most of the world's large cities and all their history" in this century or at the latest next century, if fossil fuel emissions continue at a high level). Hansen is the Director of the Climate Science, Awareness and Solutions program at Columbia University Earth Institute. For a YouTube video of Hansen summarizing the paper, see www.youtube.com/watch?v=JP-cRqCQRc8.

such as the U.S., Japan and Germany, should do more than developing countries because they have produced most of the world's greenhouse gas emissions so far."[116]

In my view, therefore, the world will probably jump over the first hurdle I noted on page 338 – accepting the proposition that action is needed – very soon, notwithstanding the idiotic denials of some US political leaders. As for the second hurdle – achieving a broad-based realization of how shifting to a natural-systems agriculture can help address climate change – I believe a robust and well-targeted informational campaign can facilitate an understanding of the key details I explained in subsection IA of Chapter 5. The key points include these:[117]

- Perennial polycultures, which form a central component of agroecological husbandry – can dramatically reduce the required amount of agricultural fertilizer and chemical pesticides, which draw heavily from fossil carbon.
- Perennial polycultures can also dramatically reduce the fossil-carbon fuels needed to power farm equipment – thus creating a further reduction in (i) the draw on non-renewable fossil-carbon deposits and (ii) the levels of greenhouse-gas emissions that contribute to climate change.
- Perennial polycultures can sequester carbon, thus contributing to climate resiliency.
- Perennial grains grown in polycultures can reduce the release of both nitrous oxide and methane, two of the most damaging of the greenhouse gases (far more potent than carbon dioxide in their impact on global climate change).
- Other elements in a package of reforms to adopt agroecological husbandry – particularly those relating to the role of livestock in farming – can further reduce the release of methane.

Once properly understood and vigorously advocated, these points should help build strong support for a Global Convention on Agroecology.

IIID. The feasibility of a duty of cooperation

I have tried in the preceding paragraphs to explain why it might not be impossible to expect that a Global Convention on Agroecology can be concluded and implemented. In particular, I have identified reasons why

---

116 Bruce Stokes, Richard Wike, and Jill Carle, *Global Concern about Climate Change, Broad Support for Limiting Emissions*, Pew Research Center, Nov. 5, 2015, www.pewglobal.org/2015/11/05/global-concern-about-climate-change-broad-support-for-limiting-emissions/ (reporting results of surveys conducted between late March and late May of 2015).

117 For details on several of the following points, see text accompanying notes 2–7 in Chapter 5, *supra*.

specific obligations that such a Convention would impose on Contracting States – relating specifically to precaution, economics, and climate change – might not in fact be considered so onerous as to dissuade most countries from participating. In each of those cases, I have asserted that instead of being considered *impossible*, such participation might be considered so *essential* as to make the creation of such a Convention feasible.

Now I turn to a fourth grounds on which, at first glance, it might be considered impossible to achieve the kinds of consensus among states that such a Convention would require. Surely, one might argue, the record of non-cooperation among the world's states and societies – indeed, the ever-increasing frictions that divide peoples one from another – shows that a comprehensive effort at global cooperation is doomed from the outset. Expressed differently, surely (the argument goes) a principle of global cooperation of the sort expressed in Principle 3 of the Convention that I described earlier is impossible in the face of news reports that abound about frictions of all sorts between people who regard themselves as fundamentally different on grounds of religion, nationality, class, gender, race, and other supposed distinctions.

To address this argument, I challenge its underlying assumption. A recent book by David Cannadine focuses on the social frictions and divisions that supposedly characterize world history. In its opening pages, Cannadine's book highlights how significant such frictions and divisions can be, leading the unsuspecting reader to assume that the book will offer up a documented narrative substantiating humanity's "divided past":

> This book sets out to explore and investigate . . . the six most commonplace and compelling forms of [defining group] identities, namely religion, nation, class, gender, race, and civilization . . . . [T]hese groupings have commanded widespread allegiance and commitment, on occasions for good, but often not, since every collective *solidarity* simultaneously creates an actual or potential *antagonist* out of the group or groups it excludes. Even if we confine ourselves to the twentieth century, there have been many such confrontations and conflicts variously described as religious wars, or national wars, or class wars, or gender wars, or race wars, or wars for civilization. [But of course numerous groupings and] conflicts have existed across the millennia and around the world, from Christians versus pagans during the later Roman Empire to the white supremacists versus anti-apartheid campaigners until 1994; and there is no reason to suppose that the twenty-first century will be free of such confrontations.[118]

Cannadine then takes a dramatically different tack – one that is in fact reflected in the book's title. Instead of writing under the title *The Divided*

---

118 David Cannadine, The Undivided Past 3–4 (2013).

Past, Cannadine writes of *The Undivided Past.* One by one, he dismantles each of the supposed antagonism-triggering group identities – religion, nation, class, gender, race, and civilization – by exploring and explaining how it is much more accurate to take a broader and more optimistic view of the human past:

> During the last half century or so, the conventional wisdom that "the history of humanity is based upon the immemorial divisions of its peoples" has been reinforced by a growing academic insistence on the importance of recognizing the "difference" between collective groups. According to the anthropologist Clifford Geertz, "difference is what makes the world go round, especially the political world"; [he and] many of his colleagues . . . [emphasize] what they varyingly describe as "difference," or "otherness," or "alterity," or "unlikeness," or "dissimilarity." Beyond doubt, such historical approaches have yielded significant work of enduring value . . .
>
> But the fact that humanity *is still here,* that no one has vanquished "us" or "them" on either side of any of these divides, . . . suggests that there is a case for taking a broader, more ecumenical, and even more optimistic view of human identities and relations – a view that . . . . recognizes affinities and discerns conversations *across* these allegedly impermeable boundaries of identity, which embody and express a broader sense of humanity that goes beyond our dis-similarities . . .
>
> [For instance,] Neil MacGregor, the director of the British Museum, has lamented the "brutally over-simplified notions of identity" that "sustain entrenched conflicts," when in reality, cultures constantly "overlap, borrow from each other and live together" in "a conversation with the whole of humanity." . . . This book [therefore] seeks to address these issues from a longer-term historical perspective [and to] . . . argue that the . . . very *categories of "us" and "them,"* whatever their particular articulation, and though proclaimed to be irreducible and absolute, frequently *reveal themselves to be unstable and ambiguous*; they often prove to be incoherent even in the thick of their confrontations with the implacable foe; and they are held together not so much by shared self-awareness as by the exhortations of leaders, journalists, activists – and by some historians, too.[119]

Complementing this historical perspective provided by Cannadine – emphasizing that the dominant theme of human history is *not* in fact one of division but rather one of "undivision" – is the perspective taken in another recent book titled *Better Angels of Our Nature.* The book's author, Steven Pinker, amasses volumes of data to support his thesis that "violence has

---

119 *Id.* at 5–8.

declined over long stretches of time, and today we may be living in the most peaceable era in our species' existence".[120] His analysis identifies six major trends showing "our species' retreat from violence", some of which have taken place over many centuries (such as "the Pacification Process" that occurred with the transition from the anarchy of hunting-gathering societies to the first agricultural civilizations with cities and governments) and some of which have taken place just over the past several decades. In the latter category, for instance, he places "the Rights Revolution" that started with the 1948 Universal Declaration of Human Rights and has been animated by "a growing revulsion . . . [to] violence against ethnic minorities, women, children, homosexuals, and animals".[121]

In examining *why* this surprising decline in violence has occurred, Pinker identifies five "exogenous forces that favor our peaceable motives and that have driven the multiple declines in violence".[122] He labels these five factors as (i) the Leviathan (that is, the creation of political states, inspired in part by the logic expressed in Thomas Hobbes' masterwork *Leviathan*), (ii) commerce (iii) feminization, (iv) cosmopolitanism, and (v) the "escalator of reason". He summarizes them as follows:

> The *Leviathan*, a state and judiciary with a monopoly on the legitimate use of force, can defuse the temptation of exploitative attack, inhibit the impulse for revenge, and circumvent the self-serving biases that make all parties believe they are on the side of the angels. *Commerce* is a positive-sum game in which everybody can win; as technological progress allows the exchange of goods and ideas over longer distances and among larger groups of trading partners, other people become more valuable alive than dead, and they are less likely to become targets of demonization and dehumanization. *Feminization* is the process in which cultures have increasingly respected the interest and values of women. Since violence is largely a male pastime, cultures that empower women tend to move away from the glorification of violence and are less likely to breed dangerous subcultures of rootless young men. The forces of *cosmopolitanism* such as literacy, mobility, and mass media can prompt people to take the perspective of people unlike themselves and to expand their circle of sympathy to embrace them. Finally, an intensifying application of knowledge and rationality to human affairs – the *escalator of reason* – can force people to recognize the futility of cycles of violence, to ramp down the privileging of their own interests over others', and to reframe violence as a problem to be solved rather than a contest to be won.[123]

---

120 Steven Pinker, The Better Angels of Our Nature: Why Violence Has Declined xxi (2011).

121 *Id.* at xiv–xxv.

122 *Id.* at xxv.

123 *Id.* at xxvi. Pinker also identifies "a few forces that one might have thought would be important" to the downward trend in violence but apparently are *not*. These include such things as (i) the

Non-violence is not exactly the same as cooperation, but the two are surely related. Even though Pinker's analysis might not provide firm support for the proposition that there is a trend toward global cooperation whose momentum will create the kind of consensus that a Global Convention on Agroecology would require, Pinker's analysis does present some grounds for optimism and some specific avenues to pursue.

For one thing, Pinker's analysis suggests that efforts toward some form of global authority over – and responsibility for – the natural world are not fundamentally misguided. Pinker notes that recognizing that each state in an international system has "a monopoly on force to protect its citizens from one another may be the most consistent violence-reducer", and it works as follows: "If a government imposes a cost on an aggressor that is large enough to cancel out his gains . . . it flips the appeal [that violence holds for] the potential aggressor, making peace more attractive than war."[124] Likewise, if an effective authority can impose a cost on an entity – be it an individual or a business association or a state government – that is large enough to cancel out any gains that the entity seeks to enjoy from committing violence against the natural world (through agricultural-pesticide runoff, for instance, or through large contributions to global greenhouse gas emissions), it flips the appeal that such action holds for the entity, making the action less likely. Hence, when Pinker identifies the aggregation of collective power into a Leviathan-type authority as an especially effective instrument in protecting persons against violence, it seems reasonable to try creating some such authority for ecological protection as well.

Granted, the infliction of violence and death on the natural world is different from the infliction of violence and death on humans alone. Still, several of the same "inner demons" that Pinker identifies as fundamentally responsible for inter-human violence also seem to apply in the case of the forms of violence that humans inflict on other elements of the natural world. Pinker identifies five such "inner demons":

> *Predatory* or *instrumental violence* is simply violence deployed as a practical means to an end. *Dominance* is the urge for authority, prestige, glory, and power, whether it takes the form of macho posturing among individuals or contests for supremacy among racial, ethnic, religious, or national groups. *Revenge* fuels the moralistic urge toward retribution, punishment, and justice. *Sadism* is pleasure taken in another's suffering. And *ideology*

---

development of ever-more-sophisticated weaponry, (ii) "resource determinism" (the theory that people inevitably fight over finite resources such as land or minerals), (iii) the increase in overall world prosperity (Pinker says that "tight correlations between affluence and nonviolence are hard to find"), and (iv) religion. None of these factors, Pinker concludes, has consistently worked to reduce violence.

124 *Id.* at 680.

is a shared belief system, usually involving a vision of utopia, that justifies unlimited violence in pursuit of unlimited good.[125]

It takes little imagination to identify in those five "inner demons" at least some that help explain why some humans act in ways that seriously degrade the environment in which we live. For instance, some people practice "instrumental violence" in exterminating some species and their habitats — prairie dogs and prairie chickens come to mind — because those species and habitats seem to interfere with a desirable short-term economic end. Accordingly, Pinker's highlighting of the violence-reducing power of Leviathan (aggregated political authority) seems consistent with an effort to create some global authority for ecological protection.

The same logic holds for the other "pacifying forces of history" that Pinker identifies — namely, commerce, feminization, cosmopolitanism, and the "escalator of reason". All four of those are on the rise in today's world. Compared with a century or even a decade ago, the integration of a global market for goods and services has thickened substantially. To draw from Pinker's words again, "technological progress [increasingly] allows the exchange of goods and ideas over longer distances and among larger groups of trading partners, [and this makes] other people . . . less likely to become targets of demonization and dehumanization". Expressed simply, commerce raises the prospects for global cooperation in general. Whether these raised prospects in general can be directed toward securing cooperation over agricultural and ecological matters specifically will turn on other factors. Among them are Pinker's other "pacifying forces of history". Consider, for instance, the factor of cosmopolitanism. As Pinker explains, "living in a more cosmopolitan society, one that puts us in contact with a diverse sample of other people and invites us to take their points of view, changes our emotional response to their well-being."[126] Certainly technological change in communications — offering at least exposure (often including visual exposure) to other cultures and indeed other species with which we share the planet — has brought virtually all humans today into a more cosmopolitan world. We might see in this trend further grounds for hope that a consensus can be reached about the need for global cooperation on agroecological affairs.

In sum, a close analysis of human history, and of the last century in particular, provides some reason to expect that a form of global cooperation is in fact possible. By viewing the past more accurately — in a way that does *not* exaggerate the divisions and frictions among various types of people based on more or less artificial differences and that does *not* exaggerate the level of violence in today's world but rather recognizes the strong

---

125 *Id.* at xxv.
126 *Id.* at 689.

historical trend of non-violence – we can gain some momentum toward collaborative planning for a shared future. Specifically, perhaps achieving consensus favoring a principle of global cooperation, of the sort that a Global Convention on Agroecology would require, should be seen not as *impossible* but as *essential*.

★ ★ ★

In the foregoing pages, I have addressed four particularly nettlesome issues that stand at the core of a Global Convention on Agroecology and its adoption: precaution, economics, climate change, and cooperation. In doing so, I have examined *practicalities*, and this question in particular: Is it most accurate or realistic to assume that such a Convention is *impossible* to conclude and implement because consensus simply cannot be reached on one or more of these four issues? For reasons I have explained here, I believe the answer is "no". If properly presented and forcefully argued, the reasons supporting the drafting, negotiation, adoption, and implementation of such a Convention can, I believe, prevail. I find the opposite assumption – that consensus is not possible – inconsistent both with historical precedent and with recent trends.

Part V
Conclusion

8 Pluralistic sovereignty and eco-states?

In Chapter 7, I explored the possibility of a new treaty – a Global Convention on Agroecology – aimed at facilitating a revolutionary change in agriculture. Now, in this final chapter of the book, I turn briefly to the concept of sovereignty, and to how it can and should be made more realistic and sophisticated in order to facilitate a change in our form of agriculture.

In doing so, I will first offer a short account of how our modern concept of sovereignty has developed, resulting in what I call "monolithic sovereignty", and then introduce a different concept, that of a "pluralistic sovereignty" that features a prominent role for "eco-states". However, on all of these topics – "monolithic sovereignty", "pluralistic sovereignty", and "eco-states" – I do not aim to offer here a comprehensive assessment. Instead, I can only sketch the general contours. The details will need to await further elaboration in a forthcoming companion book to this one.[1]

---

1 John W. Head, A GLOBAL CORPORATE TRUST FOR AGROECOLOGICAL INTEGRITY: MANAGING A NEW AGRICULTURE IN A WORLD OF ECO-STATES (forthcoming). For an abbreviated treatment of these topics, in which I have used the terms "thick sovereignty" and "thin sovereignty" in lieu of "monolithic sovereignty" and "pluralistic" sovereignty, see John W. Head, *International Law, Agro-Ecological Integrity, and Sovereignty – Proposals for Reform*, 63 THE FEDERAL LAWYER 56 (June 2016).

I. A nutshell account of "monolithic sovereignty"

IA. Bodin, Hobbes, and Grotius: building a Westphalian sovereignty

Sovereignty is both a historical and a legal concept. That is, an understanding of the concept of sovereignty as it prevails in today's world requires both (i) as a historical matter, an appreciation for how a cluster of key factors have influenced the development of international relations, especially in the Western world, and (ii) as a legal matter, an appreciation for the central position that sovereignty is thought to occupy in the world's legal framework. In the following paragraphs, I will offer my own summary of sovereignty from both a historical and a legal perspective.

In an earlier publication,[2] I have suggested that the history of what we today call international law can usefully be thought of as falling into four main chapters. Only in the fourth chapter can the term "international law" be properly used. Before that time, for reasons explained below, a more appropriate term would be "the set of legal rules governing relations between peoples of different legal and political cultures" – or, stated more simply but less accurately, "the law of nations".[3]

The first chapter in the history of the law of nations, like so much else in Western civilization, features the Greeks. In conquering and possessing other territories, establishing colonies, entering into alliances, engaging in trade and concluding peace treaties, the Greeks exemplified the functioning of "viable rules affecting the interactions of self-conscious political societies", to use the term one expert suggests.[4] For the Greeks, the most important political societies were city-states.[5]

The second chapter in the history of the law of nations features the Romans. While Rome was a republic, it entered into treaty relations with other powers, including Carthage. From those treaties a few rules and doctrines of the law of nations emerged, including the doctrine of a "just war."[6] It was when Rome became an empire, however, that it made its main

---

2   See generally John W. Head, *Supranational Law: How the Move Toward Multilateral Solutions Is Changing the Character of "International" Law*, 16 KANSAS LAW REVIEW 605 (1993).

3   Legal scholars and politicians generally used the term "the law of nations" until the late 1780s, when Jeremy Bentham coined the term "international law". See Charles S. Edwards, HUGO GROTIUS: THE MIRACLE OF HOLLAND 147 (1981). For a discussion of these two terms, see Mark W. Janis, INTERNATIONAL LAW 248–254 (5th ed. 2008). The loss of accuracy that comes from using "the law of nations" instead of my longer formulation ("the set of rules governing relations between peoples of different legal and political cultures") derives from the fact that the primacy of the nation-state in recent centuries prompts most people, especially in the USA, to equate "nation" with "state". As indicated below, however, in earlier ages the law of nations was by no means a "law of states".

4   Edwards, *supra* note 3, at 71.

5   Arthur Nussbaum, A CONCISE HISTORY OF THE LAW OF NATIONS 11–13 (1947).

6   *Id.* at 16–18. See also Malcolm N. Shaw, INTERNATIONAL LAW 13–14 (2nd ed. 1986) (discussing the early origins of the law of nations).

contribution to the development of the law of nations. Rome was, after all, an empire of many nations, if the word "nations" is understood to denote peoples or "self-conscious political societies" or, in Latin, *gens*.[7] Questions arose in the Roman Empire as to whether, and how, the laws of the city of Rome proper (*jus civile*) should apply to those nations.[8] The answer was to develop a new body of law, derived partly from those aspects of laws and customs common to Romans and non-Romans, and partly from a sense of equity and justice.[9] This new body of law became known by the term *jus gentium*, translated as the "law of nations".[10]

The third chapter in the history of the law of nations features the (Christian) Church. The Christian religion, at first considered a threat to Roman civilization, was by the fourth century CE the religion of the most influential Roman classes.[11] With the fall of the Western Roman Empire the power of the Church and its leaders increased, in part because the Church "represented a tie to the civilized, unitary past and survived the empire as the one institution seemingly capable of promoting universal values and of imposing some measure of order throughout the politically shattered West."[12] The vehicle for promoting those values and imposing that order was canon law, also called ecclesiastical law, as promulgated and administered by churchmen.[13] This canon law, "though oftentimes confused and contradictory, assumed the appearance of a law common to all those within the fold."[14] Its influence

---

7 Edwards, *supra* note 3, at 72.

8 *Id.*

9 See generally Henry S. Maine, ANCIENT LAW 49–52, 58–61 (1861) (discussing the development and features of the *jus gentium*); Nussbaum, *supra* note 5, at 19 (same). See also Edwards, *supra* note 3, at 73 (referring to the views of Sir Henry Maine).

10 Edwards, *supra* note 3, at 73. Scholars disagree over the precise logic and chronology by which the Roman judicial officials came to develop the notion and term *jus gentium*. It seems clear, however, that the term had come into general use by the 1st century BCE, when Cicero described the *jus gentium* as a common law of mankind. See *id.* at 74. For the view that the sources of *jus gentium* are "a matter of conjecture," see J. A. C. Thomas, TEXTBOOK OF ROMAN LAW 63 (1976). See also Shaw, *supra* note 6, at 15–16 (discussing confusion among Roman lawyers over the relationship between the law of nature and *jus gentium*).

11 Edwards, *supra* note 3, at 75.

12 *Id.* at 75–76.

13 *See id.* at 76. A somewhat different picture of Europe during the Dark Ages that followed the fall of the Roman Empire is drawn by René David, a leading comparative law scholar. David claims that "the reign of law had ceased" and that "society had returned to a more primitive state." René David and John E.C. Brierley, MAJOR LEGAL SYSTEMS IN THE WORLD TODAY 38 (3rd ed. 1985). The church, according to David, encouraged this view, "extoll[ing] love of one's fellow man instead of justice . . . . [L]aw itself was considered a bad thing." *Id.*

14 Edwards, *supra* note 3, at 75. For an account of the development of canon law, as well as several other aspects of the historical narrative offered here, see John W. Head, GREAT LEGAL TRADITIONS: CIVIL LAW, COMMON LAW, AND CHINESE LAW IN HISTORICAL AND OPERATIONAL PERSPECTIVE 60–62, 77–80 (2011). See also Walter Ullman, LAW AND POLITICS IN THE MIDDLE AGES 119–159 (1975).

extended increasingly beyond spiritual and ecclesiastical matters into areas that had previously been under the jurisdiction of secular authority.[15]

For approximately seven centuries after the fall of the Western Roman Empire, the political landscape of Europe presented no effective obstacles to the gradual expansion of the Church's authority. Feudalism, the prevailing form of political and social organization between the eighth and the twelfth centuries, fostered decentralization of political power into relatively small geographic areas.[16] Largely unchallenged, the churchmen clung to the notion of a universal spiritual community – Christendom – with the pope as the head.[17]

Then the political landscape of Europe changed radically. The modern state system, composed of national monarchies, swept across Europe from the thirteenth century onward.[18] Having gained power partly as a natural extension of feudalism[19] and partly as an outgrowth of the Reformation,[20] the nation-state had firmly established itself by the time Hugo Grotius came of age in the late sixteenth century.[21]

---

15 Edwards, *supra* note 3, at 76. See also Nussbaum, *supra* note 5, at 24 (explaining that "ecclesiastical law was the dominant type of universal law in the Middle Age."). By the close of the period, however, canon law was by no means the only source of "universal law" in Europe. The rediscovery of the *Digest* from Justinian's *Corpus Juris Civilis* around the end of the 11th century facilitated a revival of interest in Roman law, and a body of commercial law had been developing since the time of the Crusades. See John W. Head, *Justinian's Corpus Juris Civilis in Comparative Perspective: Illuminating Key Differences Between the Civil, Common, and Chinese Legal Traditions*, 21 MEDITERRANEAN STUDIES 91–121 (2013). See also John H. Merryman, THE CIVIL LAW TRADITION 8–9, 12–13 (2nd ed. 1985). Roman law and canon law combined to create what has been called the *jus commune* (common law) of Europe that lasted until the rise of the nation-state. *Id.* at 10–11.

16 See Edwards, *supra* note 3, at 76; Nussbaum, *supra* note 5, at 28. See also J. L. Brierly, THE LAW OF NATIONS 2–3 (1963) (describing the feudal system).

17 Edwards, *supra* note 3, at 77. Edwards points out that in 1075 Pope Gregory VII "maintained that the Roman church was divine in origin, that the Roman pontiff alone could be called universal ruler, that the pontiff could depose kings and emperors and be judged by no human powers, and that the Roman church had never erred and could not err." *Id.* at 77–78. See also Nussbaum, *supra* note 5, at 23 (noting that "the pope became in the latter part of the Middle Ages the foremost representative of unitary rule in Western civilization").

18 Edwards, *supra* note 3, at 81.

19 A central feature of feudalism, loyalty to a lord in return for protection, gradually translated into loyalty to a national monarch. Brierly, *supra* note 16, at 4.

20 See Brierly, *supra* note 16, at 1, 3; Edwards, *supra* note 3, at 82.

21 According to Brierly, "[t]he Peace of Westphalia, which brought to an end in 1648 the great Thirty Years War of religion, marked the acceptance of the new political order in Europe." Brierly, *supra* note 16, at 5. For details about the Peace of Westphalia and the change in relations among European states that followed it, see Nussbaum, *supra* note 5, at 86–96. See also Janis, *supra* note 3, at 168 ("The Peace of Westphalia legitimated the right of sovereigns to govern their peoples free of outside interference . . . . The Peace was a great property settlement for Europe, a quieting of title across the continent."). See generally Leo Gross, *The Peace of Westphalia, 1648–1948*, 42 AMERICAN JOURNAL OF INTERNATIONAL LAW 20 (1948) (discussing the character, background and implications of the Peace of Westphalia).

Thus began the fourth chapter in the history of the law of nations. One historian emphasizes how dramatically the rise of the nation–state changed the respective roles of the Church and the monarchs:

> The Christian faith cut across territorial boundaries, and the peoples within the realms were oftentimes caught in a dilemma of loyalty to new sovereign authority or loyalty to the church. In order to strengthen their positions and to promote the loyalty of their peoples, the monarchs intensified their resistance to any possible encroachments of the papacy . . . [T]he vast religious upheaval which swept all of Europe [in the sixteenth century] helped to promote the phenomenon of nationalism. By 1560 most of the basic tenets of Protestantism had been asserted, and the changing circumstances dispelled once and for all any lingering notion that Christendom was a monolithic unity directed from Rome.[22]

It was in this chaotic setting that Hugo Grotius, by all marks a genius,[23] set about writing his *De Jure Belli ac Pacis*[24] (*The Law of War and Peace*) as a manual of legal rules by which the monarchs of Europe should be guided in their relations with each other. He was not the only one to do so, nor was he the first. Machiavelli had written *The Prince* in 1513 as "a handbook of advice for a ruler on how to achieve, exercise, and retain political power."[25] A central theme of Machiavelli's book was that there was no law or set of obligations among rulers of states, since the ruler of a state could (and indeed must) be a law unto himself in order to create a stable state for the benefit of himself and his people.[26]

Contrasting with Machiavelli's view was that of Jean Bodin, whose *Six Livres de la Republique* (*The Six Books of the Commonwealth*), first published in 1576, defined sovereignty in a way that made a king, even though supreme within his kingdom and therefore not bound by laws that he makes, bound nonetheless by several external sources of law.[27] These included divine law

---

22 Edwards, *supra* note 3, at 81–82.
23 Grotius, born Huig de Groot in Delft, Holland in 1583, was declared "the miracle of Holland" at the age of fifteen. *Id.* at 1, 183 n.1. He died in 1645. Hersch Lauterpacht, writing on the three-hundredth anniversary of Grotius's death, asserted that Grotius "was one of the greatest international figures of his age – a prodigy, almost a miracle, of learning". Hersch Lauterpact, *The Grotian Tradition in International Law*, 23 BRITISH YEARBOOK OF INTERNATIONAL LAW 1, 2 (1946). See also Nussbaum, *supra* note 5, at 96–97 (describing Grotius as a "child prodigy").
24 Hugo Grotius, DE JURE BELLI AC PACIS LIBRI TRIS [THE LAW OF WAR AND PEACE IN THREE BOOKS] (Francis W. Kelsey trans., 1925) (1625).
25 Edwards, *supra* note 3, at 82. See generally Niccolo Machiavelli, IL PRINCIPE [THE PRINCE] (Ninlan H. Thomson trans., 2nd ed. 1897) (1513).
26 Edwards, *supra* note 3, at 82. See also Brierly, *supra* note 16, at 6. Brierly notes that "Machiavelli's *Prince* . . . [gave] the world a relentless analysis of the art of government based on the conception of the state as an entity entirely self-sufficing and non-moral." *Id.*
27 See Brierly, *supra* note 16, at 9; Edwards, *supra* note 3, at 83–84; Shaw, *supra* note 6, at 20. For translated excerpts from Bodin's *Six Livres de la Republique*, see Jean Bodin, ON SOVEREIGNTY:

(since sovereign power came from God), natural law (the universal precepts of nature discernible to all national beings) and the customs and traditions of the people being ruled.[28]

Of these two approaches, Machiavelli's "every state for itself" approach soon became dominant, leading to what many theorists at the time considered an appalling disregard for decency in the relations between monarchs. "Though history, like a giant hammer, had fallen upon the West, breaking it into numerous pieces, these theorists [who were opposed to Machiavelli's view] came to believe that the fragments could be held together in a meaningful whole with universal law serving as a bonding medium."[29] Many of these theorists insisted that rules governing the behavior of nations derived from two sources: (i) a law of nature and (ii) custom evidencing common consent.

Perhaps the most important of these theorists was not Hugo Grotius but Francisco Suarez, a Spanish Jesuit theologian who conceptualized a system consisting of four types of law – eternal, divine, natural and human.[30] As explained in his treatise *De Legibus ac Deo Legislatore* (*On Laws and God the Lawgiver*), published in 1612, eternal law was at the root of all law; it resided from the beginning of creation in the mind of God. It was not promulgated, however, and thus not accessible to human knowledge. Humans could, though, know divine law, which was directly promulgated by God. They also could, according to Suarez, know natural law,[31] which was based on "the natural qualities of mankind."[32] The final category, human positive law, was entirely the product of the actual customs of nations and was intended to supplement natural law.[33] Of the four types of law, Suarez said, both natural law and human positive law could be considered part of the law of nations.[34]

While Suarez and others provided theoretical foundations, Grotius built a tangible, visible structure of the rules governing relations between states. His *De Jure Belli ac Pacis* relied heavily on the work of those who preceded him – so much so, in fact, that some scholars have accused him of lifting

---

FOUR CHAPTERS FROM *THE SIX BOOKS OF THE COMMONWEALTH* (Julian H. Franklin ed. and trans., 1992).

28 See Edwards, *supra* note 3, at 84.

29 *Id.* at 85.

30 See *id.* at 86–88. Other authors give greater prominence to another Spanish theologian, Francisco de Vitoria, and to an Italian Protestant, Alberico Gentilli. See Brierly, *supra* note 16, at 26–27; Nussbaum, *supra* note 5, at 72, 75–85; Shaw, *supra* note 6, at 21–22.

31 See Edwards, *supra* note 3, at 86–88.

32 *Id.* at 106 (quoting 2 Francisco Suarez, SELECTIONS FROM THREE WORKS 271 (Gwladys L. Williams *et al.* trans., 1944) (translating bk. II, ch. XIV, sec. 7 of *De Legibus ac Deo Legislatore*, one of three works featured in the 1944 translation)).

33 Nussbaum, *supra* note 5, at 66–67. See also Edwards, *supra* note 3, at 88.

34 Edwards, *supra* note 3, at 90.

ideas directly from Suarez and others without proper attribution.[35] The difference was largely one of presentation: Grotius's treatise offered not only a comprehensive theoretical exposition, but also a practical substantive manual.[36]

The gist of Grotius's view of the law of nations is that the relations between sovereigns on behalf of their states are governed by binding rules derived from two sources: *jus gentium voluntarium* and natural law.[37] His *jus gentium voluntarium* (often translated as "the volitional law of nations") was based on consent of states as manifested in custom and usage. His law of nature was the "dictate of right reason,"[38] a set of "broad and unchanging principles of justice" derived from Christian theology.[39]

Grotius thus combined the new with the old. He acknowledged the primacy of the nation-state as the fundamental unit of political organization in Europe; nevertheless, he insisted that the state was subject in its external relations to the principles of natural law that had governed behavior under the previous regimes.

Only a short time after Grotius's great treatise appeared in 1625, Thomas Hobbes wrote his master work *Leviathan*. For Hobbes, the development of territorial states with centralized power brought the prospect of order to a politically chaotic Europe, and this development therefore deserved legal legitimization. Mark Janis offers an explanation of the seventeenth-century setting in which Hobbes (and Grotius) lived and of the grand solution that Hobbes devised. Janis focuses first on the issue of *competing allegiances*:

> The medieval era, which the bloody Thirty Years War [1618–1648] shattered, had been characterized by criss-crossing political, legal, religious, and moral allegiances. The ties of feudalism, of King and baron, of the Holy Roman Empire, of the Catholic Church, indeed of all the settled order of medieval Europe, bound men and women this way and that . . . . The conflicting allegiances of Europe had contributed to the terrible toll of confusion, death, and destruction from 1618 to 1648.

---

35 See, e.g., *id.* at 148–153. Edwards offers several explanations for Grotius's failure to cite Suarez. These include the fact that Suarez's treatise had been burned publicly by the kings of both England and France for its endorsement of a right of tyrannicide; for Grotius to have cited Suarez heavily could have invited trouble not only for Grotius's treatise but for Grotius himself, who lived in France at the time. *Id.* at 152–153.

36 See *id.* at 95–96. See also Brierly, *supra* note 16, at 29 ("Grotius's purpose was practical").

37 See Brierly, *supra* note 16, at 30; Nussbaum, *supra* note 5, at 104.

38 Grotius, *supra* note 24, at 38–39 (translating bk. I, ch. X, sec. 1).

39 Edwards, *supra* note 3, at 60–67, 105. For a comprehensive treatment of Grotius's view of the law of nature, see *id.* at 27–69. Edwards' thesis is that "even though Grotius . . . freed natural law theory from its traditional medieval tie [with divine revelation], he was not a secularist [one who takes the view that theological considerations should be excluded from the subject of natural law]." *Id.* at 47. In other words, natural law for Grotius was not grounded in the rationality of man alone, but instead had some theological foundation. See *id.*

In the mid-seventeenth century, many Europeans sought a simpler and, it was hoped, safer set of loyalties.[40]

Janis also emphasizes that the seventeenth century saw a sharp rise in nationalism, and that the outbreak of the Thirty Years War resulted in part from the claim "that a people of a certain language, society, and tradition had a right to choose their own religion and to govern themselves, free from the competing universalistic claims of Emperor and Pope".[41] By the close of the war in 1648, this claim – this demand for national autonomy and a single set of loyalties – prevailed. Janis then explains the significance of Hobbes and his theory of sovereignty:

> What 1648 most significantly inaugurated (and what Thomas Hobbes most significantly conceptualized) was the organizing principle of the state, particularly the sovereign state. Sovereignty as a concept formed the cornerstone of the edifice of international relations that 1648 raised up. Sovereignty was the crucial element in the peace treaties of Westphalia, the international agreements that were intended to end a great war [the Thirty Years War] and to promote a coming peace. The treaties of Westphalia enthroned and sanctified sovereigns, gave them powers domestically and independence externally. But what exactly did "sovereignty" mean? How did being a "sovereign" work? Because Hobbes in 1651 provided answers to these questions . . . , *Leviathan* took on the lasting importance it did.
>
> Hobbes crafted and fit a crucial puzzle piece into an emerging picture of the new Europe. Hobbes' lasting contribution was the envisioning, in his own words, of "that great Leviathan, or rather (to speak more reverently) of that *Mortall God*, to which we owe under the *Immortall God*, our peace and defence." Rather than believing in any number of loyalties, Hobbes believed that all men required "a Common Power, to keep them in awe, and to direct their actions to the Common Benefit." This Common Power, the Leviathan, required a single authoritarian state [by which persons within a territory would] "conferre all their power and strength upon one Man, or upon one Assembly of men, that may reduce all their Wills, by plurality of voices, unto one Will" . . . [so that one person or a single assembly in which that power is concentrated] "is called Soveraigne, and said to have *Soveraigne Power*; and everyone besides, his Subject".
>
> . . . So successful was the political settlement of Westphalia and so useful was Hobbes' concept of Leviathan and the sovereign state that they became deeply imbedded in the public consciousness. *It is difficult*

---

40 Janis, *supra* note 3, at 167.
41 *Id.*

now even to conceive that a world of sovereign states is an intellectual abstraction, a humanly devised creation, albeit one of tremendous force and utility for more than three centuries.[42]

The foundations for the modern concept of sovereignty, then, can be traced to a confluence of factors dating from the sixteenth and seventeenth centuries, and particularly to the influences of Bodin, Grotius, and Hobbes. As of the late seventeenth century, the developments might be summarized by highlighting these key features:

- Bodin was especially influential in his design of sovereignty as a concept, envisioning a form of control that was highly centralized but that nevertheless made a king bound by external law, including natural law.
- Grotius, like Bodin, gave special legal significance to natural law, drawn from Christian teaching, but Grotius also acknowledged the significance of the newly-emerged nation-states in the system of law. What might be called "the Grotian Solution"[43] was in essence a formula defining international law – or what was at the time referred to as "the law of nations". That formula was this: "the law of nations = *jus gentium voluntarium* + natural law". In this way the rising power of the state was "married" to the principles of natural law drawn from Christianity.
- Hobbes injected much more absolutism into the concept of sovereignty, thereby putting in place a central legal pillar for the international community – that of state sovereignty. In doing so, he demoted natural law from its place of central prominence and promoted the single human sovereign, acting under a presumed divine right of kings, to a position of sole "holder" of sovereignty.
- All of these developments came against the backdrop of a specific set of challenges arising in a peculiar historical and political environment that involved religious conflict and growing nationalism in a post-feudal political system.

Figure 8.1 offers a pair of diagrams intended to reflect in a simplified way both Grotius's view of sovereignty and Hobbes' view of sovereignty. As represented in the first diagram in Figure 8.1, the system of sovereignty envisioned by Grotius has sovereignty being "held" by God in respect of a few topics (of human relationships, that is) that are of fundamental importance. These appear within the smaller dark oval. For all other topics, though, the sovereignty would in Grotius's view be "held" by the "crowned heads of

---

42 *Id.* at 168–169 (with quoted passages from Hobbes' *Leviathan*) (emphasis added).
43 Various writers have coined the terms "Grotian Tradition", "Grotian Quest", "Grotian Moment", and so forth. See, e.g., Lauterpacht, *supra* note 23; Richard Falk, *The Grotian Quest*, in Richard Falk *et al.*, INTERNATIONAL LAW: A CONTEMPORARY PERSPECTIVE 36 (1985).

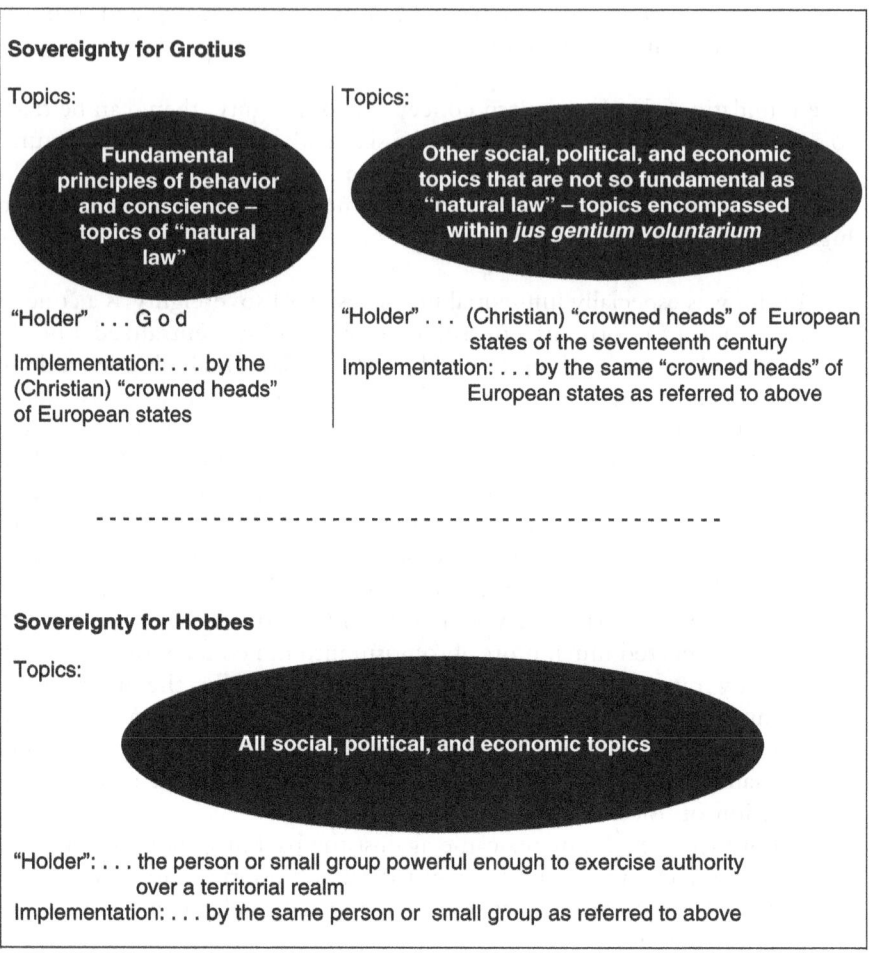

Topics:

Fundamental principles of behavior and conscience – topics of "natural law"

Topics:

Other social, political, and economic topics that are not so fundamental as "natural law" – topics encompassed within *jus gentium voluntarium*

"Holder" . . . G o d

Implementation: . . . by the (Christian) "crowned heads" of European states

"Holder" . . . (Christian) "crowned heads" of European states of the seventeenth century

Implementation: . . . by the same "crowned heads" of European states as referred to above

Sovereignty for Hobbes

Topics:

All social, political, and economic topics

"Holder": . . . the person or small group powerful enough to exercise authority over a territorial realm

Implementation: . . . by the same person or small group as referred to above

Figure 8.1 Simplified representations of sovereignty for Grotius and for Hobbes

Europe" – that is, kings and queens and other royalty – as shown in the larger dark oval. Sovereignty in Hobbes' view, by contrast, is simpler: for all topics, sovereignty is concentrated in a single person or small group.

In short, both diagrams in Figure 8.1 reflect answers to these questions:

- What are the *topics* (regarding law, governance, etc.) that are the subjects of the sovereignty at issue (the topics appear inside the dark oval shapes)?
- What types of entities (that is, persons, groups, etc.) are the *"holders"* of sovereignty in the sense that they have standing, authority, stake, voice,

or vote to make decisions on those topics – or at least have the right to have their interests represented and protected?

- What entities are most closely involved in or responsible for *implementation* of such decisions – that is, for ensuring that the rules are really followed?

In Figure 8.2, appearing in subsection IC, I will offer another pair of diagrams that will illustrate how the answers to those questions changed over the centuries following Hobbes and Grotius.

IB. Sovereignty elaborated and celebrated: the eighteenth century to today

The "Grotian Solution" – in which Christian natural law had a formal role to play in the law of nations, thus essentially making God the "holder" of sovereignty over fundamental principles of behavior and conscience – did not last long, but the Hobbesian view of sovereignty has endured and strengthened. Grotius's attempt to limit the growing strength of the nation-state by "marrying" it to the Church-based rules of proper behavior met with failure within decades of his death. A split soon reappeared between the "naturalist" school and the "positivist" school of theorists. Those in the naturalist school downplayed or denied the relevance of custom or treaty rules to a discussion of international law, focusing instead on theoretical constructions. Those in the positivist school, by contrast, largely or completely dismissed natural law as an element of international law, focusing instead on the actual practice of states.[44]

The positivist approach prevailed, partly because it reflected the growing acceptance of empiricism derived from the Renaissance[45] and partly because it offered such wide latitude for the monarchs and other governmental authorities in the various states (in Europe, that is) to do whatever their military and diplomatic power would make possible. In short, starting by the latter part of the eighteenth century and continuing throughout the nineteenth century, the dominant view of international law was that the only rules governing relations between states, as a practical matter, were voluntary in character, as manifested in custom or treaties.[46] Sovereignty as a concept

---

44 Shaw, *supra* note 6, at 23–24. See also Brierly, *supra* note 16, at 33. For discussions of the "naturalist" and "positivist" writers following Grotius in the seventeenth century, see Brierly, *supra*, at 35–37; Nussbaum, *supra* note 5, at 112–125; Gerhard von Glahn, LAW AMONG NATIONS 31 (1986).

45 See Shaw, *supra* note 6, at 24. "Empiricism as formulated by Locke and Hume denied the existence of innate principles and postulated that ideas were derived from experience." *Id.*

46 *Id.* at 24–27. Shaw writes of the nineteenth century as follows:

Positivist theories dominate this century. The proliferation of the powers of states and the increasing sophistication of municipal legislation gave force to the idea that laws were basically commands issuing from a sovereign person or body. Any question of ethics or morality was

was as robust as Hobbes could have envisioned – perhaps even more so – and generally unfettered by the types of natural-law constraints that Bodin and Grotius envisioned.

For instance, one of the most widely read international law treatises of the period was Vattel's *Le Droit des Gens* (*The Law of Nations*),[47] published just over half-way through the eighteenth century. Although Vattel recognized some theoretical importance to natural law, he claimed that only the "voluntary law of nations" – that set of rules to which states had agreed – was enforceable.[48] By emphasizing the independence of states, Vattel was responsible for "cutting the frail moorings which bound international law to any sound principle of obligation."[49] Hence, "[b]y the end of the nineteenth century, most authorities on international law conceded only the will of nations to be the source of the law, a view typical of a period in which the absolute sovereignty of states was affirmed with conviction by virtually every statesman and publicist."[50]

One other development of the eighteenth and nineteenth centuries warrants attention in this survey of the development of sovereignty in theory and practice: the rise in a few countries (most dramatically in France and the USA) of "popular sovereignty". With the political revolutions that broke out in the late eighteenth century and continued into the nineteenth, one aspect of the Hobbesian concept of sovereignty was modified: the "holders" of sovereignty increasingly came to be seen, at least as a theoretical matter, not as the *monarchs* of European or European-inspired territorial states but rather as the *populations* of those territorial states. The idea took hold that "we the people" – acting as a body, reflecting a common will, presumably through some effective way to reflect the population's shared interests – constituted the entity in which true sovereignty rested. Granted, slavery

---

irrelevant to a discussion of the validity of man-made laws. The approach was transferred onto the international scene and immediately came face to face with the reality of a look of supreme authority . . . . Since law was ultimately dependent upon the will of the sovereign in national systems, it seemed to follow that international law depended upon the will of the sovereign states.

Id. at 27. For a comprehensive discussion of the development of international law in the eighteenth and nineteenth centuries, see Nussbaum, *supra* note 5, at 126–237.

47 See Brierly, *supra* note 16, at 37. See generally E. de Vattel, Le Droit des Gens [The Law of Nations] (Charles G. Fenwick trans., 1916) (1758).

48 See Brierly, *supra* note 16, at 38 (citing Vattel, *supra* note 47, at 304–306 (bk. III, ch. XII)).

49 Brierly, *supra* note 16, at 40. According to Brierly, Vattel's "exaggerated emphasis on the independence of states had the effect . . . of reducing the natural law, which Grotius had used as a juridical barrier against arbitrary action by states towards one another, to little more than an aspiration after better relations between states . . . ." *Id.* at 38.

50 von Glahn, *supra* note 44, at 35. Contributing to this wave of positivism, and to the corresponding momentum of absolutism in sovereignty, was the widely-known view of John Austin, expressed in his 1832 book *The Province of Jurisprudence Determined*, that international law could not in fact properly be called "law" at all. See Janis, *supra* note 3, at 2–4.

was still legally accepted in some states, and voting rights for some ethnicities and for women were denied or seriously restricted in many countries. Nonetheless, the "holder" of sovereignty was substantially changed from earlier views, especially those of Hobbes, who celebrated the concentration of power into a single individual.[51] As a practical matter, the actual *implementation* of sovereignty continued to rest with the official political authorities of the state – the government – so the theoretical change had less fundamental practical effect than might at first be expected. Moreover, the rise of positivism tended to support claims of exclusive power on the part of governing regimes, so the absolutism inherent in Hobbes' views of "the sovereign" continued to predominate. By the close of the nineteenth century, therefore, state sovereignty was nearly absolute in practical terms, especially in the most powerful states, with a widespread nod of acknowledgement toward "popular sovereignty" as being the theoretical ground on which governments acted.

Thus the status of international law as of the early part of the twentieth century may be stated as follows: With the disappearance of natural law as an element of international law, the only remaining source of international law was, to use Grotius's phrase, *jus gentium voluntarium*. Although the rise of "popular sovereignty" made at least a formal or rhetorical modification to Hobbes' views, the fact remains that the state had become "absolutized"[52] to a significant degree – largely bound only by those rules that emerged from its own volition, as shown by its own actual practice. This view was reflected in one of the most widely quoted definitions of international law, the one announced in 1927 by the Permanent Court of International Justice ("PCIJ")[53] in the *Lotus* case:

> International law governs relations between independent States. The rules of law binding upon States *therefore* [that is, *because* the states are all independent] emanate from their own free will as expressed in conventions or by usages generally accepted as expressing principles of law . . . .[54]

The same view of international law – emphasizing the element of *jus gentium voluntarium* and excluding any reference to natural law – is reflected in Article 38 of the Statute of the International Court of Justice ("ICJ"). Article 38(1) is widely viewed as providing a comprehensive list of sources of international law. It reads as follows:

---

51 For a synopsis of how views of sovereignty, incorporating especially the notions of popular sovereignty, have evolved, see Eyal Benvenisti, *Sovereigns as Trustees of Humanity: On the Accountability of States to Foreign Stakeholders* 107 AMERICAN JOURNAL OF INTERNATIONAL LAW 295, 296 (2013).

52 The term "absolutize" in this context is borrowed from Burns H. Weston *et al.*, INTERNATIONAL LAW AND WORLD ORDER 1086 (2nd ed. 1990).

53 The PCIJ was established at about the same time as the League of Nations, following the First World War. For a comprehensive history of the PCIJ, see Antonio S. de Bustamante, THE WORLD COURT (Elizabeth F. Read trans., 1925).

54 S. S. "Lotus" (Fr. v. Turk.), 1927 P.C.I.J. (sec. A) No. 10 (Sept. 7), at 18 (emphasis added).

The Court, whose function is to decide in accordance with international law such disputes as are submitted to it, shall apply:

a   international conventions [that is, treaties], whether general or particular, establishing rules expressly recognized by the contesting states;

b   international custom, as evidence of a general practice accepted as law [usually labeled "customary international law"];

c   the general principles of law recognized by civilized nations;

d   . . . judicial decisions and the teachings of the most highly qualified publicists of the various nations, as subsidiary means for the determination of rules of law. [55]

None of the items listed in Article 38 reflects Grotius's view that *natural law* is a source of international law and therefore some sort of restriction on state sovereignty. Even the reference to "general principles of law" in subsection c is qualified by the words "recognized by civilized nations."[56] These general principles of law therefore are not immutable, but instead are subject to the changing and competing views of human society.[57]

In short, traditional international law had evolved by the first part of the twentieth century into a body of rules that (i) acknowledged and supported the primacy of the state endowed with a very robust form of sovereignty and (ii) consisted only of those rules that had been accepted by states, either by treaty or through practice. This framework offered little or no room for entities other than states or for rules not "emanate[ing] from their own free will," to repeat the words of the Permanent Court of Justice in the *Lotus* case mentioned on page 365. State sovereignty – construed as guaranteeing the complete legal independence of states – had become the central pillar of international law.

To a large degree, this view of sovereignty persists today. Although some voices have been raised in protest against this view,[58] and although I will

---

55 STATUTE OF THE INTERNATIONAL COURT OF JUSTICE, June 26, 1945, art. 38, 59 Stat. 1031, 1055, 1976 U.N.Y.B. 1052 [hereinafter ICJ STATUTE], art. 38, para. 1. The ICJ was established in 1945 as the judicial organ of the United Nations. Except for some introductory wording, the language of Article 38 of the ICJ Statute is identical to the wording of Article 38 of the statute of the PCIJ, which is reprinted in Bustamante, *supra* note 53, at 360.

56 For an explanation of the drafting history of the "general principles" language, see Ian Brownlie, PRINCIPLES OF PUBLIC INTERNATIONAL LAW 15–16 (4th ed. 1990). Brownlie points out that an earlier draft of the language, designed to reflect natural law concepts ("the rules of international law recognized by the legal conscience of civilized peoples") was rejected by the committee of jurists that prepared Article 38 of the PCIJ statute. *Id.*

57 For further discussion of general principles of law, see *id.* at 15–19; von Glahn, *supra* note 44, at 22–24.

58 *Id.* at 24–25. See, e.g., Brierly, *supra* note 16, at 47 (criticizing the Hobbesian view of sovereignty by arguing that "[t]o the extent that sovereignty has come to imply that there is something

explain later some of the ways in which a few important departures from this view have been made in the past 75 years or so, the fact remains that our world in the early portion of the twenty-first century largely embraces what I call a "monolithic sovereignty" concept.[59]

It is important to focus closely on two key features of this "monolithic sovereignty" concept, partly because I will soon suggest how a "pluralistic sovereignty" concept would depart from those two features. First, today's "monolithic sovereignty" is *territorial* in its conception. Within a single (usually contiguous) physical territory, the government of a state is thought to have nearly unimpeded authority – which explains in part why China is trying to "create territory" by building up more permanent sand-bars in the appropriately-named Mischief Islands.[60]

Second, the "monolithic sovereignty" concept is *national* in its assertion (or pretension) that state territorial boundaries widely reflect "nationalities", so that persons residing within State A are of one nationality and persons residing within State B are of another nationality. As explained earlier, the historical backdrop against which Hobbes drew up his vision of a Leviathan – a centralized state with a monopoly over legal and political power – was one of intense nationalism and of religious and ethnic conflict between different (perceived) nationalities.

Indeed, in both respects – territoriality and nationality – the "monolithic sovereignty" concept reflects the peculiar historical influences of fifteenth-century through nineteenth-century Europe. That is, this view of sovereignty grew out of a Europe struggling to replace (i) the chaos of cross-cutting political and religious loyalties that had produced a Thirty Years War with (ii) a more orderly political system providing largely unfettered authority to make and execute laws within territorial boundaries that at the time roughly reflected the distribution of various ethnic or "national" groups across the face of Europe. As Europe enjoyed increasing power – economic, political, military, and other forms of power – in the centuries following the Thirty

---

inherent in the nature of states that makes it impossible for them to be subjected to law, it is a false doctrine . . .").

59 For other contributions to the rich literature on sovereignty, see W. J. Stankiewicz, IN DEFENSE OF SOVEREIGNTY (1969); Hurst Hannum, AUTONOMY, SOVEREIGNTY, AND SELF-DETERMINATION 14–26 (1990) (discussing the history and limits of sovereignty and its relation to nationalism, especially for the modern nation-state); L. Ali Khan, THE EXTINCTION OF NATION-STATES: A WORLD WITHOUT BORDERS (1996) (discussing popular sovereignty, universal sovereignty and "Free State") [hereinafter EXTINCTION]; Stephen D. Krasner, ed., PROBLEMATIC SOVEREIGNTY: CONTESTED RULES AND POLITICAL POSSIBILITIES (2001); Jens Bartelson, A GENEALOGY OF SOVEREIGNTY (1995); Robert Lansing, NOTES ON SOVEREIGNTY (1921); and Harold J. Laski, STUDIES IN THE PROBLEM OF SOVEREIGNTY (1968).

60 See, e.g., *With the U.S. Distracted, China Builds 'The Mischief Islands'*, CHICAGO TRIBUNE, Mar. 15, 2015 (noting that China is literally "manufacturing a new [territorial dispute] in the South China Sea by transforming a series of lonely reefs into small islands").

Years War, this peculiar European-rooted view of state sovereignty continued to gain strength.

Additional momentum came from colonization and conquest. Beginning in the 1500s, several European states succeeded in conquering and colonizing much of the world's territory.[61] At one time or another, these European colonies covered most of North America, South America, the Indian sub-continent, Australia and the islands of the Pacific, as well as substantial portions of Asia and the Middle East. Later, when they gained political independence, the former colonies and conquered territories became legally independent states.[62] Thus through the process of colonization and decolonization the European model of sovereign independence of states, together with the traditional view of international law that this model had produced, was imposed on the rest of the world.[63] In short, the Euro-centric "monolithic sovereignty" concept, with its emphasis on territoriality and nationality, came through this blend of factors to circle the globe as we know it today.

IC. *Recent exceptions to "monolithic sovereignty"*

There are, however, some important exceptions. In several ways, sovereignty has evolved in the past 75 years or so to diverge from the Hobbesian view. Two such exceptions warrant special attention: (i) the general recognition of human rights as lying at least partly outside the scope of authority of individual states and their governments; (ii) the transfer of certain elements of sovereignty from individual nation-states to public international organizations. I will deal with each quickly in turn.

Through a series of developments that can be traced at least as far back as the 1920s with the development of the International Labour Organization – designed in large part to protect laborers from especially dangerous working conditions – international human rights have gained much prominence at the global level. The Universal Declaration of Human Rights, adopted by the UN General Assembly in 1948, purported to announce legal protections

---

61 For a survey of European colonization and conquest of the world, see GREAT LEGAL TRADITIONS, *supra* note 14, at 232–233. For these purposes, I include Britain as part of Europe, although the differences between Britain and Continental Europe (including of course their legal traditions) are quite substantial.

62 Among the first such states to emerge from European colonies was (or were) the United States of America, gaining political independence in the late eighteenth century. The decolonization of South America had yielded several new states by the mid-1800s. South and Southeast Asian colonies, such as India and Indonesia, gained independence just following the Second World War. The decolonization of Africa took place largely in the 1960s.

63 For other accounts of the "exportation" of the concept of the nation-state and the principle of sovereignty from Europe to the rest of the world, see ENCYCLOPEDIA OF PUBLIC INTERNATIONAL LAW 973–975 (Rudolf Bernhardt ed., 1992); Antonio Cassese, INTERNATIONAL LAW IN A DIVIDED WORLD 42, 47 (1986); Shaw, *supra* note 6, at 36–37.

that all states are required to extend to all persons (nationals or aliens alike) within their borders. Although some specific limitations on that requirement were written into the two general human rights treaties adopted by the same body in 1967 and later formally accepted by most countries in the world – the International Covenant on Civil and Political Rights and the International Covenant on Economic, Social, and Cultural Rights – the creation of these and other international legal instruments seems to reflect a broad consensus that individual human beings have rights at the international level.

Viewed from the perspective of sovereignty, the development of international human rights can be interpreted as representing the creation of a new category of "holders" of sovereignty: individuals. To date, most individuals lack access to any international legal forum in which to protest alleged violations of those rights. An exception exists in Europe, where the most effective regional legal framework for the protection of human rights has been built under the authority of the European Convention on Human Rights and Fundamental Freedoms. Still, it would be difficult to dispute the status of individual human beings as "holders" of sovereignty (on this single topic, that is) as I have defined it earlier. I defined "holder" for these purposes as an entity having standing, authority, stake, voice, or vote to make decisions on a specific topic – or at least that has the right to have his or her interests represented and protected.

Some observers would assert that human rights are part of natural law, in the sense that a violation of human rights (as recorded in such international legal instruments as I have mentioned above) constitutes a breach of natural law on the part of the government that commits or permits that violation. Other observers would consider any reference to natural law (or any other specific source) unnecessary; they would simply assert that human rights are governed by some universal and immutable rules that lie outside the authority or control of any government. Either way, it could be said that sovereignty for these purposes rests with – is "held" by – individual human beings.

That is the way in which I have portrayed the matter in Figure 8.2, in which I present two diagrams intended to show in a simplified manner how sovereignty has evolved somewhat since the nineteenth century. The upper diagram in Figure 8.2 represents the status of sovereignty as of the early twentieth century. The lower diagram there represents, in simplified form, the status of sovereignty today, highlighting the exceptions to the "monolithic sovereignty" concept as I have described it in this chapter. The lower diagram shows international human rights as a topic as to which sovereignty is "held" by individual persons – in the sense that it is the individual human being who has the only valid authority to determine, subject only to very limited public-safety restrictions, his or her own behavior in such protected areas as public expression, exercise of religion, political participation, and the like.

The lower diagram in Figure 8.2 also reflects, though, another exception to the "monolithic sovereignty" concept that prevailed through the early twentieth century. That exception relates to international organizations.

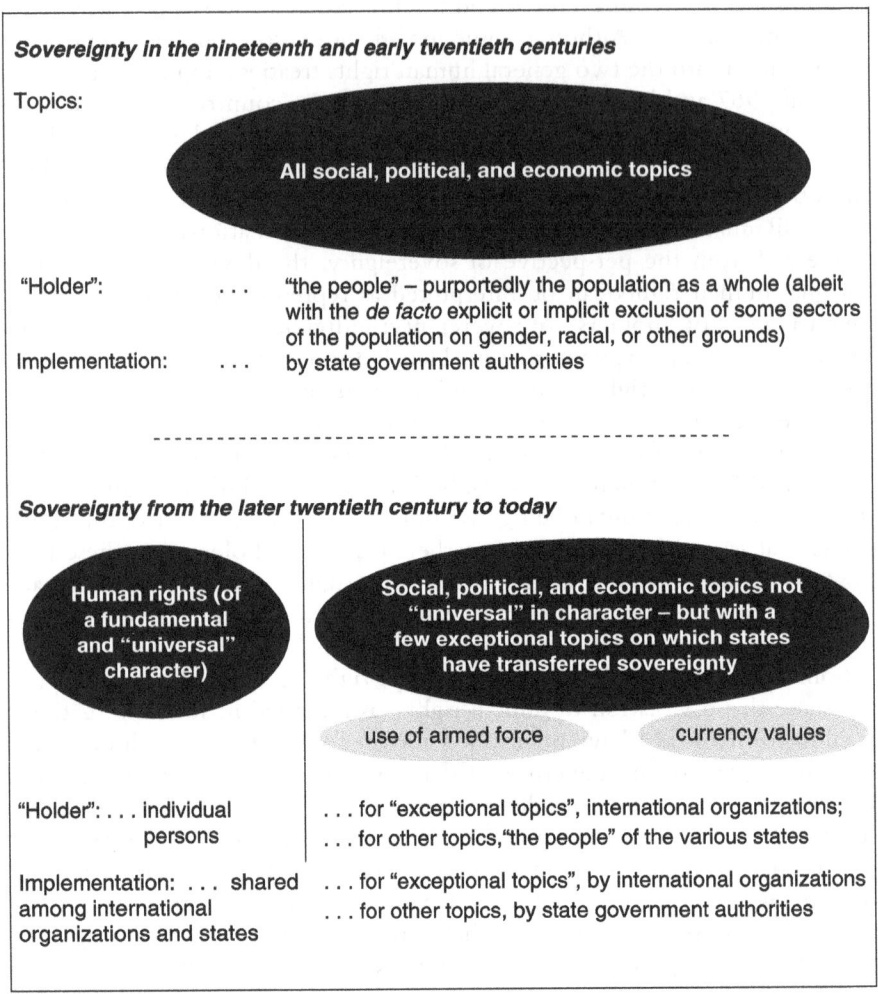

Figure 8.2 Simplified representations of sovereignty as evolved from the nineteenth
century to today

Somewhat over 20 years ago I wrote about the dramatic rise of international
organizations as new players on the international scene. After explaining that
the history of such organizations "is quite short, dating only from the middle
of the nineteenth century" with the creation of the International Telegraphic
Union and the Universal Postal Union, and mentioning the League of
Nations (founded, like the International Labour Organization, just following
the First World War), I concentrated on the 1940s:

> The turning point in the rise of international organizations . . .
> came with the end of World War II. Of the several new international

organizations formed then, the most widely known is the United Nations
. . . . It is difficult to overstate what a radical change the establishment
of the United Nations represented in 1945. Up to that point, the
traditional view of international law held that a state had virtually
unlimited legal authority to wage war in its own self-interest. Although
some attempts had been made to change the traditional view by limiting
the hasty exercise of that authority – the Covenant of the League of
Nations, for example, required a state to observe a three-month "cooling-
off period" before resorting to war – no effective means had been
established before 1945 for prohibiting the aggressive use of force by one
state against another.

The United Nations Charter sought to change all that. By including
Article 2(4) in the Charter, the states that created the United Nations
agreed to surrender the sovereign right to use force in their international
relations, except in self-defense. Instead of resting with states, the right
to use such force rests with the United Nations Security Council, which
Chapter VII of the Charter authorizes to take a range of collective
security measures . . .[64]

Having emphasized the importance of the UN and the "surrender of sovereignty" that its Charter recorded, I turned my attention to other international
organizations:

The United Nations is only one of several international organizations
formed at the close of World War II. Among the most influential of
those organizations today are the International Monetary Fund (IMF)
[and] the World Bank . . . . Like the United Nations, these organizations
are significant because they represent, either in their inception or in
their evolution, surrenders of sovereignty by states to international
organizations.

[For instance, in] the case of the IMF, the element of sovereignty that
member states surrendered was the prerogative to set currency exchange
rates as they wish.[65]

The foregoing illustrations, focusing on international organizations (such
as the UN, the IMF, and the World Bank) that were created in the wake of
the Second World War and in the anticipation of reshaping the political
world through the process of decolonization, represent what can be described
as an explosion of new entities that are fundamentally different from nation-
states, even though nation-states have created them. Such entities cover a

---

64 Head, *Supranational Law, supra* note 2, at 623–624.
65 *Id.* at 627. I also referenced the General Agreement on Tariffs and Trade, which has had its
   institutional status amended and expanded by the World Trade Organization.

range of issues concerning political, economic, environmental, military, research, diplomatic, and other forms of cooperation and authority – again, reflecting among other things a political reordering that emerged from decolonization. Most of those institutional arrangements, moreover, involve transfers of sovereignty by their member states to the institutions themselves. A particularly noteworthy illustration of this appears in Europe, where member states of the European Union have transferred significant portions of their sovereignty to a variety of EU agencies.[66]

As of today, therefore, the concept of sovereignty has evolved to a position that has a very few crucial topics "carved out" from the authority of states, in two separate ways. First, some types of behavior, such as the treatment of human beings by governments of the territories where they live, are governed not by rules enacted by state authorities but instead by rules and principles that are widely regarded as universal and immutable. Put more simply, sovereignty has been *removed* from states in respect of fundamental human rights. Second, a further "carving out" of certain topics from the authority of states has occurred when sovereignty over those topics has been *transferred* voluntarily by the states themselves in order to gain individual benefits through collective action. These topics include the use of armed force – as recorded, for instance, in the UN Charter – and selected economic behaviors as specified in the IMF Charter.[67]

Because sovereignty has either been removed or transferred in these various ways, the types of "holders" of sovereignty in today's version of that concept have increased significantly. Now it is not only the populations of states ("we the people") or the political leaders of those states who are the "holders" of sovereignty. Instead, (i) the "holders" of sovereignty in respect of those (few) rights that have been recognized thus far as immutable (in the areas of human rights and humanitarian law) are individual human beings; and (ii) to some extent international organizations (formed themselves by the treaty-making action of states) are "holders" of those elements of sovereignty that have been transferred to them by the states that held those elements before they were transferred.[68]

---

66 For a discussion of these transfers of sovereignty, see Richard E. Levy, *The Law and Economics of Supranationalism*, 40 European Journal of Law and Economics 1 (2015), http://link.springer.com/article/10.1007/s20657-015-9508-x.

67 The IMF Charter was amended in the 1970s to negate in part the "surrender of sovereignty" that its member states had originally engaged in over currency values, but in numerous important ways such "surrender of sovereignty" remains in place and still limits the authority of an IMF member state over the value of its currency. For details, see John W. Head, Losing the Global Development War: A Contemporary Critique of the IMF, the World Bank, and the WTO 114–115 (2008) [hereinafter Head-2008].

68 What I refer to here as "elements of sovereignty" are often portrayed as sticks in a bundle, with state authorities frequently making policy decisions about which "sticks" to retain and which to transfer elsewhere in order to best serve their interests.

Figure 8.2 is designed to reflect these developments. As noted on page 369, the upper diagram there represents, in simplified form, the status of sovereignty as of the early twentieth century. The lower diagram there represents, in simplified form, the status of sovereignty as of today, with the sorts of exceptions I have explained above. Textual material in both diagrams summarizes the three key features – that is, (i) the *topics* at issue (these topics appear inside the dark oval shapes), (ii) what types of entities (that is, persons, groups, etc.) are the "*holders*" of sovereignty in the sense that they have standing, authority, stake, voice, or vote to make decisions on those topics – or at least have the right to have their interests represented and protected – and (iii) what entities are most closely involved in or responsible for *implementation* of such decisions, to ensure that the rules are really followed. In the lower diagram, the introduction of international organizations as "holders" of some elements of sovereignty (as transferred to them by their member states) is represented by the two light-shaded ovals. For this purpose, only two such topics – use of force and currency values – have been used as illustrations.

II. "Pluralistic sovereignty": blended authority in global eco-states

I have offered in section I of this chapter a simplified summary of how the concept of sovereignty has evolved over about the last four centuries. One overarching constant is that the state has, throughout that period, remained the predominant "holder" of sovereignty. I have emphasized that certain narrow exceptions and restrictions have been introduced, especially in the past 75 years or so, so that certain other "holders" of sovereignty have reduced the power of the state, at least theoretically. Still, as a practical matter, the entities that are most closely involved in or responsible for *implementation* of the decisions emerging from an exercise of sovereignty are almost exclusively state government authorities; the only significant exceptions apply where international organizations have implementation authority. However, international organizations are themselves created by and ultimately responsible to states.

I have also emphasized that the concept of sovereignty that prevails in today's world, which I have referred to as the "monolithic sovereignty" concept, reflects in two respects – territoriality and nationality – the peculiar historical influences of fifteenth-century through nineteenth-century Europe. These elements also reflect the overarching importance of the state in the current concept of sovereignty.

In drawing this book to a close, I will offer the basic outlines of a different concept – that of "pluralistic sovereignty" – with the expectation that I will leave for later work the task of developing a more complete examination of this "pluralistic sovereignty" concept and its implementation by a new international institutional structure. I start with a "punch line" by making this assertion: a pluralistic sovereignty concept, designed to supplant the "monolithic sovereignty" concept that has prevailed for several centuries, might be based in part on "eco-zone clusters" or "eco-states" (both of which terms I will define in the following pages), instead of being based overwhelmingly on the type of states have that dominated the global political scene for several centuries. In the next few paragraphs I offer a survey of the "eco-state" notion, and then I will set forth several grounds favoring the adoption of such a pluralistic sovereignty concept.[69]

IIA. *A survey of ecozone clustering*

Many efforts have been made to "map" the world on the basis of ecoregions and biomes,[70] and these efforts date back nearly two centuries.[71] Recall that these two terms – ecoregions and biomes – are the ones used by the World Wildlife Fund ("WWF") in applying a simplified classification system to the world's enormous diversity of climates, soils, land cover, species distribution, and other environmental factors.[72] Under the WWF classification system,

---

69 Several elements of the following discussion draw from John W. Head, Kate Marples, and Jon Simpson, *Mediterranean Agriculture, Ecology, and Law: Creating a New Non-State Actor to Counteract Agroecological Collapse in the Mediterranean Basin*, 24 MEDITERRANEAN STUDIES (forthcoming 2017).

70 One early US-based illustration of such an effort – although based on watersheds, not ecoregions or biomes – can be found in the proposal made by John Wesley Powell that the boundaries of western US states should be based on watersheds of rivers in that region. See Rhett B. Larson, *Interstitial Federalism*, UCLA LAW REVIEW 908, 916–917 (2015), citing Wallace Stegner, BEYOND THE HUNDREDTH MERIDIAN: JOHN WESLEY POWELL AND THE SECOND OPENING OF THE WEST 322 (1954). For an example of a recently-prepared map of the "United Watershed States of America", see http://communitybuilders.net/wp-content/uploads/2013/11/Watershed_States_NoBorderlo.jpg.

71 For example, the Köppen climate classification system referred to earlier represents a late-nineteenth-century effort, itself reflecting the work of Alexander von Humboldt in the early part of that century. See text accompanying note 114 in Chapter 3, *supra*. See also note 62 and accompanying text in Chapter 4, *supra*.

72 For information about the WWF classification system, see text accompanying note 62 in Chapter 4. For details about other classification systems, and in particular their definitions of "biome", see, e.g., (i) US Fish and Wildlife Service, *Ecosystem Conservation Glossary*, www.fws.gov/midwest/EcosystemConservation/glossary.html and (ii) UN Environment Programme, *Global Environment Outlook: Environment for Development (GEO-4) 515 (Glossary)* (2007), www.unep.org/geo/geo4/report/Glossary.pdf. There are many complexities in the identification and labeling of biomes. For a brief discussion of these, see John W. Head, GLOBAL LEGAL REGIMES TO PROTECT THE WORLD'S GRASSLANDS 20–24 (2012). As noted there, these complexities have been addressed in different ways by the UN Environment Programme, the FAO, the US Geological Survey, the Land Cover Working Group of the Association of Remote Sensing, the Asian Institute of Technology, and the UNEP Environment Assessment Programme for Asia and the Pacific.

there are 867 terrestrial ecoregions,[73] fitting within 14 terrestrial biomes. Those 14 terrestrial biomes[74] are:

- Tropical and subtropical moist broadleaf forests (Biome 1)
- Tropical and subtropical dry broadleaf forests (Biome 2)
- Tropical and subtropical coniferous forests (Biome 3)
- Temperate broadleaf and mixed forests (Biome 4)
- Temperate coniferous forests (Biome 5)
- Boreal forests/taiga (Biome 6)
- Tropical and subtropical grasslands, savannas, and shrublands (Biome 7)
- Temperate grasslands, savannas, and shrublands (Biome 8)
- Flooded grasslands and savannas (Biome 9)
- Montane grasslands and shrublands (Biome 10)
- Tundra (Biome 11)
- Mediterranean forests, woodlands, and scrub (Biome 12)
- Deserts and xeric shrublands (Biome 13)
- Mangroves (Biome 14).

Much of the world's land surface currently devoted to agriculture falls within Biome 4, Biome 8, and Biome 12 – especially for production of corn (maize), wheat, and soybeans. Rice production occurs most intensely in Biome 1. The concentration of production of those four principal crops in those four biomes will be obvious by comparing Figure 8.3, Figure 8.4, Figure 8.5, and Figure 8.6 (showing where the specific ecoregions falling within each of those four biomes are found around the world) with the maps provided in Figures 1.2, 1.3, and 1.4 in Chapter 1 (showing where maize, wheat, and rice are produced around the world).

Other "maps" of the world, drawn on the basis of ecological and other environmental factors, have been prepared by an entity called Esri. In collaboration with the US Geological Survey, Esri released in 2014 what it refers to as an "ecological land units map of the world".[75]

---

73 For details, see the pages for *What is an Ecoregion?* and *About Global Ecoregions* on the website of the World Wildlife Fund, wwf.panda.org/. The latter of these explains that "[t]he aim of the Global Ecoregions analysis is to ensure that the full range of ecosystems is represented within regional conservation and development strategies, so that conservation efforts around the world contribute to a global biodiversity strategy."

    For further background on ecoregions, see David M. Olson, D. Eric Dinerstein, Eric D. Wikramanayake, Neil D. Burgess, George V. N. Powell, Emma C. Underwood, Jennifer A. D'amico, Illanga Itoua, Holly E. Strand, John C. Morrison, Colby J. Loucks, Thomas F. Allnutt, Taylor H. Ricketts, Yumiko Kura, John F. Lamoreux, Wesley W.Wettengel, Prashant Hedao and Kenneth R. Kassem, *Terrestrial Ecoregions of the World: A New Map of Life on Earth*, 51 Bioscience 933 (2001), www.worldwildlife.org/science/ecoregions/delineation.html.

74 A fifteenth category – not truly a biome – is labeled "rock and ice".

75 See http://blogs.esri.com/esri/esri-insider/2014/12/09/the-first-detailed-ecological-land-unitsmap-in-the-world/. For further information on this Esri map, see http://ecoexplorer.arcgis.

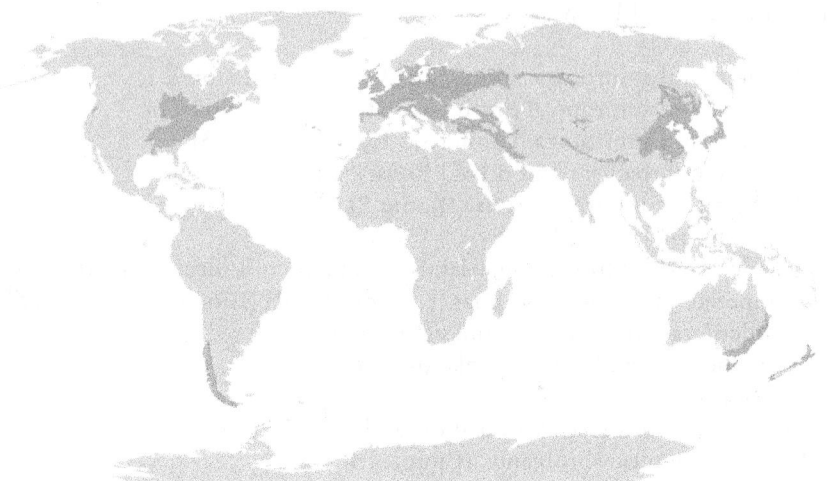

Figure 8.3 Ecoregions in the "temperate broadleaf and mixed forests" biome (Biome #4 in WWF classification)[76]

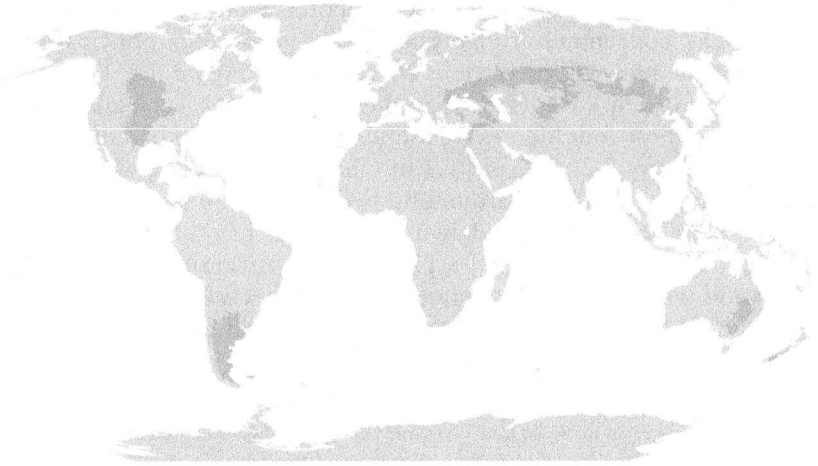

Figure 8.4 Ecoregions in the "temperate grasslands, savannas, and shrublands" biome (Biome #8 in WWF classification)

com/eco/. Esri is an entity headquartered in Redlands, California; it released in 2012 the ArcGIS Online, a cloud-based mapping system. For details, see www.esri.com/about-esri.

76 The maps in Figures 8.3, 8.4, 8.5, and 8.6 are all drawn from Wikimedia Commons, the free media repository (listed author: Terpischores), and are available at https://commons.wikimedia.org/wiki/File:Biome_map_04.svg, at https://commons.wikimedia.org/wiki/File:Biome_map_08.svg, at https://commons.wikimedia.org/wiki/File:Biome_map_12.svg, and at https://commons.wikimedia.org/wiki/File:Biome_map_01.svg, respectively.

Figure 8.5 Ecoregions in the "Mediterranean forests, woodlands, and scrub" biome
(Biome #12 in WWF classification)

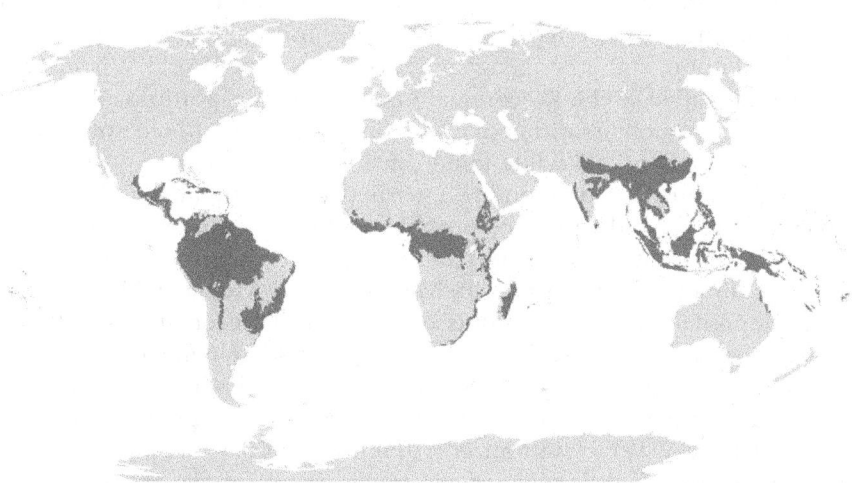

Figure 8.6 Ecoregions in the "tropical and subtropical moist broadleaf forests" biome
(Biome #1 in WWF classification)

Similarly, the Royal Society has prepared a world map of climates, divided
into the following 12 categories: warm temperate dry, warm temperate
moist, cool temperate dry, cool temperate moist, polar dry, polar moist,
boreal dry, boreal moist, tropical montane, tropical west, tropical moist, and

tropical dry.[77] The Nature Conservancy has also prepared (or in some cases revised or reproduced) global ecoregions maps, both for terrestrial and for maritime areas.[78] Likewise, the Food and Agriculture Organization of the UN ("FAO") has prepared a map of the world's "global ecological zones".[79]

The various maps showing terrestrial biomes and ecoregions, including those appearing as Figures 8.3, 8.4, 8.5, and 8.6, all are aimed at offering divisions and categorizations of the world's land-mass based on a common set of ecology-related factors. The factors involve, as noted on page 374, climate, soil type, land cover, species distribution, and the like. Taken together, the maps might be regarded as presenting an image of the world based on what I will call "ecozone clusters".

Having introduced that term – "ecozone clusters" – I will explain briefly what I mean by it and how it differs from other related terms. Although the terminology might vary depending on context, I generally adhere to this set of definitions and distinctions:

- A "biome", as described in the preceding four pages, is a particular category of dominant land cover, such as "Tropical and Moist Broadleaf Forests" or "Temperate Grasslands, Savannas, and Shrublands". The World Wildlife Fund has identified 14 in the terrestrial portions of the Earth.

- An "ecoregion", as also described above, is a specific territorial tract displaying a distinctive array of climate, soil type, land cover, species distribution, and other geographical features. The World Wildlife Fund has identified over 800 ecoregions in the terrestrial portions of the Earth. Each ecoregion fits within one (and only one) of the 14 territorial biomes under the World Wildlife Fund categorization system.

- A "composite ecoregion" – a term that is sometimes used but that I typically avoid to reduce possible confusion with "ecoregion" – is a set of adjacent or nearly-adjacent ecoregions within a particular region. For instance, the "European–Mediterranean Montane Mixed Forests composite ecoregion" consists of eight specific ecoregions in southern Europe and North Africa that encompasses some temperate coniferous forest ecoregions (in Biome 4 under the WWF classification system) and some temperate broadleaf and mixed forest ecoregions (in Biome 5 under the WWF classification system).

- An "ecozone" is the *physical and territorial manifestation of a biome* as it is found in a particular part of the world – such as the set of ecoregions in, say, North America that fall within the "Temperate Grasslands, Savannas, and Shrublands" biome – and it differs from a biome in the sense that whereas a biome is technically a *category*, an ecozone constitutes

---

77 See http://rsif.royalsocietypublishing.org/content/royinterface/9/71/1105/F2.large.jpg.
78 See http://maps.tnc.org/gis_data.html.
79 See www.fao.org/docrep/006/ad652e/ad652e10.htm.

all of the physical *areas* in a particular part of the Earth that fall within one of the 14 terrestrial biomes (again, using the World Wildlife Fund classification system)

- An "ecozone cluster" is the set of *all* ecozones fitting within a particular biome category, so that all those portions of the Earth that fall within, say, the "Temperate Grasslands, Savannas, and Shrublands" biome, wherever they are located on the Earth's surface, would constitute an ecozone cluster. Hence the Temperate Grasslands, Savannas, and Shrublands ecozone cluster would include all the shaded areas appearing in the map in Figure 8.4 – that is, all those territories on various continents that fall within ecoregions classified under the Temperate Grasslands, Savannas, and Shrublands biome (Biome 8 under the WWF classification system).

- An "eco-state" would (in my formulation) be the *political and legal manifestation of a biome,* and it would be a legal entity in roughly the same way that a territorial "state" is a legal entity in the system of international relations that has prevailed in the world for several centuries. For the past several decades, numerous international organizations have been created also as legal entities, with international legal personality. The territory of an eco-state would comprise the territory encompassed in its corresponding ecozone cluster, such that all ecozones (whether in North America or in South America or elsewhere in the world) in, say, the "Temperate Grasslands, Savannas, and Shrublands" ecozone cluster would fall within the jurisdiction of a single eco-state. The extent to which such an eco-state has *sovereignty,* and what kind of sovereignty it might be, is the subject I examine briefly in subsection IIB.

Before turning to that issue of eco-state sovereignty, however, consider further the various maps I have referred to previously, including those appearing as Figures 8.3, 8.4, 8.5, and 8.6. Two key features of all those maps are of fundamental significance. First, the boundaries between the ecozone clusters (falling within the categories of "biomes", to use the WWF terminology) are almost entirely different from the *political* boundaries that exist in the states that define today's world. Second, the maps show that virtually all of the world's ecozones "repeat" – that is, they appear in several portions of the world. For instance, ecoregions falling within what the WWF classification system calls the "Temperate Grasslands, Savannas, and Shrublands" biome appear on all of the continents except Antarctica.[80] Similarly, ecoregions falling within what the WWF classification system calls the "Tropical and Subtropical Grasslands, Savannas, and Shrublands" biome (which is

---

80 Using the WWF's system of "biogeographical realms" – largely equivalent to continents – those temperate grassland ecoregions appear in the Nearctic, Neotropic, Palearctic, Afrotropic, and Australasia "realms"; they are absent only from the Oceania "realm" and the Indo-Malay "realm".

not shown in any of the maps here but has also been largely converted now into agricultural use) appear on all of the continents except Europe and Antarctica.[81]

Curiously, the manifestations of one of the WWF-classification biomes – namely, the Mediterranean Forests, Woodlands, and Scrub biome – appears only in five places in the world: in the Mediterranean Basin itself, of course, and also in limited regions of south-central and southwestern Australia, in the fynbos of southern Africa, in the Chilean matorral, and in certain limited areas of California (including around the Napa Valley, which is famous, as are those other areas, for wines and olives). Mediterranean Forests, Woodlands, and Scrub ecoregions are characterized by hot and dry summers, while winters tend to be cool and moist. Although the habitat is globally rare, it features an extraordinary biodiversity of uniquely adapted animal and plant species, which can adapt to the stressful conditions of long, hot summers with little rain.[82] It is because of the relative rarity of this biome in the world, together with the overwhelming significance economically, politically, and historically of the Mediterranean Basin, that I have undertaken a companion research project into that area.[83]

IIB. *Pluralistic sovereignty: a tentative account and Metaphor B*

In the preceding few paragraphs I have explained the efforts that have been made to draw world maps on ecological grounds rather than historical-political grounds. I have dwelt on that point in order to introduce this assertion: Making an effective transition from modern extractive agriculture to agroecological husbandry – and realizing thereby the advantages I have explained in Chapters 4 and 5 of this book – will almost surely require a restructuring, sharing, and "layering" of authority over global ecological protection and agricultural production. I have summarized earlier in this Chapter 8 the "monolithic sovereignty" concept as it has developed over the past several hundred years and that prevails in today's world – with some narrow exceptions emerging in recent decades. I believe that a new "pluralistic sovereignty" concept should now be developed and embraced, carefully but urgently, in order to create a new form of blended authority in

---

81 Using the WWF's system of "biogeographical realms" – largely equivalent to continents – those tropical grassland ecoregions appear mainly in the Neotropic, Afrotropic, and Australasia "realms", but a few of them exist also in the Nearctic realm (just along the Gulf Coast), the Indo-Malay "realm" (just south of Nepal), and the Oceania "realm" (in the Hawaiian Islands). They are absent only from the Palearctic "realm".

82 For further details, see the WWF account of this biome at www.worldwildlife.org/biomes/mediterranean-forests-woodlands-and-scrubs.

83 See John W. Head, Eco-Crisis in the Mediterranean Basin: A Legal and Institutional Framework for Regional Agroecological Integrity (forthcoming).

ecozone clusters – or, to refer to them now not in physical, territorial terms but rather in legal terms, in "eco-states".[84]

There is, after all, a strong element of artificiality in the theory that within a certain territory, all matters of law and administration are subject to the control of a single authoritative entity. The real world has countless examples of territories that are subject to layered authority, with those layers of governmental regulation determined by *topic*. In the case of the region of Trentino-Alto Adige in northern Italy, for instance, a special layering of authority allocates most legislative and administrative powers to the two largely self-governing provinces that make up the region – namely, Trentino and Südtyrol (South Tyrol). Hence, even though the roughly one million people of the region are Italian citizens, and Trentino-Alto Adige is unmistakably part of Italy, the exercise of sovereign control cannot be said to be strictly territorial.[85] Instead, it is "topical" (my term for it) in that sovereignty is allocated on the basis of topics or aspects of administration and operation.

The world is replete with other illustrations of the same reality. The USA presents an obvious illustration: its federal-state composition involves extensive "layering" of legislative and administrative authority defined in constitutions and frequently erupting into disputes. Likewise, the European Union involves a layering or blending of sovereignty. A conference conducted in Trento in 2009 highlighted many other such illustrations of layered or blended sovereignty, in which the allocation of administrative and other forms of control over certain territories has been made on the basis of the topics at issue – trade, currency, education, taxation, defense, and so forth.[86] Such illustrations appear in Catalonia (in its relations with Spain), in Quebec (Canada), in Aceh (Indonesia), in Sabah & Sarawak (Malaysia), in Südtyrol (Italy), in Wales (UK), and elsewhere in the world. In all these cases, and

---

84 The term "eco-state" does not originate with me. See, e.g., James Meadowcroft, *From Welfare State to Ecostate*, appearing as Chapter 2 in John Barry and Robyn Eckersley, eds., THE STATE AND THE GLOBAL ECOLOGICAL CRISIS (2005). See also Karin Bäckstrand and Annica Kronsell, eds., RETHINKING THE GREEN STATE: ENVIRONMENTAL GOVERNANCE TOWARDS CLIMATE AND SUSTAINABILITY TRANSITIONS (2015). These works examine the notion of "eco-state" (or, in the case of the second one, "green state") in ways substantially different from what I am describing here. Both concentrate mainly on enhancing environmental-protection efforts within existing states; for instance, introductory comments in the book edited by Barry and Eckersley note that it aims to explore "to what extent it might be possible to 'reinstate the state' as a facilitator of progressive environmental change rather than environmental destruction".

85 For details about the administrative and legislative powers of Trentino-Alto Adige, see the website for that region, at www.regione.trentino-a-adige.it/. The special statutory foundation for these powers is found in the *Statuto Speciale di Autonomia per il Trentino-Alto Adige* at www.regione.trentino-a-adige.it/Moduli/933_STATUTO%202015.pdf.

86 See REGIONAL SELF-GOVERNMENT, CULTURAL IDENTITY AND MULTINATIONAL INTEGRATION: COMPARATIVE EXPERIENCES FOR TIBET (Roberto Toniatti and Jens Woelk, editors) (forthcoming 2016). A principal purpose of the conference was to provide a range of models from which China might draw in considering appropriate relations between China and Tibet.

numerous others, sovereignty is not territorially exclusive and absolute but rather layered, blended, shared, mixed. It is the set of characteristics that prompts me to use the term "pluralistic sovereignty".

There is also a strong element of artificiality in the theory that the states in today's world are "national" in character. In a book published shortly before his death, Patrick Glenn emphasized the fact that there is not and never has been such a thing as a "nation-state".[87] Instead, all states are "cosmopolitan" in character.

> [My] argument is that *all* states are cosmopolitan in character, often in spite of themselves and of the choices they may have made. Most people do not presently think that this is the case . . . but this is principally because the western theory of states has been formulated in terms of necessary equality and uniformity within states. It is the notion of a "nation-state" that has largely prevailed . . . .
>
> [That notion] began as an essentially romantic idea in eighteenth century France – the idea that "*la nation*" should have its own legal and political structures . . . [but it] is an idea that has failed. We have been trying to make it succeed for over two centuries now, but it has not succeeded. It is simply not the case that there has ever been a nation-state, and we may now therefore draw the conclusion that there never will be.[88]

In short, the international community is replete with diversity not only between various political cultures but also within them. The international community consists of many power centers, and we have legitimized some of them legally on a largely historical basis. The "monolithic sovereignty" concept pretends that this legal legitimization has been done (i) on the basis of territoriality, with exclusive authority in a particular territory being exercised by only one sovereignty entity, and (ii) on the basis of nationality, with territorial boundaries reflecting national identities. In fact both of those assertions are largely false in a world that is characterized by multiple layers of authority and by massive relocations and interbreeding of peoples.

Expressed differently, ours is a world in which sovereignty – that is, the exercise of power widely accepted as legitimate – is, as a practical matter, allocated along topical lines and placed with a variety of entities. Those entities that are states are not, for the most part, based today on nationality but are instead cosmopolitan in character. The folklore of "monolithic sovereignty", emphasizing territoriality and nationality, tells us something

---

87 See generally H. Patrick Glenn, THE COSMOPOLITAN STATE (2013). For a shorter overview of his thesis, see H. Patrick Glenn, *The Cosmopolitan State*, 61 KANSAS LAW REVIEW 735 (2013) [hereinafter Glenn-KLR].

88 Glenn-KLR, *supra* note 87, at 737–738.

that is misleading in terms of the legitimate and appropriate exercise of authority.

From a legal perspective, then, a central challenge of our day is to design some form of "pluralistic sovereignty" that more accurately reflects the political reality of the world and that will change the character and theory of sovereignty in ways that can provide protection for the ecosphere, with special attention to agriculture. I will explore these issues in another book, but for now I will offer a collection of six observations that I consider central to meeting the challenge of designing a "pluralistic sovereignty".

First, consider the matter in terms of "keystone species". A "keystone species" is a plant or animal that plays a unique and crucial role in the way an ecosystem functions.[89] For well over a hundred million years, dinosaurs served as Earth's keystone species – or, more accurately keystone "clade", which is a large group of genera with a common ancestor.[90] Then the dinosaurs lost their "keystone" status. By some accounts, the Earth's keystone species for several thousand years has been *Homo sapiens.*[91] What happened to rob the dinosaurs of this power of geo-domination and to transfer it to humans? Some combination of the processes of extinction of the dinosaurs and evolution of humans – processes that might at some point likewise rob humans of their position as a keystone species and award it instead to some set of insects.[92]

In like fashion, from a political perspective, the world has awarded the position of "keystone species" for the past 400 years or so to a particular type of political entity. At first it might have been accurate to consider it the territorial nation-state, but for reasons explained briefly earlier it is simply, in today's world, the state – rooted in the peculiarities of European history,

---

89 See *Keystone Species,* on the website of the National Geographic, at http://education. nationalgeographic.com/education/encyclopedia/keystone-species/?ar_a=1. That account also defines "foundation species", "umbrella species", and "indicator species". The same article notes that the theory that the balance of ecosystems can rely on one keystone species was first established in 1969 by American zoology professor Robert T. Paine. For a discussion of "cultural keystone species", see Ann Garibaldi and Nancy Turner, *Cultural Keystone Species: Implications for Ecological Conservation and Restoration,* 9 ECOLOGY AND SOCIETY no. 3 (2004). The authors explains that according to some observers, "just as there is a biosphere, i.e., the region of the earth's crust and atmosphere occupied by living organisms, there is an ethnosphere, defined as '. . . the sum total of all thoughts, beliefs, myths and institutions made manifest today by the myriad cultures of the world.'" *Id.*

90 According to the entry for "dinosaur" on Wikipedia (an adequate source for this purpose), there are likely more than 500 genera of dinosaurs, the earliest of which appeared during the Triassic period, 231.4 million years ago. They "were the dominant terrestrial vertebrates for 135 million years, from the beginning of the Jurassic [period] (about 201 million years ago) until the end of the Cretaceous [period] (66 million years ago)", when they started becoming extinct.

91 Some observers would dispute this. For an overview of competing perspectives, see Amy Frietag, *Are Humans a Keystone Species?,* at www.southernfriedscience.com/?p=5330.

92 Ali Khan also uses the notion of extinction in his assessment of the nation-state and sovereignty. See Ali Khan, EXTINCTION, *supra* note 58.

regarded (somewhat artificially) as having exclusive sovereign control over the governance of a defined region that at one time might have been home to persons who shared a particular nationality or ethnicity. Recent decades have seen reality shift, though, so that the status of the state as a "keystone species" is increasingly under question.

Indeed, if we were to explore world history, or at least Western history, from the perspective of identifying "keystone species" of political structures, we might include in the chronology the following "species": tribe-state, city-state (as in Greece), empire (as in Rome), feudalism (as after the fall of the Western Roman Empire through the early Middle Ages), and the nation-state (as recognized formally in the Peace of Westphalia, 1648). What I am suggesting is that the status of the state as a "keystone species" now seems in the process of changing.

In the face of the various forms of unsustainability that I have surveyed in Chapters 2 and 3 of this book, I believe it would be sensible for us to welcome these changes. On economic grounds, on ecological grounds, and on social and other grounds, the existing political structure that features the state as the "keystone species" has shown serious failings. These failings warrant, in my view, intense effort to identify, at least for ecological issues, a new "keystone species" of political structure that can do a better job at arresting and reversing the damage already done in a wide range of aspects of human life – of which the aspect on which this book focuses is agriculture.

Second, consider what should be the criteria for the legitimization of a new "keystone species" of world-stage entities – again, with particular attention to agroecological integrity and protection. In the course of my earlier studies of international organizations, especially international economic institutions,[93] I have identified several central features that such institutions should have and that most contemporary institutions lack, at least in part. These include extensive transparency, broad and representative participation, legality (establishing clear rules and following them), competence (as a legal matter and among the staff), accountability (both of the institution, through a form of external "judicial review", and of the member entities), distributional justice, and others.[94] In providing legal legitimization to entities that would succeed or supplement the state, special attention would need to be given to the fiduciary, trusteeship-like duties that such entities must discharge in protecting the interests of the beneficiaries whom they are to serve. Those beneficiaries would include humans, of course, but also other species and elements of the ecosphere.

One possible entity that would succeed the state as a "keystone species" in the global political order is the eco-state as I described it near the end of subsection IIA. Perhaps a structure of governance could be created that

---

93 See generally Head, Losing the Global Development War, *supra* note 67.
94 See *id.* at 276–310.

would place legal authority with eco-states in respect of ensuring agroecological integrity[95] – thus facilitating the implementation of the Global Convention on Agroecology described in Chapter 7.

Third, consider whether international organizations in their current formulation could be regarded as a new "keystone species" in the global political order. Although the significance and authority of some international organizations, such as the United Nations, the World Bank, the WTO, and the IMF, might support an argument that a shift of this sort is occurring, the fact remains that those organizations remain subservient in most respects to the state in its current formulation. More specifically, all of those institutions are in greater or lesser extent dominated by Western, especially US, interests and influence – even to the extent in the IMF and the UN, for instance, of having explicit provisions in their charters permitting the USA an effective veto over any major actions proposed for those institutions, including a proposal to amend the charter provisions that provide such veto powers.

Having said that, perhaps a new and modified form of international organization could be designed that would *not* have this same dependency-relationship to the state. For instance, if some form of global constitutional convention were conducted in a manner generally regarded as fair and effective in registering the desires of various elements in the international community, and if such a convention were to "charter" a new form of international institution that had representative responsibility not to *states* as constituents but rather to other entities – individuals, private corporate entities, public NGOs, charitable foundations, etc. – then it is conceivable that the international institution emerging from such a process would have legitimacy and authority largely independent from the state, and therefore be insulated against the sort of "capture" that undermines the effectiveness of some of the existing international organizations.

Fourth, consider the parallel between the challenges that face the world today and those that faced Europe in the first half of the 1600s. Recall that it was in that half-century that Hugo Grotius and Thomas Hobbes both offered legal and political solutions to the chaos they saw afflicting Europe in the form of the Thirty Years War. Some modern authors have referred to that era as a "Grotian Moment", to signify that it presented a time of crisis in which bold action was both demanded and permitted.[96]

---

95 The special significance that agroecological integrity has for guarding against a collapse of human civilization is reflected in several of the Principles that I described above in outlining a proposed Global Convention on Agroecology. See subsection IIA2 in Chapter 7, *supra*.

96 One set of authors, citing Richard Falk and others, calls the "Grotian Moment" 'a period of normative uncertainty in which [an old structure] . . . of international relations is being superseded, but not yet fully or in any precisely defined direction or manner". Weston *et al.*, *supra* note 52, at 1086–1097. More recently, Michael Scharf explored the notion of a "Grotian Moment", defining it as "a transformative development in which new rules and doctrines of customary international law emerge with unusual rapidity and acceptance". Michael P. Scharf,

The seventeenth-century "Grotian Solution" as I summarized it previously was to offer a new definition of the law of nations (international law) that acknowledged the power of a then-new "keystone species" (the nation-state) while calling on the crowned heads of Europe to hew also to the key principles of natural law that their shared Christianity allegedly bound them to follow. For Grotius, that is, the Grotian Solution could be expressed as a formula as "the law of nations = *jus gentium volutarium* + Christianity-based natural law".

If we were to view the world today as being in a new, twenty-first-century "Grotian Moment", what might be the new "Grotian Solution" that we could posit? Perhaps in the place of Christianity-based natural law, we would opt for a set of principles aimed at and crucial for ecospheric health. In this regard, some insight might be gained from the so-called "Gaia Hypothesis" or "Gaia Principle", attributable largely to James Lovelock. As posited by Lovelock, the Earth can usefully be regarded as being similar to a super-organism that has complex mechanisms for self-regulation,[97] rather than as a mere collection of interdependent but relatively separate sets of processes and systems. One conclusion that can flow from viewing the Earth in this way is that an essential role of humans is to avoid interfering with the Earth's mechanisms for self-regulation. While the Gaia Hypothesis has many detractors,[98] and while I do not endorse it myself, perhaps something akin to it could be elaborated that would encompass a set of principles of the sort I referred to previously – that is, a set of principles giving primacy to *ecospheric health*, so that absolutely nothing would have higher priority than guarding against the degradation of the Earth. If such a set of principles were developed and enunciated, we might call that set of principles "Gaian-like natural law" or (more generically and less inelegantly) "ecospheric natural law". [99]

Seizing the "Grotian Moment": Accelerated Formation of Customary International Law in Times of Fundamental Change, 43 CORNELL INTERNATIONAL LAW JOURNAL, 439, 444 (2010) (citing work by Saul Mendlovitz and Merav Datan).

97 This simple statement of the Gaia hypothesis appears on Wikipedia: "The Gaia hypothesis, also known as Gaia theory or Gaia principle, proposes that organisms interact with their inorganic surroundings on Earth to form a self-regulating, complex system that contributes to maintaining the conditions for life on the planet." For further descriptions of Gaia theory, see *Gaia Theory: Model and Metaphor for the 21st Century*, www.gaiatheory.org/overview/.

98 See, e.g., Toby Tyrrell, ON GAIA: A CRITICAL INVESTIGATION OF THE RELATIONSHIP BETWEEN LIFE AND EARTH (2013). Like numerous other criticisms, this one challenges the scientific rigor and usefulness of the Gaia principle. "I believe Gaia is a dead end. Its study has, however, generated many new and thought provoking questions. While rejecting Gaia, we can at the same time appreciate Lovelock's originality and breadth of vision, and recognise that his audacious concept has helped to stimulate many new ideas about the Earth, and to champion a holistic approach to studying it." *Id.* at 209. For a survey of criticisms, see the relevant Wikipedia entry, https://en.wikipedia.org/wiki/Gaia_hypothesis#Recent_criticism.

99 One of the most important such principles that might fall within the ambit of ecospheric health (or ecospheric resilience, sustainability, and preservation) and that has particular pertinence to agroecology might be the "law of return" that I discussed in Chapters 3 and 6 of this book. For

The late J. Stan Rowe has offered a perspective that resembles the Gaia Hypothesis in some ways but in my view improves upon it. Rowe, a geo-ecologist and environmentalist who worked as a research forester with Forestry Canada for 19 years and also served as Professor of Plant Ecology at the University of Saskatchewan, described himself as "[n]ot a misanthrope, but a defender of Earth against the excesses of anthropes".[100] Rowe's writings urge us to stop thinking about "organisms as possessing life" and start thinking instead of "life as possessing organisms".[101] He emphasizes the importance of changing our viewpoint in a way that (1) does *not* focus on an individual *organism* – whether it is a single-celled paramecium or a pronghorn antelope – and regard that particular entity as a separate object that comprises a collection of atoms and molecules and that also *has life in it* (along with many non-living ingredients) but that (2) focuses instead on the entire *ecosphere* that the Earth comprises and regard it as *a form of life that has organisms in it*. If we were to change our viewpoint in this way, we would be changing to what might be called an "ecospheric worldview".[102]

Such an ecospheric worldview also resembles traditional Native American values. The late John Mohawk, a Seneca historian, writer, and social activist, has offered these views about the natural world and natural law:

> The natural world is our bible. We don't have chapters and verses; we have trees and fish and animals . . . . The Indian sense of natural law is

---

explanatory references to the "law of return", see text accompanying note 117 in Chapter 3, *supra*, and text accompanying note 33 in Chapter 6, *supra*. Others might emphasize the need for the close knowledge of nature that "bioregionality" implies, the necessity of species diversity, the dangers of a "commodification" of land, and other principles discussed earlier in this book. More generally, the principles of a "Gaian-like" natural law or an "ecospheric natural law" would surely need to acknowledge the responsibilities of humans to recognize and protect the interests of other species (reflecting the principle of interspecies equity) as well as the interests of both human and non-human species in the future (reflecting the principle of intergenerational equity). Some of these principles and others related to them were featured in my earlier discussion in Chapter 4 of an agroecological future.

100 See biographical information about Stan Rowe on the homepage for the Ecocentrism website, www.ecospherics.net/pages/aboutauthors.html#rowe. Other pages of that website provide copies of Rowe's works *The Living Earth and Its Ethical Priority* (2003) and *Biological Fallacy: Life = Organisms* (1992).

101 Rowe offers these observations about the relationship between life, the ecosphere, and organisms:

> For thousands of years, humans have been viewers immersed in the Ecosphere . . . [and thus unable to see clearly that the Ecosphere is in fact an entity that itself has life.] It is not a superorganism; it is supraorganic: a higher level of organization than plants and animals, including people. The lively Ecosphere gives the lie to those who see the world's reality as little more than a competitive arena, for without compliant cooperation among its multitudinous parts the diversifying creativity of the planet could not have evolved nor could its overall homeostasis continue.

> J. Stan Rowe, *Biological Fallacy: Life = Organisms* (1992), www.ecospherics.net/pages/RoBiolFallacy.html.(first published at 42 BIOSCIENCE no. 6, 1992).

102 For similar views espoused by Wes Jackson, see note 34 in Chapter 7, *supra*.

that nature informs us and it is our obligation to read nature as you would a book, to feel nature as you would a poem, to touch nature as you would yourself, to be a part of that and step into its cycles as much as you can.[103]

Now let me draw more clearly the parallel or analogy that I am suggesting between the seventeenth-century "Grotian Solution" and one that might be possible in the twenty-first century. In subsection IA, I suggested that a more comprehensive and less historically-driven term for what we now call international law, and what was earlier referred to broadly as the law of nations, would be "the set of legal rules governing relations between peoples of different legal and political cultures".[104] Grotius drew from Christianity-based natural law in offering his definition of the law of nations. Christianity-based natural law holds little practical relevance in today's world, in which most people are not Christian. If broad consensus could be achieved as to the importance of ecological protection, then a new Grotian Solution might draw from that consensus, just as Grotius drew from the consensus he saw regarding Christian natural law, so that a new Grotian Solution might be this: "the set of legal rules governing relations between peoples of different legal and political cultures = *jus gentium voluntarium* + ecospheric natural law."

In such a formula, the term *jus gentium voluntarium* would have a broader content than the content Grotius had in mind in his use of the term. Instead of referring to the voluntary action of territorial, politically distinct *states* (which Grotius had in mind), the term *jus gentium voluntarium* in a new twenty-first-century Grotian Solution could refer to the voluntary action engaged in through a decision-making process involving entities that would be characterized by the values I enumerated earlier: transparency, participation, legality, competence, accountability, distributional justice, and the like.[105] Some of these entities might be states similar to the ones we are accustomed to, but other entities would have different structures and constituencies.

Let me make one further comparison between (i) the substance of what Grotius was attempting in designing his seventeenth-century Grotian Solution and (ii) the substance of what would be involved in designing a twenty-first-century Grotian Solution. Grotius was intent on facilitating a transition from feudalism to the nation-state as the fundamental political unit in Europe. In order to do that, Grotius emphasized foundational precepts – drawn from Christianity-based natural law – that concerned *inter-human*

---

103 John Mohawk, *A Native View of Nature – Interview by Charlene Spretnak* (Jan. 26, 2005), http://lists.ibiblio.org/pipermail/livingontheland/20050126/000517.html.

104 See text accompanying note 3, *supra*.

105 See text accompanying note 94, *supra*.

relationships and the relationships of humans to a Christian God in which his readers professed belief or were expected to believe.

By contrast, a twenty-first-century Grotian Solution of the sort I have been describing here would aim at facilitating a transition from the state to some other form of dominant political entity as a "keystone species" of political organization, which I have suggested might be an "eco-state". In order to do that, it might be possible to emphasize foundational precepts – drawn from something I have tentatively referred to here as "ecospheric natural law" drawing elements from the worldviews urged by such writers as Lovelock, Rowe, and Mohawk, summarized in the preceding three pages[106] – that concern not inter-human relationships but *inter-species relationships* and the relationship of humans to the ecosphere. Stated in an overly simplistic way, (i) God would be replaced by the Earth viewed as an ecosphere (a "supraorganic" entity, in Rowe's words, "Gaia" in Lovelock's, the natural world in Mohawk's), (ii) the nation-state would be replaced by the eco-state as a "keystone species" of political organization, and (iii) inter-human relationships of the sort addressed by Christian natural law would be replaced by inter-species relationships, which ecospheric natural law would require humans to manage and protect in the interest of long-term planetary sustainability.[107]

Fifth, consider what would be required to instill a new set of values of the sort I have alluded to by using the term "ecospheric natural law". Assuming it would be *desirable* to instill such a new set of values, would it be *possible* to do so? For Grotius, of course, the inclusion of Christianity-based natural law in his "Grotian Solution" did not involve *educating* his readers about natural law or the Christianity from which he drew it. He engaged merely in *reminding* his readers about natural-law principles which their shared Christianity already encompassed. By contrast, I believe most humans now are deeply ignorant of, and divorced from, the processes and systems of the natural world. With now more than half of the world's human population living in urban areas rather than rural areas,[108] a massive education process would need to be undertaken to instill a new set of values concentrating on the Earth and on humans' proper role in its preservation.

---

106 Some similar themes are present in the Encyclical *On Care for Our Common Home* as recently released by Pope Francis, who asserts there that (i) all humans are interconnected, (ii) all the rest of Nature is interconnected, (iii) humans and Nature are interconnected. See generally http://w2.vatican.va/content/francesco/en/encyclicals/documents/papa-francesco_20150524_enciclica-laudato-si.html.

107 For some observations about the meaning of the adjective "long-term" in this context, see note 34 in Chapter 1, *supra*.

108 For a discussion of the fact that now about 54 percent of the world's population lives in urban areas, see a posting on the UN website, *World's Population Increasingly Urban With More than Half Living in Urban Areas* (July 10, 2014), www.un.org/en/development/desa/news/population/world-urbanization-prospects-2014.html. The same posting reports that the proportion "is expected to increase to 66 per cent by 2050." *Id.*

The challenge posed by this massive education process is more difficult than the challenge faced by Grotius in several other ways as well. For one thing, Grotius could count on his readers sharing not only a common religion but also a single language – Latin. By contrast, there is no common language of the world today: by some estimates, only about 1.2 billion people speak English either as a first language or as a second language, ranking slightly behind Mandarin (at 1.35 billion) and ahead of Hindi (540 million), Spanish (460 million), and Arabic (390 million).[109] In other words, no single language is spoken by more than about a fifth of the world's 7 billion people.

Moreover, there is no common "cultural language" or shared narrative for ecospheric values: there is no settled authoritative text purporting to set forth those values, as in the case of the Christian Bible; there is no broadly shared hero recognition, as in the case of the Christian worship of Jesus.[110] Furthermore, ecospheric values do not have associated with them any temples or other physical structures in which a recognized priesthood can preach about those values to a congregation, as the seventeenth-century Christians had in their vast distribution of churches and clergy across the face of Europe. Moreover, Grotius had a limited audience to convince – the crowned heads of Europe and the members of their courts. With today's more egalitarian worldview, a spokesperson for ecospheric values (that is, someone serving in the role of a new Grotius) would need to reach and persuade huge numbers of people in order to instill those Earth-based values of ecospheric natural law widely enough to make a difference.

Sixth, and finally, consider these topics specifically from the perspective of agriculture. From this perspective, another parallel construction comes to mind. The analogy I have speculated on in the preceding paragraphs draws parallels between (i) what Grotius attempted in the seventeenth century in order to facilitate a transition away from an old political system (feudalism, with Church teachings providing a kind of "glue") to a new political system (the Westphalian nation-state system with sovereignty viewed as being strictly territorial and national in character) and (ii) what would be required to facilitate in our present day a transition away from what is now the old state system (with "monolithic sovereignty" at its core) to a new political system (with some other entities, perhaps "eco-states" at its core). Now I wish to propose a different analogy, one that I believe might usefully be drawn between (i) the journey of a tribe or band of travelers seeking food security and (ii) the journey of a tribe or band of travelers seeking political

---

109 These figures are drawn from the Wikipedia page for "List of languages by total number of speakers" (probably adequate for these purposes), relying mainly on details set forth in ETHNOLOGUE (2013, 17th edition) and earlier studies by George H. J. Weber.

110 Incidentally, these various elements – a shared authoritative text (or set of texts), a shared hero, and a shared language (at least written language) – were present in dynastic Chinese culture, contributing to traditional China's cohesion for many centuries. The same could be said of several of the world's major religions.

security through law. In this regard, recall Figure 4.3, which appeared in Chapter 4 to illustrate a metaphor – Metaphor A – that I introduced there to help explain the significance of agroecological husbandry. I now repeat Figure 4.3 and I also introduce a new Figure 8.7, as a means of illustrating a different metaphor – Metaphor B.

As should be clear by now, the tribe or band whose journeys are represented in both Figure 4.3 and Figure 8.7 is the human species. I have explored in this book how the group whose journey is represented in Figure 4.3 can now blaze a trail through a process of bushwhacking toward a different destination of food and ecological sustainability. My analysis has included

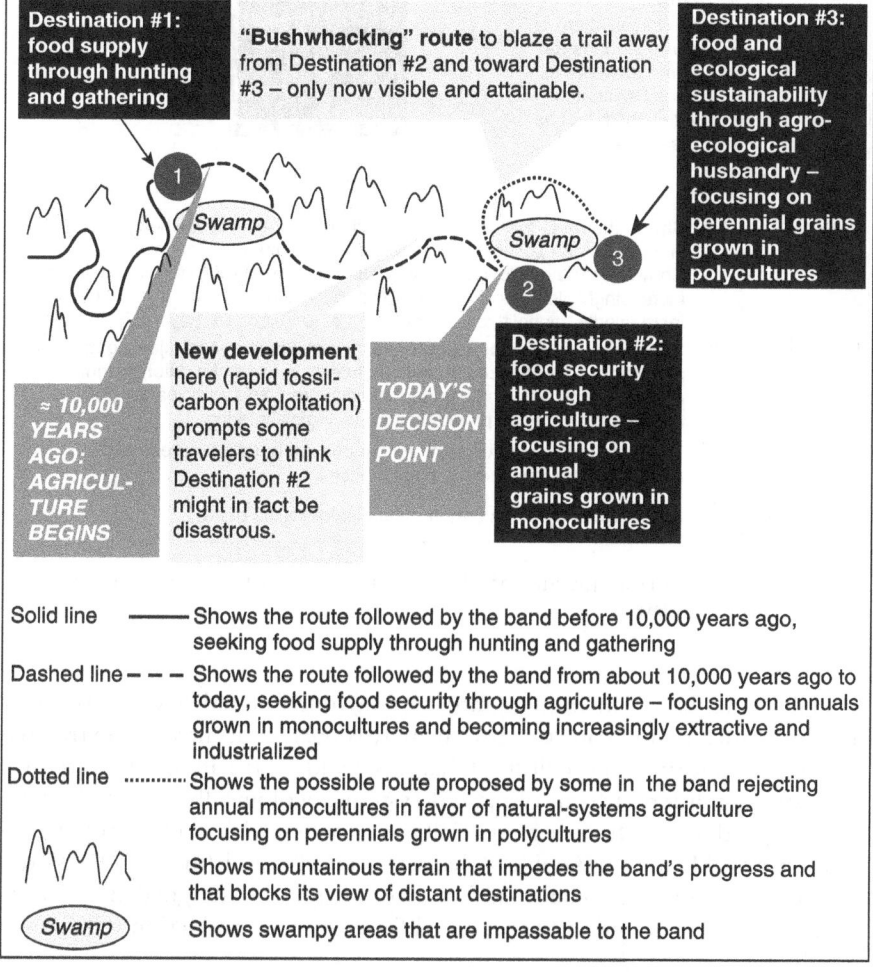

Figure 4.3 (repeated) (Metaphorical) journey of a band of travelers seeking food supply, security, and sustainability

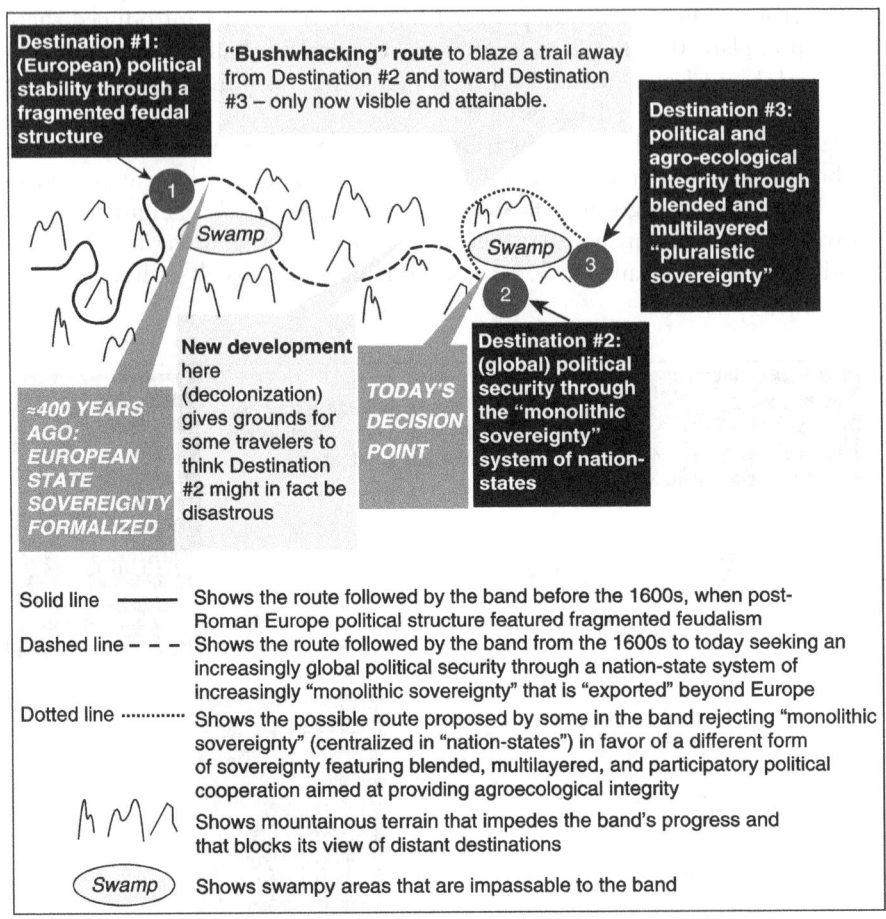

Figure 8.7 (Metaphorical) journey of a band of travelers seeking political security through law

several elements: (i) an inquiry into certain fundamental flaws in the path that humanity is now on, toward an increasingly concentrated, extractive, industrialized form of agriculture (Chapters 2 and 3); (ii) an inquiry into how a profoundly different form of natural-systems, ecologically-oriented system of food production, focusing especially on grains and legumes, might now be both possible and preferable (Chapters 4 and 5); and (iii) a study of how a national and international legal framework might be designed and built to facilitate a global transition to that novel form of food production (Chapters 6 and 7).

It is for that third component – the legal framework – that the metaphorical narrative I have introduced also has explanatory power. Although any metaphor has imperfections and limitations and cannot be pressed too far without breaking, I believe both Metaphor A (on agriculture) and Metaphor B (on law) invite some useful insights. Just as humanity has travelled for roughly 10,000 years toward a particular destination of *food* security based on annuals grown in monocultures, only to find now that this destination is in fact deeply unattractive, likewise humanity has traveled for the past four centuries toward a particular destination of *political* security based on a system of states monopolizing political power that also now seems deeply unattractive. I believe that just as recent scientific advances now make it possible to blaze a trail to a new destination for food production consistent with ecological protection, experience of the past century or so makes it possible now to recognize defects in the international legal system and to work toward correcting those defects – at least in terms of ecospheric health in general and the task of feeding ourselves in particular. Some of this effort involves more sophisticated and proven forms of cooperation among states and non-state actors, and some of it involves examining and effectively updating the concept of sovereignty.

The analysis that I have introduced in these last two chapters concentrates on why a "bushwhacking" exercise is needed in order to arrive at a different destination of political and ecological stability through new forms of eco-regional multilateral action and through a blended and multilayered "pluralistic sovereignty" – one that is in fact already practiced in some settings around the world and is ripe for more nuanced development and application in this context as well. I have tried to explain both (i) how such a "pluralistic sovereignty" concept (I have also referred to it as a "sovereignty of restraint") should be more realistic than "monolithic sovereignty" is in reflecting the character of the international community as it exists in today's post-decolonization world, and (ii) how it would contribute to a new definition of international law – creating, in essence, a new "Grotian Solution" that would serve the same sort of purpose as the one devised by Hugo Grotius in the early 1600s. I have offered various perspectives on this effort to develop a "pluralistic sovereignty" concept. In several cases I have tried to present provocative analogies and rather daring propositions – some of which I do not fully accept or endorse myself – with the intention of exploring separately, in the companion book to this one, the details of how a "pluralistic sovereignty" should be designed.[111]

For now, my principal aim has been to emphasize that the *legal* initiatives that will be needed in order to facilitate a transition from modern extractive

---

111 See John W. Head, A Global Corporate Trust for Agroecological Integrity: Managing a New Agriculture in a World of Eco-States (forthcoming). In that book I will compare the pluralistic sovereignty concept with various other formulations, including the "trustee sovereignty" discussed by Eyal Benvenisti, under which sovereigns – while owing fiduciary duties to "their own people" – must *also* take into account the interests of foreign stakeholders. Benvenisti, *supra* note 51, at 296, 314.

agriculture to agroecological husbandry extend beyond the type of treaty law that I have proposed in Chapter 7 in the form of a Global Convention on Agroecology. Although developing such a treaty will, in my view, be a necessary step, it will not be a sufficient step. Further legal changes will be needed, and they will involve a fundamental reorientation of sovereignty.

III. Closing observations

At several points in the preceding pages I have drawn special attention to the fact that difficulties, especially political difficulties, will be encountered if the transition I discuss here – from modern extractive agriculture to agroecological husbandry – is attempted. Some of the difficulties revolve around cultural, legal, and constitutional rules and values that are deeply held and broadly accepted. For instance, some of the legal reforms that I have discussed in Chapters 6 and 7 could run contrary to perceived private property interests in land. Similarly, much of the discussion in Chapter 7 about a new treaty system, and in this chapter about an updated and improved concept of sovereignty, definitely runs counter to the perceived short-term national interests of many countries, including some that are especially powerful.

My general view regarding such difficulties has two elements to it. First, I believe the seriousness of the ecological distress that humans are impos-ing on the Earth is profound. Much of my attention in this book has focused on the degradation to the Earth's soil – its "thin skin of life stretched over a rock" – but the other three forms of eco-crisis are also affected by modern extractive agriculture. Those other three forms of eco-crisis affect (i) the Earth's air, weather, and climate, (ii) the Earth's water, both supplies and quality, and (iii) the Earth's biodiversity. If I am correct in this assessment, and assuming I am not in a miniscule minority in this regard,[112] then it makes sense to at least *consider*, and try to *design*, what seem to be radical changes in existing norms and practices. Indeed, I would go further than that: it would be irresponsible *not* to do so.

Second, I believe that some of those existing norms and practices are in fact not as worthy as they are made out to be. By that informal language, I mean this: a historically and culturally astute observer will realize that private property rights, especially in respect of land, have varied dramatically from one time and culture to another; indeed, conceiving of "property" as being the thing itself, rather than just a right in the thing, is a relatively recent development that reflects in part the emergence of markets and business transactions as being central to the identity of some especially influential

---

112 As noted above, public sentiment is running strongly now in favor of acknowledging the dangers of climate change and supporting official action to address it. See text accompanying note 116 in Chapter 7, *supra*.

societies. An astute observer will also realize that the "monolithic sovereignty" concept is also relatively recent and had its roots in a cluster of specific historical circumstances that no longer exist and that in any event were confined just to one very small portion of the globe. And he or she will realize that there have in fact been a number of exceptions to and departures from *all* of these norms and practices as circumstances have required, so that proposing and designing alterations to them is hardly as heretical as it may at first appear.[113]

In short, I see two conflicting perspectives. On the one hand, proposals for a fundamental transformation of agriculture as an institution, along with the changes that such a transformation would involve for other aspects of our legal and cultural landscape, seem very problematic substantively. On the other hand, the problematic character of such proposals is almost surely overestimated. Indeed, if global ecological circumstances deteriorate quickly in coming decades, as seems likely, proposals of the sort discussed in these pages might seem too mild and timid – not too bold or overreaching – in short order.

From my own personal perspective, these issues bear directly on the family farm that I described in the opening pages of this book. As I explained there, our family farm participates in the system of modern extractive agriculture, as do the other farms in that region and radiating out in every direction. How realistic is it to envision a transition from that system to a radically different system – namely, a system of agroecological husbandry centered on perennials grown in polycultures? Would Will Bier, the neighbor who farms that land now and who occasionally invites me to join in the corn harvest, embrace such a change? Could the people of that area, and in farming regions all around the world, adopt a renewed land ethic and a revised economic system and a novel international legal framework fashioned along the lines I have described in this book?

This all remains to be seen. However, humans have shown themselves to be remarkably inventive and adaptable, especially in their development of agriculture and of the increasingly sophisticated civilization that it supports. That inventive, adaptive character of our species gives me hope that the crisis the world now faces as a result of the particular evolutionary path

---

113 In examining, for instance, the issue of private property in the context of climate change, a distinguished environmental law professor at the University of Utah has asserted that "[o]ver-attentiveness to private rights at the expense of public needs . . . can impede effective adaptation to climate change" and that one important change in legal and cultural norms "will be the reassertion of community and public values, even if they come at the occasional expenses of individual property rights and liberties". Robin Kundis Craig, *Becoming Landsick: Rethinking Sustainability in an Age of Continuous, Visible, and Irreversible Change*, in Jessica Owley and Keith Hirokawa, eds., Rethinking Sustainable Development to Meet the Climate Change Challenge 56–57 (2015). Craig identifies several doctrines in US law "that underscore the occasional primacy of public and community values", including public nuisance law, the public trust doctrine, and the public necessity doctrine. *Id.* at 56–58.

that agriculture has followed – toward the extractive, industrial version that dominates my family farm and the rest of the world – can in fact be reversed. To express the matter differently by invoking the metaphors I have offered here: I believe the "band of travelers" depicted in both Figure 4.3 and Figure 8.7 *can* decide to change direction and to "bushwhack" its way toward a new destination. Whether we *will* do so depends importantly on public consciousness and political will, and on how they both can be developed to build new legal frameworks at the highest levels to facilitate a transition to a new agriculture . . . agroecological husbandry.

Appendix
A "bare-bones legal and policy brief"

Building worldwide legal foundations for a new agriculture

The following few pages offer an outline of a brief – similar in style to a legal brief – that aims to summarize the principal points made in the main text of this book, but expressed more in the form of advocacy than of explanation. It is worth noting that many of the propositions presented here are not persuasive in this "stand-alone" format without detailed explanations or citations to authority. For those explanations and citations, see the corresponding discussions in the main text. References are provided below [in brackets] to the chapter, section, and subsection in which each point in this "bare-bones brief" is discussed.

Proposition #1. *The form of "extractive" agriculture that humans have developed over about 10,000 years presents a cluster of problems, especially in its most modern "industrial" form. These problems are economic, ecological, and social in character, and they are so substantial as to conclude that modern extractive agriculture has failed.*

1   Modern extractive agriculture is economically unsustainable. [Ch.2]

   A   Farming presents **unusually serious economic perils**. From the perspective of the farmer, it includes very high entry costs (particularly in developed economies). As an economic sector it is unusual in that the predominant production input is land, which is neither mobile nor fungible. It involves heavy dependency on factors beyond the farmer's control. These include markets (the farmer typically "sells wholesale but buys retail" for inputs) as well as nature, in the form of adverse weather, pests, and disease. Moreover, farming typically involves the merger of family and business interests in ways that distinguish it from most other occupations and ways of life. [Ch.2: IA]

   B   Some **government efforts** have been made to counter those intimidating economic realities of farming operations. During the economic turmoil of the Great Depression that gripped the world in the 1930s, many developed countries tried to implement economic initiatives within their borders and also to export their agricultural surpluses – thereby adding sometimes to the distress of farmers in other lands. [Ch.2: IB]

C The USA attempted both of those approaches, and as consequence brought an unprecedented shift in its agricultural sector, characterized by **enormous concentration**, in two respects: (i) concentration of ownership in the hands of a small minority of the population, and (ii) concentration in the types of grains and other agricultural products. US influence elsewhere in the world has brought these same types of concentration (especially the second type) to other countries. One result is that now 90 percent of the world's food comes from 30 crop species, even though about 7,000 crop species exist. This shift in the agricultural sector toward concentration represents not only a dramatic drop in diversity but a fundamental change in farming that has (i) driven most people out of farming and (ii) left the remaining individual farmers facing even greater economic challenges. [Ch.2: IC]

D **Systemic sustainability?** It is implausible to suggest that even though some farmers find farming unprofitable, the *system* of modern agriculture as a whole is economically sustainable. The suggestion is implausible because (i) it is actually the smaller farms with diversified land ownership and crops (not the huge ones with concentrated operations) that have higher productivity and greater poverty-reduction potential; (ii) even for large agribusiness enterprises, modern agricultural techniques are quite wasteful and inefficient when true costs are properly accounted for; and (iii) even with heavy subsidization by some governments, modern concentrated agribusiness has not succeeded in increasing crop yields as much as promised and as much as needed. [Ch.2: II]

E Likewise, it is disingenuous to suggest that **world food demand** (and the need to feed a growing global population) is a "trump card" in justifying modern extractive agriculture despite its economic unsustainability. In fact, the system of food supply and distribution resulting from modern agriculture – even with the impressive results of the so-called "Green Revolution" – continues to fall short. One illustration is that the number of malnourished people has stayed stubbornly above 800 million for over 40 years. While this malnourishment problem is attributable not just to agricultural practices but also to issues of food distribution, food waste, political frictions, and other influences, the claim that modern extractive agriculture's value is proven by its ability to meet global human food supply needs is unpersuasive: it has not done so and shows no long-term capacity to do so. [Ch.2: III]

F In short, modern agriculture of the sort promoted by US farm policy and spread around much of the world involves **a litany of**

economic problems. These include unacceptably high risk and cost, a deleterious degree of concentration, and economic unsustainability not only for small farmers but for large concentrated operations as well – and it cannot be justified on grounds of increased crop yields or as an adequate response to world food demand. [Ch.2: IV]

2   Modern extractive agriculture is ecologically unsustainable. [Ch.3: I]

A   It creates substantial **habitat loss and degradation**. This is true worldwide, in all cropland settings. North America provides a potent illustration: at one time virtually all of the acres (nearly a billion) currently used for farming in the USA were relatively undisturbed habitat. Compared with earlier times, or even 50 years ago, modern American farms are increasingly silent and sterile. [Ch.3: IB1]

B   It creates massive **soil erosion**, even with "low-till" or "no-till" farming techniques. For instance, although topsoil can be replenished at a rate of less than one inch in 200 years, current rates of soil erosion in the USA (even with aggressive soil-conservation efforts in some locations) run 12 times higher than soil formation rates. Moreover, soil loss problems in many regions elsewhere in the world are much worse, so that by one estimate 75 billion tons of soil are lost to erosion worldwide each year. [Ch.3: IB2]

C   In addition to soil erosion, modern extractive agriculture also results in serious **soil degradation** – that is, in its fertility, its resilience, its organic matter, and other elements of its quality. Particularly troubling in this regard is the initiative of the past half-century to use massive amounts of outside synthetic chemical inputs that kill or injure countless microbes, worms, insects, and other participants in the soil's architecture of life. [Ch.3: IB3]

D   Moreover, modern extractive agriculture creates enormous dead zones and other forms of aquatic poisoning and contamination because nitrate, phosphorus, and other substances emitted from agricultural operations are transported downstream. Similarly, emissions of ammonia are transported downwind in the air, inducing species destruction and stress from acid raid. As a consequence, both **terrestrial and aquatic ecosystems** (including of course wildlife relying on them) are degraded. [Ch:3: IB4]

E   The damage to habitats highlighted above – including both terrestrial habitat and that of waterways and oceans – is creating an unprecedented reduction in **biodiversity**, which puts at risk the "ecosystem services" that such biodiversity provides, and ultimately presents food-security risks as well. [Ch.3: IB5]

F   Special concerns arise in the case of **pesticides** used in modern extractive agriculture: aside from the risk they might pose to human health (addressed in paragraph 3, below), the massive use of chemicals in agriculture (especially in the USA and other developed economies) can contribute to the destruction of beneficial species, increases in pest resistance, reduction in pollination, crop losses, ground and surface water contamination, and more. [Ch.3: IB6]

G   Spoiling of the natural environment brings direct injury to humans also in the form of severe degradation of **water quality and air quality**. Beyond releasing fertilizers (noted above in sub-paragraph D), modern farming operations also release a range of inorganic and organic matter that can contribute to salinization, transmission of infectious disease, bioaccumulation (in aquatic species more than terrestrial species), and other forms of degradation. Moreover, the sheer volume of water involved in modern agriculture (by one estimate, 70% of fresh water worldwide) is cause for concern. [Ch.3: IB7]

H   Modern extractive agriculture causes even greater damage – and substantially adds to an existential planetary threat – by its direct and indirect contribution to **global climate change**. Anthropogenic climate change started in a significant way with the advent of agriculture thousands of years ago, and the past two centuries have seen a dramatic increase in it – and much of this is attributable still today to agriculture. Overall, roughly 13 percent of worldwide greenhouse gas emissions come directly from agricultural activities (including livestock operations, but not counting such indirect sources as the conversion of tropical rain forests to soybean production), and most of these emissions are of nitrous oxide and methane, which are much more potent than carbon dioxide in their impact on global climate change. [Ch3: IB8]

3   Modern extractive agriculture poses undue risks to human health. [Ch.3: II]

A   Although (i) **agricultural chemicals** have long life cycles and (ii) the long-term effects of their concentrated accumulation over time is not yet known, they are still used widely in modern agriculture and in many countries are not made subject to rigorous handling restrictions. [Ch.3: IIB]

B   Although **genetic technology** can bring important benefits – and no injury seems likely to result from the genetic modifications that have occurred thus far in terms of *direct* implications for human health – modern extractive agriculture has permitted such technology to be used inappropriately in two ways: it has permitted undue privatization of the technology, and it has taken an unduly narrow and anthropocentric view of how to evaluate and regulate genetic-modification developments in agriculture. [Ch.3: IIC]

C Moreover, in a limited sense modern extractive agriculture contributes to serious **food–borne illnesses.** Haste in handling food products (especially meat, and particularly in the context of concentrated animal feedlot operations) increases the transmission of pathogens. Even more troubling is the over-prescription and improper use of antibiotics in agricultural livestock operations without adequate caution to guard against the emergence of pathogens that are resistant to those antibiotics. [Ch.3: IID]

D A particularly disgusting result of modern extractive agriculture is its contribution to **obesity**. The key problem relating to agriculture's role in that epidemic is that the structure of agricultural subsidies (like that of fossil-carbon-related subsidies) is deeply flawed, in a way that encourages production and sale of unhealthful products and discourages the production and sale of healthful food. [Ch:3: IIE]

4 Modern extractive agriculture runs historically and socially "against the grain" of human development in several ways. [Ch.3: III]

A It pays far too little attention to "**bioregionalism**" – the obvious reality that different ecoregions (around the world but sometimes also separated only by a few miles) have different characteristics and therefore require different and specialized agricultural approaches. It also largely disregards the "**law of return**": instead of the nutrient cycling that was common until the mid-nineteenth century, modern American agriculture relies almost exclusively on exogenous inputs – especially synthetic chemical fertilizers and pesticides. [Ch.3: IIIA, IIIB]

B Modern agriculture also ignores the principle of "**restraint**" that earlier agricultural societies practiced. Instead of aiming for single-crop yield maximization through concentrated inputs, they emphasized diversity, restrictive cultivation, moisture conservation, and selectivity. Moreover, it has abetted the development of a "consumption ethic" and changed our relationship to land by "commodifying" land. And it has had a similar effect on our relationship to community and shared resources and destinies. [Ch.3: IIIB, IIIC]

C What some observers would consider the crowning success of modern agriculture – the so-called **Green Revolution** – should be questioned more broadly. On reflection, the Green Revolution should be regarded (despite its laudable intentions) as the most visible manifestation of an "industrial" form of agriculture that has done deep and lasting harm to society by transforming farming, rural life, and food production in ways (i) that unwisely discard some values and efficiencies that for thousands of years have been central not only to our production of food but also to our role and identity within the ecosphere and (ii) that have, by linking agriculture with industry (and the fossil carbon on which industry depends), made agriculture today economically unsustainable, environmentally unsustainable, and dangerous to human health. [Ch.3: IIIC]

Proposition #2. *A fundamentally different form of food production and rural life – agroecological husbandry – is possible, and it is highly preferable to modern extractive agriculture, particularly in terms of producing grains and legumes that account for the largest portion of human caloric intake.*

5   Agroecological husbandry constitutes a dramatic departure from modern extractive agriculture in several ways. [Ch.4]

A   It focuses on **"husbandry"** in the sense of a way of life on land, and as part of the natural world, that encompasses conservation, frugality, economy, and the prudent use and nurturing of resources in the interests of the long-term viability of an ecosystem for its own sake and because of its own value. [Ch.4: IA, IC]

B   It takes the **ecosystem** as its standard. That is, it starts from the assumption that nature's economy – and particularly the economy and architecture of the native grasslands that constitute the setting for a great deal of agricultural production – should provide the guidance for a "natural-systems" agriculture that focuses on perennials grown in polycultures, not annuals grown in monocultures. [Ch.4: IB]

C   This emphasis on "natural systems" and "husbandry" implies a rejection of the emphasis that modern extractive agriculture places on production at all costs, including a rejection of the form of disruptive "clear-cutting" that modern agriculture (relying on annual crops) represents. Instead, agroecological husbandry reflects a **land ethic** that recognizes the seamless connection between healthy soil, healthy ecosystems, and vibrant human communities. Accordingly, it traces its roots to earlier advocates (Liberty Hyde Bailey, Sir Albert Howard, and others) who emphasized the importance of farming in concert with nature. [Ch.4: IC, ID]

D   One specific effort to provide the scientific wherewithal to make agroecological husbandry a viable substitute for modern extractive agriculture is reflected in the work of **The Land Institute**, where researchers have worked for several decades to develop perennial grains (as distinct from annual grains) that would be grown in polycultures (as distinct from monocultures). [Ch.4: II]

E   Viewed more broadly, agroecological husbandry plays a central role in a larger picture of **humans' place in nature** and its method of serving as the keystone species responsible for guarding the Earth's natural systems. [Ch.4: III] This has implications for land husbandry [Ch.4: IIIB], for biodiversity [Ch.4: IIIC], for global human population [Ch.4: IIID], for the production and use of energy [Ch.4: IIIE], and for economic theory and practice [Ch.4: IIIF].

6 A transition to agroecological husbandry holds strong promise for improving dramatically on – and offsetting the numerous disadvantages of – modern extractive agriculture. [Ch.5]

A Agroecological husbandry addresses the **economic, ecological, and health-related problems** associated with modern extractive agriculture. [Ch.5: IA]. More broadly, it brings food production **closer to nature** – a feature to be valued as a practical matter because many problems with modern extractive agriculture stem ultimately from the fact that it *rejects* nature in a variety of ways. [Ch.5: IB]

B Fortunately, agroecological husbandry is not a pipe-dream but instead has **solid prospects for success** – including prospects for commercial success. Although it is unconventional, and therefore outside the typical orbit of research institutions funded by agribusiness interests, it has been gaining wide acceptance in the scientific world. To date, substantial progress has already been made (on relatively meager research resources thus far) not only in (i) the development of perennial varieties but also in (ii) the understanding of polycultures. [Ch.5: II]

Proposition #3. *New legal initiatives and designs, both at the national level and at the global level, can facilitate a shift from modern extractive agriculture to agroecological husbandry – and the necessary steps in this direction should be taken immediately.*

7 Numerous changes should be made in substantive law at the *national* level – especially in the USA as the principal "engine" of modern extractive agriculture, but also elsewhere in the world as possible and necessary – in order to facilitate such a transition to agroecological husbandry. [Ch.6: I]

A In order to address the ways in which modern extractive agriculture is unsustainable as a matter of **economics**, legal and regulatory changes should be made that focus specifically on (i) reducing the high entry costs and other hurdles to small farmers and beginning farmers eager to engage in agroecological husbandry, (ii) encouraging an increase in the size and diversity of farms and rural populations through support for the overall infrastructure involved in agroecological husbandry and the communities that would rely on it, and (iii) encouraging a dramatic increase in the diversity of crops – to reverse the trend that has resulted in maize, wheat, and rice accounting for 89 percent of all cereal grain production (as of 2012) – through a wholesale reorientation of agricultural subsidies. [Ch.6: IA]

B   In order to address the ways in which modern extractive agriculture is unsustainable as an **ecological** matter, legal and regulatory changes should be made (i) to expand dramatically the ongoing scientific research into perennial species of food grains and legumes that can gradually supplant the annual crops that dominate today's agriculture, (ii) to expand, in like fashion, the ongoing scientific research into foodcrop polycultures, (iii) to reorient agricultural subsidies in many ways, including tying them largely to ecological performance rather than crop yields, (iv) to remove fossil-carbon subsidies as gradually as necessary to avoid chaos but as quickly as necessary to blunt the worst effects of global climate change, (v) to stiffen agriculture-specific anti-pollution protections to offset and reduce ecological damage from agriculture as currently practiced and force a shift to agroecological husbandry, and (vi) to give special attention to other means of reducing greenhouse gas emissions from agricultural operations, including livestock production. [Ch.6: IB]

C   In order to counter some of the **human–health risks** of modern extractive agriculture, legal and regulatory changes should be made (i) to adopt the anticipatory approach of the Precautionary Principle as practiced in Europe and as reflected in numerous international legal instruments (and thereby reject the "reactive" approach taken in many countries, including the USA, under which restriction on the production or use of agricultural chemicals typically occurs only *after* a specific and clear injury has been shown), (ii) to place the overall supervisory responsibility for and control over GM/GE research in the public domain instead of predominantly in private hands, (iii) to impose special regulations on transgenic manipulation, which poses peculiar risks, and (iv) to reduce agriculture's role in food-borne illnesses and in obesity. [Ch.6: IC]

D   Several of the legal and regulatory changes enumerated above will help counter the fact that modern extractive agriculture goes **"against the grain"** of human development. Beyond those legal steps, however, are some more fundamental legal and institutional reforms that would require action outside the realm of the nation-state and therefore should be addressed at the international level – as summarized in paragraphs 9–10. [Ch.6: ID]

8   In the particular context of changes in laws and regulation in the USA, overriding emphasis should be placed on the enactment and implementation of a "50-year farm bill", with elements drawn also from numerous other proposals that have been offered to transform American agriculture. [Ch.6: II]

A.   A 50-year farm bill, in sharp contrast to the usual short-term agricultural legislation that characterizes US government action in this sector, would announce a **national agricultural policy that is**

based on ecological principles. Its specific aims would be to address biodiversity and ecosystem health, soil degradation and erosion, water pollution from agricultural run-off, agriculture-pesticide dangers, fossil-carbon dependence, greenhouse-gas emissions, carbon sequestration, and the restoration of farm and rural communities. [Ch.6: IIA1]

B   Consistent with those aims, the **substantive provisions** of a 50-year farm bill would include provisions to put in place the specific types of requirements, restrictions, and initiatives enumerated above in paragraph 7 – addressing economic, ecological, human-health, and other problems of modern extractive agriculture, and facilitating a transition to agroecological husbandry. [Ch.6: IIA2]

C   A 50-year farm bill would also specify an **implementation schedule** to assure that (i) acreages devoted to food crops other than annual monoculture commodity grains would start increasing within five years after the legislation's enactment, (ii) appreciable changes in overall nationwide allocations among annual and perennial foodcrops would occur, and designs of workable polycultures would be in final stages, within ten years after enactment, (iii) new perennial grains would be in widespread use, some of them in polycultures, by 20 years after enactment, (iv) nationwide allocation of cropland acreages to perennials would be 60% (much of it in polycultures) by forty years after enactment, and (v) by the end of the 50-year period, perennials would have gained dominance over annual grains, ecology-smart agricultural subsidies would have reduced various forms of farm-based pollution and largely arrested soil degradation and erosion – with a corresponding reduction in US agriculture's contribution to greenhouse gas emissions and a big increase in US agriculture's carbon-sequestration capacity – and rural depopulation would be have reversed and a high degree of regionalization and localization of food markets achieved. [Ch.6: IIA3]

D   The **policy prescriptions found in numerous other proposals** would supplement the ones noted above. These proposals focus on such matters as training a new generation of farmers, making specific amendments to the Clean Water Act of 1972 (and other statutes), and of course revising the system of agricultural subsidies. Some of them also recommend ways to "agriculturize" urban areas and ways to improve marketing of local and regional food. [Ch.6: IIB]

9   A new treaty system could facilitate a transformation away from modern extractive agriculture and toward agroecological husbandry. [Ch.7: I, II]

A   The **principles announced in a Global Convention on Agroecology** would, among other things, (i) emphasize the need for a reintegration of humans with the rest of the natural world,

(ii) acknowledge the responsibility that humans have to restore the Earth's ecological integrity, (iii) unambiguously adopt the Precautionary Principle, (iv) recognize a duty of cooperation, as a legal matter, among humans, (v) announce the parties' endorsement of a new form of agricultural life and production, (vi) specify that the specific objectives of such a new form of agricultural life and production are to reverse the economic, ecological, and other deficiencies of modern extractive agriculture, (vii) assert that certain "safeguard rights" of all people (basic civil and political rights) will be protected during the transition to a new form of agriculture, and (viii) note the special urgency of the need to address the issues of climate change, ecological preservation, and food security. [Ch.7: IIA]

B   The **obligations of contracting states** to the treaty would be to facilitate the implementation of the principles enumerated above, particularly through legislation, regulation, education, and research. Special attention would be given to the latter two of these, and financial support would be provided (for economically less-developed states) as needed under a form of "special and differential treatment" of the sort found in other modern treaty systems. [Ch.7: IIB, IIC]

C   The adoption and implementation of some of the principles of the Global Convention on Agroecology will require fundamental and difficult **reorientation of views** on several issues. These include (i) the proper balance to be struck between the importance of precaution (as in the Precautionary Principle), the role of technology, and the aims of research, (ii) the role that growth has in a global economic theory, (iii) the relationship between global climate change and agriculture, and (iv) the feasibility of a legal "duty of cooperation" among states and ethnic groups. On all these issues, though, modern thinking provides adequate grounds for optimism that a reorientation of views, though difficult, is by no means impossible and in any event necessary. [Ch.7: III]

10   The long-term success of a new treaty system – and of the efforts to transform agriculture – depends also, however, on a **reorientation of sovereignty** as a central concept in international relations. [Ch.8: I, II]

A   The traditional concept of sovereignty – a so-called "**monolithic sovereignty**" – emerged in the political setting of sixteenth-century and seventeenth-century Europe. Although it proved useful in that context, and although it has evolved some since then (to reflect at least the rhetoric, for instance of "popular sovereignty"), this "monolithic" version of sovereignty has limited relevance to today's world. [Ch.8: I]

B   Intense effort should be devoted now to the development of a new version of sovereignty – a so-called "**pluralistic sovereignty**"

– that is more appropriate to today's world. Given the essential significance of global ecological protection, and the role that agriculture must play in achieving that protection, such a "pluralistic sovereignty" concept might rest on a global political framework in which "eco-states", not "nation-states" (which are largely artificial as a practical matter now anyway), would be the planet's fundamental political units. [Ch.8: II]

Selected bibliography

In writing this book, I have surveyed countless sources – books, journal articles, websites, and miscellaneous other sources (and have relied on my research assistants to do the same) – in hopes of gaining adequate understanding of the law and literature relating to the points of interaction between international law and agriculture. I list here some of those materials that I found most helpful and depended on most often. Readers wishing to explore beyond the limits of this book might find this list useful. Of course, further information about selected topics may also be found in the many works that are specifically cited in footnotes found throughout the text.

Works are listed primarily in alphabetical order by author or editor (or institutional author/editor). Books, treatises, and monographs (as well as the names of journals in which articles appear), are designated by LARGE AND SMALL TYPEFACE; titles of law journal articles, book chapters, and other less extensive works are usually *italicized*. Names of authors or editors are all listed in plain roman type (upper- and lower-case) for simplicity, with surnames shown in **bold** for quick reference.

Books

Mary Jane **Angelo**, Jason J. **Czarnezki** and William S. **Eubanks** II, FOOD, AGRICULTURE, AND ENVIRONMENTAL LAW (Environmental Law Institute, 2013)

Karin **Bäckstrand** and Annica **Kronsell**, eds., RETHINKING THE GREEN STATE: ENVIRONMENTAL GOVERNANCE TOWARDS CLIMATE AND SUSTAINABILITY TRANSITIONS) Routledge/Earthscan, 2015)

Sanjay Kabir **Bavikatte**, STEWARDING THE EARTH: RETHINKING PROPERTY AND THE EMERGENCE OF BIOCULTURAL RIGHTS (Oxford University Press, 2014)

Janine **Benyus**, BIOMIMICRY: INNOVATION INSPIRED BY NATURE (HarperCollins, 1977)

Klaus **Bosselmann** and J. Ronald **Engel**, eds., THE EARTH CHARTER: A FRAMEWORK FOR GLOBAL GOVERNANCE (KIT Publishers, 2010)

J. L. **Brierly**, THE LAW OF NATIONS (Clarendon Press, 1963)

Lester R. **Brown**, WORLD ON THE EDGE: HOW TO PREVENT ENVIRONMENTAL AND ECONOMIC COLLAPSE (Norton, 2011)

Ian **Brownlie**, PRINCIPLES OF PUBLIC INTERNATIONAL LAW (Oxford University Press, 4th ed. 1990)

David **Cannadine**, THE UNDIVIDED PAST (Knopf, 2013)

Willa **Cather**, DEATH COMES FOR THE ARCHBISHOP (Knopf, 1927)

Jason **Clay**, WORLD AGRICULTURE AND THE ENVIRONMENT: A COMMODITY-BY-COMMODITY GUIDE TO IMPACTS AND PRACTICES (Island Press, 2004)

George Cameron **Coggins**, RESTORATION: AN IMMODEST BLUEPRINT FOR FEDERAL PUBLIC LAND AND RESOURCES LAW REFORM IN THE 21ST CENTURY (KU, 2011)

Committee on Twenty-First Century Systems Agriculture, National Research Council, TOWARD SUSTAINABLE AGRICULTURAL SYSTEMS IN THE 21ST CENTURY (2010)

John R. **Ehrenfeld** and Andrew J. **Hoffman**, FLOURISHING: A FRANK CONVERSATION ABOUT SUSTAINABILITY (Stanford Business Books, 2013)

FAO, PERENNIAL CROPS FOR FOOD SECURITY: PROCEEDINGS OF THE FAO EXPERT WORKSHOP (2014)

Cary **Fowler** and Patrick R. **Mooney**, SHATTERING: FOOD, POLITICS, AND THE LOSS OF GENETIC DIVERSITY (University of Arizona Press, 1990)

Adam **Frank**, ABOUT TIME: COSMOLOGY AND CULTURE AT THE TWILIGHT OF THE BIG BANG (Free Press, 2011)

H. Patrick **Glenn**, THE COSMOPOLITAN STATE (Oxford University Press, 2013)

Hurst **Hannum**, AUTONOMY, SOVEREIGNTY, AND SELF-DETERMINATION (University of Pennsylvania Press, 1990)

R. M. **Harrison**, R. E. **Hester**, Joe **Morris**, Karl **Ritz** and Mark G. **Kibblewhite**, ENVIRONMENTAL IMPACTS OF MODERN AGRICULTURE (RSC Publishing, 2012)

Paul **Hawken**, Amory **Lovins** and L. Hunter **Lovins**, NATURAL CAPITALISM: CREATING THE NEXT INDUSTRIAL REVOLUTION (Little, Brown and Company, 1999)

John W. **Head**, GLOBAL LEGAL REGIMES TO PROTECT THE WORLD'S GRASSLANDS (Carolina Academic Press, 2012)

John W. **Head**, GREAT LEGAL TRADITIONS: CIVIL LAW, COMMON LAW, AND CHINESE LAW IN HISTORICAL AND OPERATIONAL PERSPECTIVE (Carolina Academic Press, 2011)

John W. **Head**, LOSING THE GLOBAL DEVELOPMENT WAR: A CONTEMPORARY CRITIQUE OF THE IMF, THE WORLD BANK, AND THE WTO (Nijhoff, 2008)

Richard **Heinberg**, THE END OF GROWTH: ADAPTING TO OUR NEW ECONOMIC REALITY (New Society Publishers, 2011)

Tim **Jackson**, PROSPERITY WITHOUT GROWTH: ECONOMICS FOR A FINITE PLANET (Earthscan, 2009)

Wes **Jackson**, BECOMING NATIVE TO THIS PLACE (Counterpoint, 1994)

Wes **Jackson**, CONSULTING THE GENIUS OF THE PLACE: AN ECOLOGICAL APPROACH TO A NEW AGRICULTURE (Counterpoint, 2010)

Wes **Jackson**, NATURE AS MEASURE – THE SELECTED ESSAYS OF WES JACKSON (Counterpoint, 2011)

Wes **Jackson**, NEW ROOTS FOR AGRICULTURE (Bison Books, 1980, 1985 reprint)

Ali **Khan**, THE EXTINCTION OF NATION-STATES: A WORLD WITHOUT BORDERS (Kluwer Law International, 1996)

Andrew **Kimbrell**, THE FATAL HARVEST READER: THE TRAGEDY OF INDUSTRIAL AGRICULTURE (Island Press, 2002)

Frederick L. **Kirschenmann**, CULTIVATING AN ECOLOGICAL CONSCIENCE: ESSAYS FROM A FARMER PHILOSOPHER (Counterpoint, 2010)

Aldo **Leopold**, A SAND COUNTY ALMANAC (Ballantine Books 1966, 1949)

Charles C. **Mann**, 1491: NEW REVELATIONS OF THE AMERICAS BEFORE COLUMBUS (Knopf, 2011)

Gregory **McIsaac** and William R. **Edwards**, eds., SUSTAINABLE AGRICULTURE IN THE AMERICAN MIDWEST: LESSONS FROM THE PAST, PROSPECTS FOR THE FUTURE (University of Illinois Press, 1994)

Bill **McKibben**, DEEP ECONOMY: THE WEALTH OF COMMUNITIES AND THE DURABLE FUTURE (Macmillan, 2008)

Donella H. **Meadows**, Dennis L. **Meadows**, Jørgen **Randers** and William W. **Behrens** III, THE LIMITS TO GROWTH (Universe Books, 1972)

Daniel **Miller**, CONSUMPTION AND ITS CONSEQUENCES (Polity, 2012)

David R. **Montgomery**, DIRT: THE EROSION OF CIVILIZATIONS (University of California Press, 2007)

R. P. C. **Morgan**, SOIL EROSION AND CONSERVATION (Wiley-Blackwell, 3rd ed. 2005)

Arthur **Nussbaum**, A CONCISE HISTORY OF THE LAW OF NATIONS (Macmillan, 1947)

Steven **Pinker**, THE BETTER ANGELS OF OUR NATURE: WHY VIOLENCE HAS DECLINED (Penguin, 2011)

Peter P. **Rogers**, Kazi F. **Jalal** and John A. **Boyd**, AN INTRODUCTION TO SUSTAINABLE DEVELOPMENT (Earthscan, 2008)

William F. **Ruddiman**, PLOWS, PLAGUES, AND PETROLEUM: HOW HUMANS TOOK CONTROL OF CLIMATE (Princeton University Press, 2005)

Susan A. **Schneider**, FOOD, FARMING, AND SUSTAINABILITY (Carolina Academic Press, 2011)

Ernest L. **Schusky**, CULTURE AND AGRICULTURE: AN ECOLOGICAL INTRODUCTION TO TRADITIONAL AND MODERN FARMING SYSTEMS (Bergin & Garvey, 1989)

Malcolm N. **Shaw**, INTERNATIONAL LAW (Cambridge University Press, 7th ed. 2014)

John **Thackara**, HOW TO THRIVE IN THE NEXT ECONOMY: DESIGNING TOMORROW'S WORLD TODAY (Thames & Hudson, 2015)

Mark B. **Tauger**, AGRICULTURE IN WORLD HISTORY (Routledge, 2010)

Gerhard **von Glahn**, LAW AMONG NATIONS (Macmillan USA, 1986)

Burns H. **Weston** Richard A. **Falk** and Anthony A. **Damato**, INTERNATIONAL LAW AND WORLD ORDER (West, 2nd ed. 1990)

Mary Christina **Wood**, NATURE'S TRUST: ENVIRONMENTAL LAW FOR A NEW ECOLOGICAL AGE (Cambridge University Press, 2013)

Donald **Worster**, THE WEALTH OF NATURE: ENVIRONMENTAL HISTORY AND THE ECOLOGICAL IMAGINATION (Oxford University Press, 1993)

Angus **Wright**, THE DEATH OF RAMÓN GONZÁLEZ (University of Texas Press, rev. ed. 2010)

Journal articles and book chapters

Mary Jane **Angelo**, *Corn, Carbon, and Conservation: Rethinking U.S. Agricultural Policy in a Changing Global Environment*, 17 GEORGE MASON LAW REVIEW 593 (2010)

Kenneth **Boulding**, *The Economics of the Coming Spaceship Earth*, in VALUING THE EARTH 297, 297–309 (Herman Daly and Kenneth Townsend, eds., 1993)

John W. **Head**, *Supranational Law: How the Move Toward Multilateral Solutions Is Changing the Character of "International" Law*, 16 KANSAS LAW REVIEW 605 (1993)

Wes **Jackson** and Wendell **Berry**, *A 50-Year Farm Bill*, THE NEW YORK TIMES, Jan. 4, 2009

H. **Lauterpact**, *The Grotian Tradition in International Law*, 23 BRITISH YEARBOOK OF INTERNATIONAL LAW 1, 2 (1946)

Charles C. **Mann**, *1491*, THE ATLANTIC MONTHLY, March 2002

David M. **Olson**, D. Eric **Dinerstein**, Eric D. **Wikramanayake**, Neil D. **Burgess**, George V. N. **Powell**, Emma C. **Underwood**, Jennifer A. **D'amico**, Illanga **Itoua**, Holly E. **Strand**, John C. **Morrison**, Colby J. **Loucks**, Thomas F. **Allnutt**, Taylor H. **Ricketts**, Yumiko **Kura**, John F. **Lamoreux**, Wesley W. **Wettengel**, Prashant **Hedao** and Kenneth R. **Kassem**, *Terrestrial Ecoregions of the World: A New Map of Life on Earth*, 51 BIOSCIENCE 933 (2001) (see www.worldwildlife.org/science/ecoregions/delineation.html)

John **Opie**, *Ecology and Environment*, appearing as the third chapter in THE GREAT PLAINS REGION (Amanda Rees, ed., 2004), a contribution to THE GREENWOOD ENCYCLOPEDIA OF AMERICAN REGIONAL CULTURES

Michael **Pollan**, *Farmer in Chief*, THE NEW YORK TIMES: THE FOOD ISSUE (Oct. 12, 2008)

J. B. **Ruhl**, *Farms, Their Environmental Harms, and Environmental Law*, 27 ECOLOGY LAW QUARTERLY 263, 265 (2000)

Carl **Sauer**, *The Agency of Man on the Earth*, in CARL SAUER – SELECTED ESSAYS 1963–1975 (Bob Callahan, ed.; Turtle Island Foundation, 1981)

World Bank, *Turn Down the Heat: Why a 4°C Warmer World Must Be Avoided* (2012)

Other resources

Grasslands Conservation Council of British Columbia, *Overview of the World's Grasslands*, at www.bcgrasslands.org/library/world.htm

Pope Francis, Encyclical *On Care for Our Common Home* (2015)

World Wildlife Fund, *Terrestrial Ecoregions Database* (showing details for all 800-plus terrestrial ecoregions within the WWF classification), available via link from the WWF webpage for "Terrestrial Ecoregions", www.worldwildlife.org/science/ecoregions/item1267.html

Index